BUS
TORRES STRAIT
CAPE YORK
Wessel Islands
CAPE ARNHEM
Groote Eylandt
GULF OF CARPENTARIA
CAPE YORK PENINSULA
PRINCESS CHARLOTTE BAY
Cooktown
Sir Edward Pellew Group
Mornington Island
Cairns
Atherton
BARKLY TABLELAND
Gregory River
Gilbert River
GREAT
SOUTH PACIFIC OCEAN
Flinders River
GREAT
Townsville
BARRIER
Mt Isa
Burdekin River
QUEENSLAND
DIVIDING
REEF
Diamantina River
Thomson River
Barcoo River
Rockhampton
Alice Springs
RANGE
Fraser Island
Creek
Cooper
Warrego River
Condamine River
Culgoa
Brisbane
CAPE BYRON
SOUTH
Lake Eyre
River
McIntyre
USTRALIA
Namoi River
RANGE
Lake Torrens
FLINDERS RANGES
BARRIER RANGES
Darling River
NEW SOUTH WALES
Lord Howe Island
DIVIDING
Lachlan River
Sydney
Adelaide
Murrumbidgee River
MALLEE
Murray River
Canberra
GREAT
Kangaroo Island
VICTORIA
Bega
Melbourne
GIPPSLAND
TASMAN SEA
GRAMPIANS
BASS STRAIT
Wilsons Promontory
King Island
Furneaux Group
TASMANIA
Hobart
Maatsuyker Island

A Field Guide to the Mammals of Australia

A Field Guide to the Mammals of Australia

by Peter Menkhorst
Illustrated by Frank Knight

253 Normanby Road, South Melbourne, Australia

Oxford University Press is a department of the University of Oxford.
It furthers the University's objective of excellence in research,
scholarship, and education by publishing worldwide in

Oxford New York
Athens Auckland Bangkok Bogotá Buenos Aires Cape Town
Chennai Dar es Salaam Delhi Florence Hong Kong Istanbul
Karachi Kolkata Kuala Lumpur Madrid Melbourne Mexico City
Mumbai Nairobi Paris Port Moresby São Paulo Shanghai Singapore
Taipei Tokyo Toronto Warsaw
with associated companies in Berlin Ibadan

First published 2001

National Library of Australia
Catalogue-in-publication data:

Menkhorst, P. W.
A field guide to the mammals of Australia.

Bibliography.
Includes index.
ISBN 0 19 550870 X.

1. Mammals—Australia—Identification. I. Knight, Frank,
1941– . II. Title.

599.0994

Edited by David Meagher
Cover and text designed, and typeset by Derrick I Stone Design
Printed in Hong Kong through the Bookmaker International Ltd

Foreword

Australia's birth as a continent was a most protracted event, from rift zone earthquakes 120 million years ago that shook dinosaurs and strange egg-laying ancestors of modern platypuses off their feet, to the final collapse into the sea, perhaps 45 to 38 million years ago, of the last dry land that linked Tasmania and Antarctica. For the next 30 million years, in splendid isolation apart from the comings and goings of bats and marine mammals, Australia's mammals became a spectacularly weird mob.

Then, rent by cycles of profound climate change from icehouse to greenhouse conditions and back again, and driven by competition with extraordinary reptiles like tree-climbing crocodiles (mekosuchines) and such birds as the gigantic 'demon ducks of doom' (dromornithids) and invasions perhaps 5 million years ago of the ancestors of our native rodents, wave after wave of evolutionary change swept across the land.

As the first fingers of aridity crept into the continent's heart and grasslands began to spread, perhaps no longer ago than 3 million years, the scene was being set for perhaps the greatest challenge to our native mammals: the arrival of wave upon wave of humans, the last being Europeans committed to reshaping the land with axes and armies of alien animals and plants.

Which brings us to this moment, which, for many of our four-footed furry friends, has become the most hazardous crossroad of rapidly changing circumstances. With 65 per cent of the continent now under the thumb of unsustainable agriculture focused on introduced species and land degradation costing an estimated $5 billion accumulating every year, the future of most of Australia's mammals is in doubt.

We live in the shadow of a growing mountain of scientific literature documenting so many long- and short-term declines that the brightest view in sight is a distinctly backwards one. But what worries me even more than the declines or losses of particular species is the failure to focus on conserving the capacity of evolution to replace these extinctions. Too gradual for most to see but catastrophically fast on a geological scale, the presumption of conservation is giving way to a panic of preservation. As the evolutionary future for wild lineages dims, we struggle now just to keep things the way they are—which is a folly because nothing about our natural environments or where they currently sit will stay the same.

Is it too late to turn the situation around for Australia's increasingly threatened mammals? I, for one, am incorrigibly optimistic because I refuse to have to explain to the next generation why Numbats, Wombats, Orange Leaf-nosed Bats, or any other kind of 'bat' vanished while our generation, far more aware than any before us of the dark forces at work, minded the shop.

What are required to turn current trends around are trials of new and sometimes challenging conservation strategies. In addition to such traditional strategies as protected areas and ecotourism, we must be prepared, like many other countries around the world, also to trial sustainable harvesting of wildlife, programs of species reintroduction to secured areas of former range, and even native animals as pets—strategies based above all on valuing our precious wildlife and rediscovering sustainable ways to put people and natural environments back together for the benefit of both. Without value-driven conservation strategies, I fear struggle for survival of both will be lost before the end of this century.

This field guide, a marvellous mirror on our mammals of the moment, will become a critical factor in increasing our sense of their value and increasing commitment in the younger generation to lose no more. For a spectacular job, Peter Menkhorst and Frank Knight deserve a mountain of praise. Certainly there has been a long list of noble predecessors to this book, such as the overviews of John Gould, Frederick Wood Jones, David Ride, Ron Strahan, Meredith Smith and Rosemary Ganf, and field guides, such as those by Ellis Troughton, Basil Marlow, Chris Watts and Heather Aslin, and Sue Churchill, but none cover all species of Australian mammals. Besides being completely up to date, in one portable, beautifully illustrated and succinctly written book, every furry beast of Oz is in the spotlight, its fate now squarely in your hands.

Mike Archer
Director, Australian Museum
Professor of Biological Science, University of New South Wales

Contents

Foreword by Mike Archer v
Acknowledgments ix

Introduction **1**
How to use this field guide **1**
Geographic coverage 1
Identifying a mammal quickly 1
Classification and names 1
Field characters 2
Identification keys 2
Species accounts 3
Plates 7
Distribution maps 7
Field techniques for identifying mammals **8**
Finding mammals 8
Field characters 11
How mammals are measured 12
Submitting records 12
References 12

Checklist of Australian Mammals **14**

Identification Keys to Australian Mammals **23**
Notes on using the keys **23**
Keys to orders and genera of small mammals 23
Key to species of marine mammals 23
Index to keys 23
Identification keys 24

Habitat types **39**

Abbreviations **43**
Key to distribution maps 43

Species Accounts and Plates **44**
Monotremes (plate 1) 44
Carnivorous marsupials (plates 2–16) 46
Bandicoots and bilbies (plates 17–19) 76
Koala and wombats (plate 20) 82
Cuscuses (plate 21) 84
Possums and gliders (plates 21–28) 84
Bettongs, rat-kangaroos and potoroos (plates 29–32) 100
Kangaroos and wallabies (plates 32–46) 106
Fruit bats (plates 47–49) 136
Insectivorous bats (plates 50–64) 142
Rodents (plates 65–81) 172
Introduced carnivores (plate 82) 206

Introduced herbivores (plates 83–87) 208
Seals (plates 88–90) 218
Dolphins, porpoises and Killer Whale (plates 91–94) 224
Whales (plates 95–103) 232
Dugong (plate 104) 250

Addendum **252**

Further Reading **253**

Glossary **255**

Index to Common Names **261**

Index to Scientific Names **267**

Acknowledgments

Frank Knight has harboured a dream of illustrating a field guide to Australian mammals for some thirty years. I am extremely grateful to Frank for his tenacity in pursuing that dream, and for offering me the opportunity to team with him to write the words to accompany his superb illustrations. From the beginning our endeavours have been truly collaborative, a true partnership. We both owe a great debt to Andrew Isles for bringing us together.

Many colleagues in Australian mammalogy have contributed directly to this work in a variety of ways, and all who have published studies on Australian mammals have contributed indirectly. Space limitations in a field guide do not allow acknowledgment of all references consulted during the preparation of this work. However, I wish to record my gratitude to all Australian mammalogists and field naturalists over the past two centuries who have undertaken often arduous field work, then carefully documented their studies and observations in the scientific and natural history literature. Without their efforts, there would be no body of knowledge to summarise in a book such as this. The reviews provided by Ron Strahan and his team of authors (Strahan 1995), Sue Churchill (Churchill 1998), and the series of action plans for Australian mammals produced by Environment Australia were particularly helpful sources of information.

People who generously made direct contributions were: reviewers of draft species accounts—Gordon Friend, Lindy Lumsden, Graham Ross, Peter Shaughnessy and John Woinarski; people who answered my many questions—Sue Churchill, Rohan Clarke, Greg Connors, Lawrie Conole, Peter Copley, Matthew Crowther, Murray Ellis, Mike Fleming, Tony Griffiths, Norm Mackenzie, Harry Parnaby, Greg Richards, John Seebeck and Steve Van Dyck; collection managers and curators at museums—Lina Frigo, Mark Darragh and Joan Dixon (Melbourne Museum), Sandy Ingleby (Australian Museum), Cath Kemper and Martine Long (South Australian Museum), John Wombey (Australian National Wildlife Collection); and Ellen Menkhorst who helped type the manuscript. These people pointed out numerous errors and deficiencies in the drafts and I have attempted to accommodate their suggestions. However, I take full responsibility for any remaining errors or omissions.

The key to marine mammals was developed entirely by Graham Ross and very generously made available for inclusion in this work. It is reproduced here with permission from Environment Australia. Professor Mike Archer kindly provided the Foreword.

At Oxford University Press the project was enthusiastically supported by a very professional team which included Peter Rose, Heather Fawcett, and Derrick Stone. David Meagher's editorial hand was always thoughtful, knowledgeable and constructive.

Lastly, I must thank my family—Barbara Gleeson and our daughters Ellen and Cassia—for tolerating (or perhaps enjoying) my reclusive behaviour in the evenings over the past three years, and for having far more than my fair share of computer time.

Peter Menkhorst

I must first thank Peter Menkhorst, the guide's author. He has been a dream to work with: the perfect collaborator. Always ahead of me but patient. I am also impressed by his practical yet passionate commitment to the Australian fauna and its environment, which I like to think I share. However, unlike Peter, my idea of a good holiday is not sitting on Lady Julia Percy Island in January looking at fur seals.

Second only to Peter, I am most grateful to Andrew Isles who recognised that we might make a team to see the mammal guide through, getting us together over lunch in order to discuss its prospects. For this and other support, I thank him.

I am also grateful to those prominent Australian zoologists who set aside time from their busy lives to try to help me get a much earlier, and ultimately unsuccessful, version of this guide off the ground. These include the late John Calaby, Mike Archer, Kath Kemper, Greg Richards, the late Peter Aitken and Merv Griffiths.

Hugh Tyndale-Biscoe was always positive at times when the guide looked to be at very long odds indeed. Likewise Mike Archer. I intend to live long enough to see his genetically reconstructed Thylacine living and breathing. David Ride spent time reflecting on the earlier guide and his thoughts helped to see it re-emerge in its present form.

OUP's editors, particularly Peter Rose and Heather Fawcett, have been great, as has Derrick Stone, our designer. As have the team at the Australian National Wildlife Collection, particularly Dick Schodde and John Wombey but also latterly Terry Chesser and Robert Palmer.

Cath Kemper of the South Australian Museum kindly let me check specimens, as did Sandy Ingleby in the Australian Museum, Sydney, Lina Frigo at the Melbourne Museum and Steve Van Dyck at the Queensland Museum. Specimens were borrowed from the West Australian Museum through Nora Cooper.

I would like to thank Geoff Sharman who, when I was his technician in the early sixties, allowed me time to develop skills in pen drawing in the preparation of some slides portraying Australian mammals that first gave me the sense that natural history illustration might be something I could do. He also paid good money for an early painting.

Finally, thanks to Heather for being around all these many years.

Frank Knight

Introduction

How to use this field guide

Geographic coverage

The field guide covers wild mammals recorded from the Australian continent, its continental islands, and the seas over its continental shelf. It does not include species restricted to outlying political territories such as Christmas Island, Cocos–Keeling Islands, Macquarie Island, Heard Island or the Australian Antarctic Territories—these areas are not biogeographically part of Australia. Introduced species that have established self-sustaining wild populations are included, as are species that are thought to have become extinct since European settlement began in 1788.

Identifying a mammal quickly

The main part of this book follows the preferred field guide layout: species text and distribution maps are on the left-hand pages, and colour illustrations of each species are in the same sequence on the facing pages. Where appropriate, a diagram of a key characteristic is included beneath the distribution map.

In most cases readers will know the broad type of mammal they have seen—for example, kangaroo, wombat, possum, bat, seal, dolphin. In such cases it is best to turn directly to the relevant illustrations, and the adjacent distribution maps, to narrow down the options before studying the species accounts.

For more difficult groups, especially small mammals, a series of dichotomous keys is provided to determine firstly the taxonomic order to which the specimen belongs, and then its genus. The appropriate illustrations, texts and distribution maps can then be examined to determine the species. In many cases the characters needed to identify small mammals require that the animals can be closely examined, preferably 'in the hand'. Thus it is not always possible to reach a definitive identification of small mammals, nor even for some very large mammals such as whales.

Classification and names

The systematic arrangement and species nomenclature broadly follow that of Strahan (1995). However, the sequence of genera and species has been varied where necessary to place close together those species or groups with similar appearance and potentially overlapping distributions. In the case of seals, those species that occur in Australian waters only as vagrants are placed together. This is a purely pragmatic arrangement to assist the identification process.

There is no current comprehensive checklist of Australian mammals that provides an agreed list of both scientific and common names. Furthermore, there are numerous taxonomic studies in progress, and changes to species arrangements and nomenclature will continue for some time. This is particularly true for bats, rodents and dasyurids.

Nevertheless, a checklist is presented of all currently recognised species of mammal recorded from Australia, and described in this book. Publication of the first descriptions of the Rusty Antechinus *Antechinus adustus*, Subtropical Antechinus *A. subtropicus* and Tan False Antechinus *Pseudantechinus roryi* came too late for them to be illustrated in colour and included within the species accounts for antechinus. However, a brief account of each is included in the addendum on page 252.

Names used generally follow Strahan (1995) except where subsequent taxonomic changes have been accepted. For whales and dolphins the nomenclature of Bannister, Kemper and Warneke (1996) is followed, except for the inclusion of two species of Bottlenose Dolphin and two species of Minke Whale. Alternative common names in current use are also listed. Synonyms for scientific names are not listed due to space limitations, but they have been listed by Walton (1988). In some cases we have used different common names that we believe have advantages over those suggested by Strahan (1995). Three major principles have been followed in developing alternative common names:

1 Simplicity and shortness are desirable.
2 Aboriginal names are preferred when there is one agreed name that clearly applies to that taxon.
3 International usage is followed where there is an agreed name for extralimital species.

Field characters

The species accounts are necessarily brief to conform to the field guide layout. Therefore, species descriptions are usually limited to those external characters necessary for making an identification. The level of detail necessary to arrive at an identification varies enormously across mammalian groups. Most large terrestrial mammals can be readily identified if a reasonable view is obtained and key features noted. Ideally, the species accounts will have been studied before entering the field so that the observer has prior knowledge and can quickly scan for key features when an animal is sighted. In contrast, small mammals are rarely seen unless trapped or otherwise captured. Even when in the hand, many rodents, bats and small dasyurids can be difficult to identify to species level, and precise measurements of diagnostic body parts may be needed. In such cases we have described or illustrated detailed features of external morphology needed for identification. In rare cases, such as the broad-nosed bats (genus *Scotorepens*) identification may not be possible without biochemical analyses.

Identification keys

For the groups of small mammals that are difficult to identify in the field, simple dichotomous keys are provided to help determine the order, and then the genus. The plates, distribution maps and species accounts of the genus can then be consulted to identify the species.

The first key distinguishes the three orders that include small Australian terrestrial mammals (up to quolls, small gliding possums and bettongs in size)—rodents, polyprotodont marsupials and diprotodont marsupials. Keys to the genera in each of these three orders follow. Bats are not included in the Key to Orders because it is assumed that readers will know that a furred animal with a flight membrane supported by greatly elongated fingers is a bat (order Chiroptera). Instead, a key to the families of bats is included, followed by keys to the genera within each family.

Marine mammals can also be hard to identify, despite their large size. The inclusion of Graham Ross's key to species of stranded marine mammals greatly improves the utility of this guide.

Species accounts

The field guide layout we have adopted presents all the information about each species within one double-page spread—species description, distribution map, drawing of key feature, and the colour illustration. For each species the recommended common name is given, followed by the current scientific name, then any alternative common name(s) in current use. Then follow values for standard measurements of adult animals that are helpful for identifying a particular mammal group. Measurements given are the range and, where appropriate, indicative average (in parentheses) for each parameter. For rodents and some dasyurids, the number of pairs of teats in the inguinal, thoracic and pectoral regions of females are also presented.

The species accounts follow a standard format. The main identifying features of the species are described first. Where space permits, a more detailed description of external morphology is presented. The descriptions should be used in conjunction with the colour illustration and the diagram beneath the map. The descriptions are based upon our field experience, examination of museum specimens, photographs, and the literature, particularly the original descriptions of many species. Terms used to describe body parts and patterns of pelage colouration are illustrated in Figure 1. The standard measurements used in this book are illustrated in Figure 2.

Abbreviations used in the species accounts are explained on page 43.

Further information is given under three headings: Distribution, habitat and status; Behaviour; and (where appropriate) Similar species. For species whose vocalisations are a useful aid to identification, a section on voice is also included. For groups in which the arrangement and form of teeth is helpful in identification, this information is presented using the dental formula, a standard shorthand method of summarising the number and arrangement of teeth. It consists of a list of the numbers of each of the four types of teeth in one half of the upper and lower

Figure 1
Names used to describe body parts and colour patterns **a** body parts of terrestrial mammal, (continued on next page).

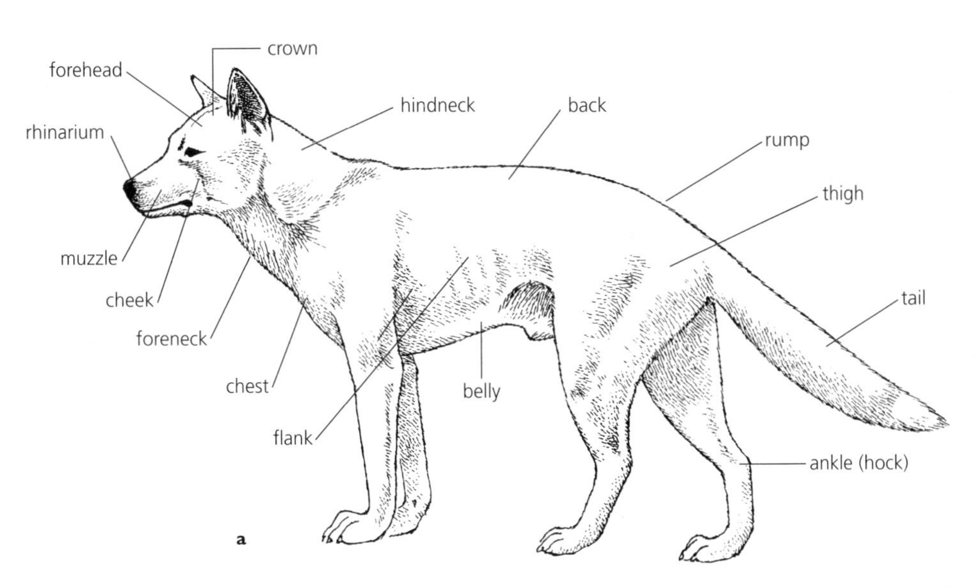

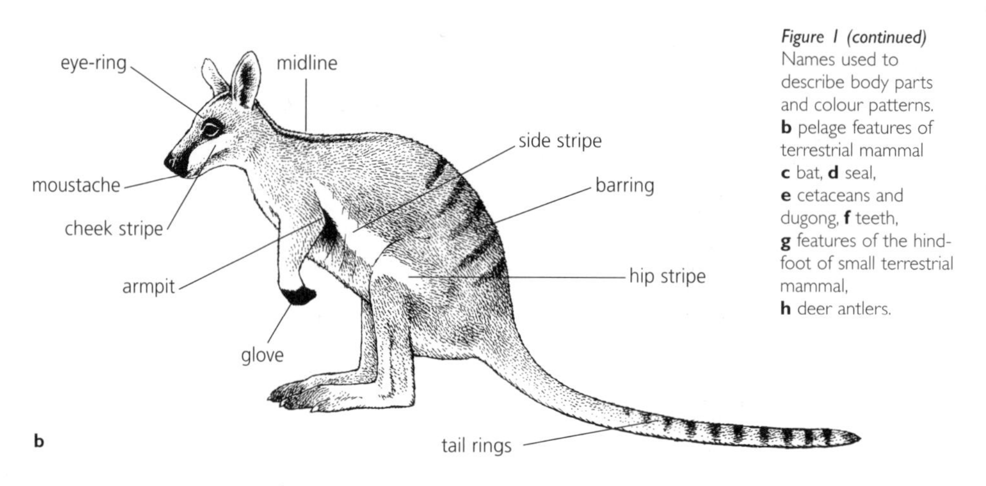

Figure 1 (continued) Names used to describe body parts and colour patterns. **b** pelage features of terrestrial mammal **c** bat, **d** seal, **e** cetaceans and dugong, **f** teeth, **g** features of the hind-foot of small terrestrial mammal, **h** deer antlers.

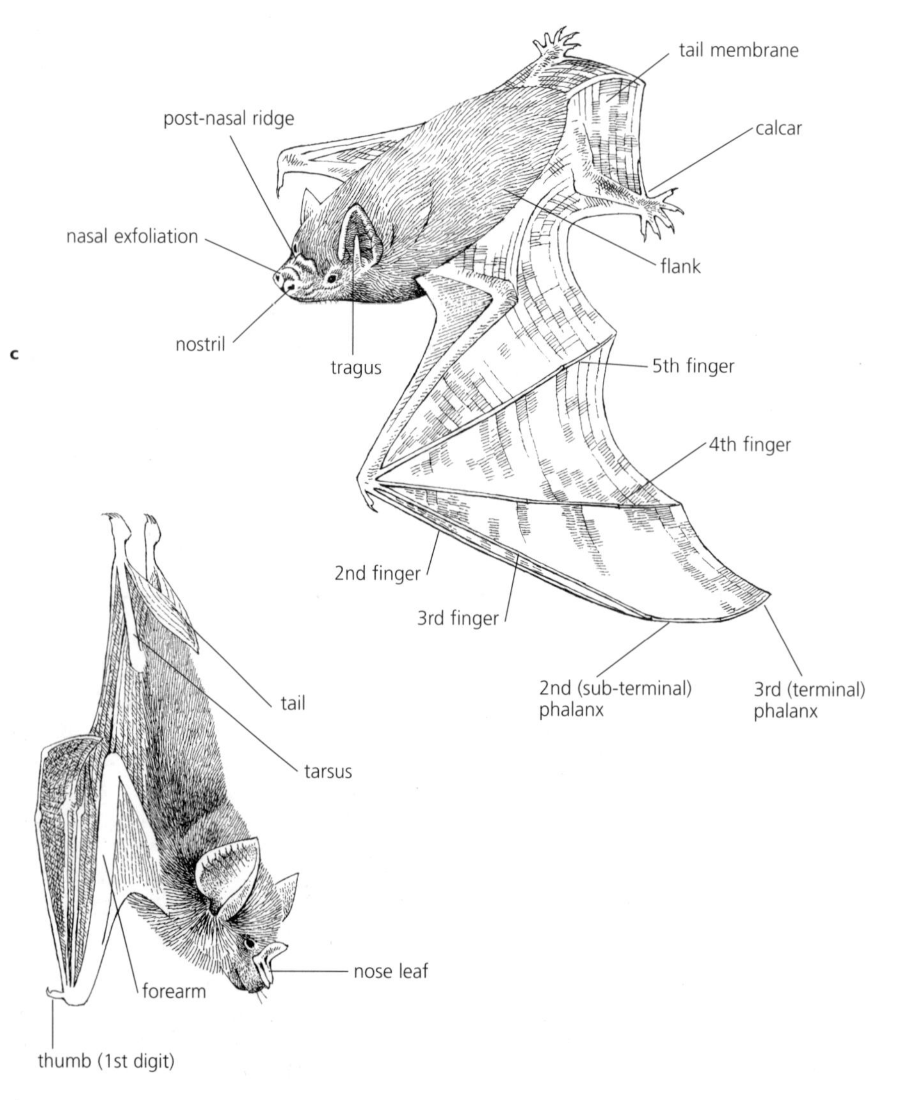

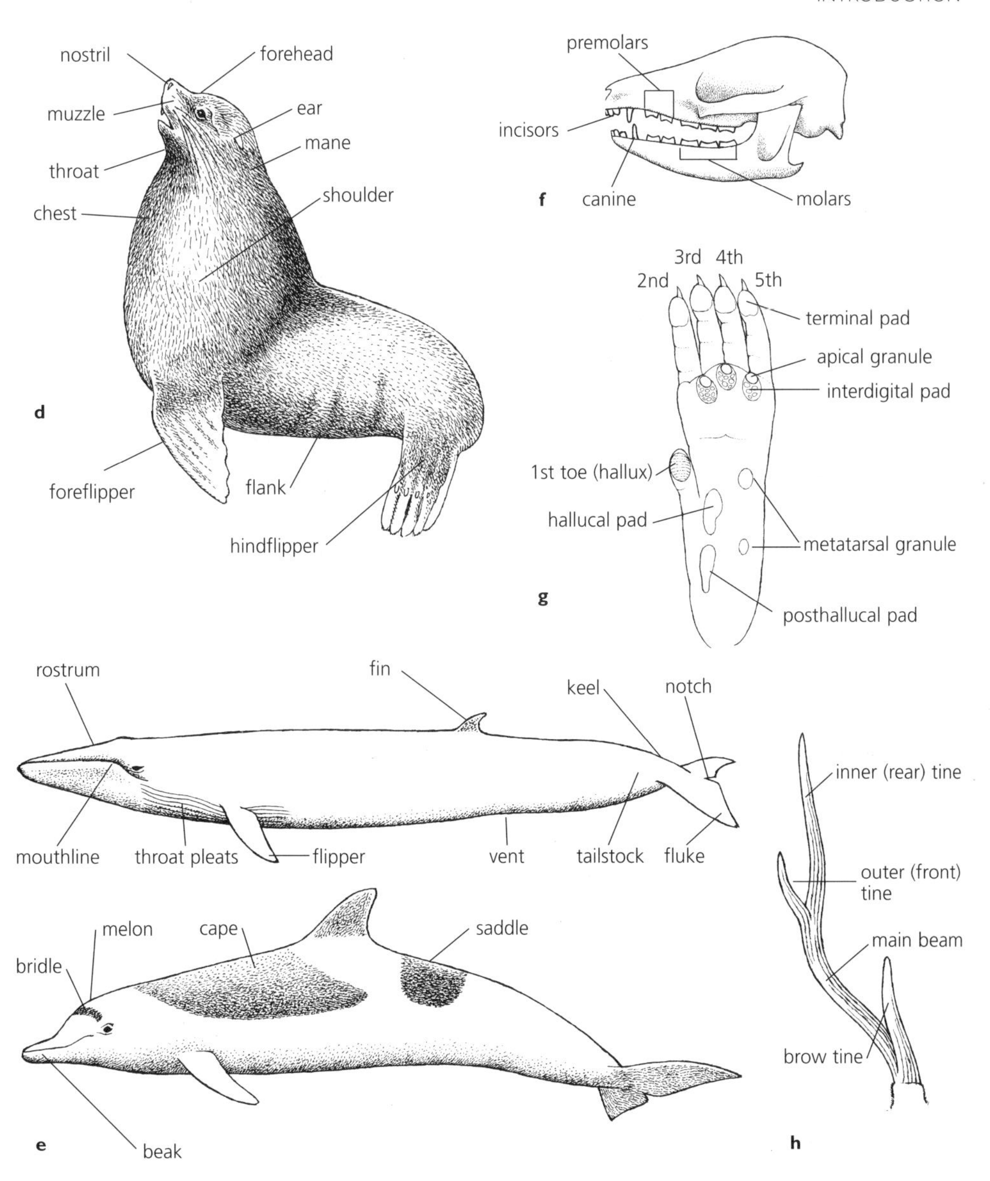

jaws, in a given sequence: incisors (I), canines (C), premolars (P) and molars (M), followed by the total number of teeth in the mouth. Thus, most macropods have the formula I 3/1 C 0/0 P 1/1 M 4/4 = 28. (I 3/1, for example, indicates three incisors in each side of the upper jaw, and one in each side of the lower jaw.) Individual teeth are referred to by the letter abbreviation and a number for position: a superscript for the upper jaw and a subscript for the lower jaw. For example, I^3 is the third upper incisor, and P_2 is the second lower premolar.

Distribution, habitat and status

This provides a concise statement of the geographic range of each species and the broad habitat types it occupies. Photographs of the major broad habitat types are shown on pages 39–42. Where possible, limits of the

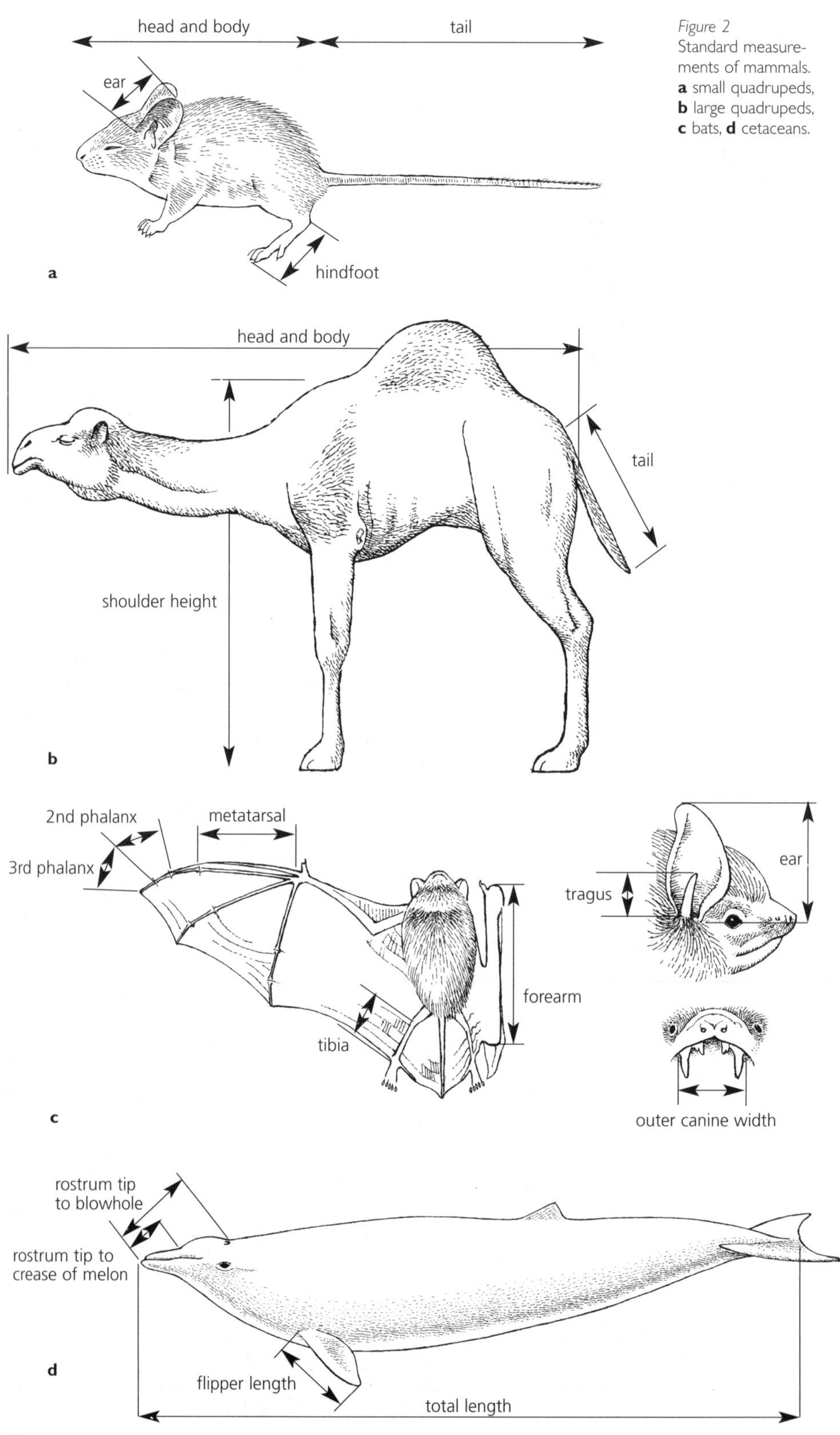

Figure 2
Standard measurements of mammals.
a small quadrupeds,
b large quadrupeds,
c bats, **d** cetaceans.

known distribution are documented to help identify new distributional information. A conservation status category is given for species classified as threatened by ANZECC (2000) or in the action plans for bats (Duncan, Baker and Montgomery 1999) and seals (Shaughnessy 1999).

Behaviour
Behavioural characteristics that may be useful in making an identification are described. Where space permits, brief descriptions of diet and breeding habits are also provided.

Voice
For species whose vocalisations are helpful in making an identification, a description of the sounds is provided, usually with an attempt at a phonetic rendering—a highly subjective process that may or may not give a good description of the sounds for other listeners.

Similar species
No field guide is complete without a list of similar species. Once you have established that your animal resembles a particular species, you will need to know what other species are similar in appearance, and possibly present at the site, before a correct identification can be assumed. Species that might be confused with each other are listed and the key differences are highlighted.

Plates

Every species is illustrated in a full-colour plate on the page opposite the species account. The plates attempt to portray an average example of each species in a typical pose. In most cases the poses depicted across members of a genus can be assumed to apply to other members of that genus. Together, they indicate some of the variety in stances adopted by species within the genus. Where there is significant geographic, age or gender variation in external appearance, or distinct colour morphs, we have attempted to depict the range of variation. In producing these plates we have drawn upon a wide range of sources of information, including our field experience, museum specimens, photographs, and published descriptions. However, as for the written descriptions, we are only too aware that for some species the information available is scant, and we would greatly appreciate feedback from readers.

Distribution maps

The distribution maps were produced by examining printouts or publications from State and Territory fauna mapping schemes where such data are readily available (Vic, NSW, Qld, SA, Tas, NT), and by examining the literature on Australian mammals. Information on the occurrence of species on islands comes mostly from Abbott and Burbidge (1995), but space does not allow the listing of all island occurrences—only major islands or island groups are mentioned. The maps are necessarily small in scale and are intended to provide only a rough indication of actual distribution. Thus, they can be valuable to provide an indication of whether a preliminary identification is plausible or unlikely. However, we emphasise that knowledge of the distribution of some species, particularly some insectivorous bats, is rudimentary and that the maps will require modification as new knowledge is gained. Where a species has undergone a marked range contraction, the former distribution is also shown. Areas in which a species occurs intermittently, or very sparsely, are also indicated.

Field techniques for identifying mammals

Finding mammals

Unlike Africa, parts of Asia and the Americas, mammals in Australia tend not to be an obvious and visible part of the fauna. This is mainly because Australia has a relatively low abundance and diversity of large ground mammals, and also because most Australian mammals are nocturnal, rarely venture from dense cover, and are often arboreal. So to see and identify many of Australia's mammals can require more effort and more complex skills than are usually needed to identify, for example, Australian birds. Nevertheless, many species can be observed by day, or at night with a spotlight, and the satisfaction gained from close observation is all the more pleasing because of the difficulty involved. As elsewhere, most of the small mammals are rarely sighted and require specific techniques to capture them before sure identification can be made.

Large mammals

Larger terrestrial mammals are often best seen from a vehicle—in many situations they will tolerate a closer approach by car than by foot. And for herbivorous mammals, roadsides often provide attractive feeding sites because of the lusher plant growth promoted by water runoff and greater light penetration. In dry conditions many mammals need to drink, so a hide set up at a bore tank or waterhole can provide exciting viewing at dusk or early morning. Areas regularly used by larger mammals always provide some signs of that use—scratches on tree trunks caused by the claws of possums or the Koala, tree hollows with the entrance rubbed smooth and clean, paths worn by the regular commuting of macropods, shallow scrapes or patches of bare ground where kangaroos lie up during the heat of the day, wallows and rubbed trees left by rutting stags, scats left at a latrine site by quolls or wombats, or topsoil churned over by feeding pigs. These, and many other signs, can be used to identify places where careful stalking and observation is likely to be rewarding.

For nocturnal species, slightly more specialised techniques are required. The simplest is stag-watching. This involves locating trees with hollows that appear suitable for use by possums and gliders, then watching the hollow entrance at dusk, preferably silhouetted against the sky, to see if any mammals emerge. These can then be illuminated with a spotlight.

Spotlighting involves walking or driving slowly through the habitat with a powerful light (50 watts is recommended). By holding the light close to one's eyes and systematically sweeping the beam through the tree tops or across the ground, one can easily detect the reflected eye-shine of mammals (and other animals, including nocturnal birds, reptiles and spiders). Most species, including possums, gliders and macropods, will remain still when caught in the spotlight beam and stare back at the light, allowing prolonged viewing through binoculars. However, some mammals, notably phascogales, usually turn quickly away from the beam and move into shadow. Many species have surprisingly bright eye-shine of a characteristic colour, visible from more than 80 metres away. When spotlighting in dense vegetation it is important to move quietly and stop frequently, with the light switched off, to listen for the sounds of animals rustling in the canopy or forest litter. Once a mammal is located, switching to a red-filtered light can allow closer observation of behaviour because many mammals cannot detect red light. The best spotlights are powered by a rechargeable battery: usually 12 volt, 10 ampere-hour sealed lead–acid batteries. These can weigh several kilograms and are best carried in a backpack.

Small mammals

In Australia opportunities to observe and identify small mammals without catching them are rare. Live trapping is the most common and informative mammal survey technique. It can be undertaken only under a permit issued by the state wildlife agency and is usually undertaken by groups such as field naturalist clubs, zoological societies, university researchers, and professional zoologists. It requires special equipment and a degree of knowledge about animal handling and animal care. Those interested in participating in such surveys should contact their state wildlife agency or local field naturalist club for information about opportunities to participate in organised and approved mammal trapping surveys.

In mammal surveys, ground-dwelling and tree-dwelling mammals are often captured in cage or box traps. These comprise a wire mesh or sheet-aluminium box, with a bait container and spring-loaded or gravity-drop door (Figure 3). The door is released by the animal tugging at the bait or treading on a floor treadle. Because the target mammal may weigh as little as 10 grams, the release mechanism must be finely tuned and carefully set. Placement of the trap is also critical to trapping success. It must be in a position where it is likely to be encountered by the target species—look for dense vegetation cover close to the ground, runways, fallen timber, rocks and scats. Traps set in the open, away from cover, are unlikely to be successful. Each trap should be securely placed on the substrate so that it does not wobble and the door closure mechanism is not obstructed. To capture arboreal species, traps can be placed in trees by tying them to large branches, or by using special brackets.

For most Australian small mammals a standard bait is used—a mixture of peanut butter, honey and rolled oats with a moist consistency. This may be rolled in a piece of gauze to hold it in the bait receptacle, or simply placed at the back of a treadle trap. Other baits which have been successfully used include bacon or sardines for Water Rats, and fruit, raisins or walnuts for possums.

Traps should be checked early in the morning, and closed during the day, especially in hot or cold weather. Wire mesh traps should have a plastic cover if rain is anticipated, and material such as shredded paper can be placed inside for insulation and nest building.

Another highly successful trap for small ground mammals is the pitfall trap combined with a drift fence. This deceptively simple technique comprises a series of pits, about 30–40 cm deep and 20 cm diameter, usually lined with a plastic or metal can, or length of PVC pipe. The pits are spaced along a low fence (20 cm high) made of nylon insect screen which passes across the centre of each pit. The base of the fence is buried to prevent animals passing easily beneath it. Any small animals which encounter the drift fence turn and run along it until they fall into a pit. This technique is particularly suited to open country such as grassland and arid areas, but can be used wherever pits can be dug. It readily captures species which are too small to trigger cage traps, or are reluctant to enter them, such as dunnarts, ningauis and pygmy possums, as well as a wonderful array of reptiles and invertebrates.

Insectivorous bats are sampled by a variety of techniques including sophisticated electronic devices which detect and characterise bat echolocation calls. In Australia, this is a rapidly developing field of bat

Figure 3
Types of traps commonly used to capture small mammals

detection, but is still in the developmental stage and in the realm of the specialist. Therefore, the technique is not included in this discussion, or in the species accounts. More commonly, bats are captured using harp traps, mist nets or trip lines. Harp traps comprise a frame holding an array of taut, vertical monofilament nylon lines, at about 20 mm spacing. Beneath the frame is a canvas catch bag. The monofilament lines are too fine to be detected by the echolocation capacity of bats. When the trap is suspended across a flight path, bats hit the lines, flutter downwards into the catch bag, then climb up the sides and collect beneath a plastic flap. Harp traps, which can have a catch area of 6 m^2 or more, can be placed low to the ground, or suspended in the canopy or over a stream.

Mist nets of the type used by bird banders can also successfully capture bats as they leave a roost site or come to water to drink. However, some bat species quickly chew holes in mist nets so this technique can become expensive. One technique which requires no expensive equipment is trip lining over water. This involves finding a body of water used by bats for drinking or foraging for insects low over the water's surface. Fire dams in forest, or pools in arid country, are often perfect. Several nylon fishing lines are stretched across the water body about 2 cm above the surface and several metres apart. Bats skimming across the surface are tripped up by the line, crash into the water, and breaststroke to shore where they can be easily collected by hand.

Many indirect methods of detecting the presence of mammals are available, such as tracks, scats and feeding sign. These are well described by Triggs (1996). Tracks of a selection of species are illustrated on the endpapers. One particularly informative method is to search for the roosting sites of owls (Barn Owl and Masked Owl are particularly useful in this regard), collect pellets of undigested material that owls regurgitate and identify skull, bone and hair fragments contained within them.

Field characters

The characters useful for identifying mammals vary between mammalian groups. For many small mammals there are no obvious morphological features or distinctive colour patterns that are useful for field identification. In these cases identification requires a specimen in the hand and use of the sorts of characters emphasised in the keys. However, many larger species can be identified on the basis of general morphology and colour patterns. The following points are important to note when observing a mammal:

- Examine the overall form of the animal, its size compared to a nearby object, body shape, mode of movement and body carriage.
- Note the relative proportion of major body parts, e.g. length of tail compared to head–body, shape of ears, length of forelimb compared to hindlimb.
- Check for obvious colour patterns, e.g. presence of dark midline along neck and back, hip stripe, facial colour pattern, patches below and behind ears, colour of ears (inside, rim and outside).
- Note the relative length of fur, especially on the tail.

For living whales and dolphins, the following points are important:

- overall length
- shape and placement of the dorsal fin along the back
- shape of flippers and of the flukes
- shape of head, e.g. presence or absence of a melon, angle at which the melon rises from the beak or mouth, presence of a beak, curvature of

the mouthline, position and number of blowholes, number of ridges along top of head
- dive sequence: are top of head and fin visible simultaneously or sequentially, are flukes raised above water when diving, and pattern and height of blow if visible
- behaviour, e.g. indulgence in breaching, flipper slapping, spy hopping, lob tailing or bow riding
- colour pattern, but note that skin colour often fades rapidly after death and is generally not a useful feature for identifying dead specimens.

For beachcast whales and dolphins, the form and number of teeth or baleen plates are important features.

How mammals are measured

The standard measurements used in this book are illustrated in Figure 2. Measurements of small mammals need to be accurate to within 1 mm, and should be taken with a pair of calipers. Total length and tail are usually measured with a steel rule, preferably with a butted end. For larger animals a tape measure is most useful and 1 cm accuracy is often sufficient.

Submitting records

Government wildlife agencies in all Australian states and territories are keen to learn of interesting mammal observations. Most state agencies maintain computer databases of mammal distribution records to which members of the public can contribute. If you believe that you have found a species that is rarely recorded, or is outside the distribution depicted in the map, the appropriate wildlife agency would be interested to learn of your observation. Written information, including date and precise locality, should be passed to the local regional office of the agency, or to the agency's wildlife research section, which is usually located in the capital city.

State museums are the most appropriate repository for scientifically interesting mammals found dead or as skeletal material. Specimens should be wrapped in plastic bags and frozen until transfer to the museum is arranged. Each specimen must be accompanied by full details of the date, precise location and circumstances of collection. Without these data the scientific value of the specimen is severely compromised. Collectors should be aware, however, that possession of such specimen material may be technically in breach of state wildlife regulations. Although there can be no guarantee, there should be no difficulties as long as reasonable attempts are being made to transfer the material to the appropriate authorities (state museum or wildlife agency).

References

Abbott, I. and Burbidge, A. 1995. The occurrence of mammal species on the islands of Australia: a summary of existing knowledge. CALMScience 1(3): 259–324.

ANZECC 2000. *Australian and New Zealand Environment and Conservation Council Threatened Fauna List, May 2000.* Environment Australia, Canberra.

Banister, J.L., Kemper, C.M. and Warneke, R.M. 1996. *The Action Plan for Australian Cetaceans.* Australian Nature Conservation Agency, Canberra.

Churchill, S. 1998. *Australian Bats.* Reed New Holland, Sydney.
Duncan, A., Baker, G.B. and Montgomery, N. (eds) 1999. *The Action Plan for Australian Bats*. Environment Australia, Canberra.
Shaughnessy, P. 1999. *The Action Plan for Australian Seals*. Environment Australia, Canberra.
Strahan, R. (ed.) 1995. *The Mammals of Australia.* Reed Books, Sydney.
Triggs, B. 1996. *Tracks, Scats and Other Traces: A Field Guide to Australian Mammals.* Oxford University Press, Melbourne.
Walton, D.W. (ed.) 1988. *Zoological Catalogue of Australia. Volume 5—Mammalia.* Bureau of Flora and Fauna, Canberra.

Checklist of Australian mammals

A total of 379 species of mammal have been recorded from Australia and its continental waters since European settlement, and are included in this field guide. The numbers of species of indigenous and introduced mammals in major taxonomic groups are shown in Table 1, followed by a checklist of these species. Within each family, genera are arranged alphabetically, as are species within each genus.

Group	*Indigenous*	*Introduced*	*Total*
Monotremes	2	–	2
Polyprotodont marsupials	89	–	89
Diprotodont marsupials	68	–	68
Marsupial moles	2	–	2
Fruit bats	12	–	12
Insectivorous bats	64	–	64
Rodents	64	5	69
Seals	10	–	10
Placental land carnivores	1	2	3
Ungulates	–	13	13
Lagomorphs	–	2	2
Cetaceans	44	–	44
Dugong	1	–	1
Totals	357	22	379

Table 1 Numbers of species of indigenous and introduced mammals in major taxonomic groups in Australia and its waters.

Class Mammalia
Subclass Prototheria (monotremes)

Order Monotremata (platypus and echidnas)
Family Ornithorhynchidae (Platypus)
Ornithorhynchus anatinus (Shaw 1799) — Platypus
Family Tachyglossidae (echidnas)
Tachyglossus aculeatus (Shaw 1792) — Short-beaked Echidna

Subclass Marsupialia (marsupials)

Order Dasyuromorphia (carnivorous marsupials)
Family Thylacinidae (Thylacine)
Thylacinus cyanocephalus (Harris 1808) — Thylacine
Family Dasyuridae (dasyurids)
Antechinomys laniger (Gould 1856) — Kultarr

Antechinus adustus (Thomas 1923)	Rusty Antechinus
Antechinus agilis Dickman, Parnaby, Crowther & King 1998	Agile Antechinus
Antechinus bellus (Thomas 1904)	Fawn Antechinus
Antechinus flavipes (Waterhouse 1838)	Yellow-footed Antechinus
Antechinus godmani (Thomas 1923)	Atherton Antechinus
Antechinus leo Van Dyck 1980	Cinnamon Antechinus
Antechinus minimus (Finlayson 1958)	Swamp Antechinus
Antechinus stuartii (Macleay 1841)	Brown Antechinus
Antechinus subtropicus Van Dyck & Crowther 2000	Subtropical Antechinus
Antechinus swainsonii (Waterhouse 1840)	Dusky Antechinus
Dasycercus cristicaudata (Krefft 1867)	Mulgara
Dasycercus hillieri (Thomas 1905)	Ampurta
Dasykaluta rosamondae (Ride 1964)	Kaluta
Dasyuroides byrnei (Spencer 1896)	Kowari
Dasyurus geoffroii Gould 1841	Western Quoll
Dasyurus hallucatus Gould 1842	Northern Quoll
Dasyurus maculatus (Kerr 1792)	Spot-tailed Quoll
Dasyurus viverrinus (Shaw 1800)	Eastern Quoll
Ningaui ridei Archer 1975	Wongai Ningaui
Ningaui timealyi Archer 1975	Pilbara Ningaui
Ningaui yvonneae Kitchener, Stoddart & Henry 1983	Mallee Ningaui
Parantechinus apicalis (Gray 1842)	Dibbler
Phascogale calura Gould 1884	Red-tailed Phascogale
Phascogale tapoatafa Meyer 1793	Brush-tailed Phascogale
Planigale gilesi Aitken 1972	Giles' Planigale
Planigale ingrami (Thomas 1906)	Long-tailed Planigale
Planigale maculata (Gould 1851)	Common Planigale
Planigale tenuirostris Troughton 1928	Narrow-nosed Planigale
Pseudantechinus bilarni (Johnson 1954)	Sandstone False Antechinus
Pseudantechinus macdonnellensis (Spencer 1895)	Fat-tailed False Antechinus
Pseudantechinus mimulus (Thomas 1906)	Carpentarian False Antechinus
Pseudantechinus ningbing Kitchener 1988	Ningbing False Antechinus
Pseudantechinus roryi Cooper, Aplin & Adams 2000	Tan False Antechinus
Pseudantechinus woolleyae Kitchener & Caputi 1988	Woolley's False Antechinus
Sarcophilus harrisi (Boitard 1841)	Tasmanian Devil
Sminthopsis aitkeni Kitchener, Stoddart & Henry 1984	Kangaroo Island Dunnart
Sminthopsis archeri Van Dyck 1986	Chestnut Dunnart
Sminthopsis bindi Van Dyck, Woinarski & Press 1994	Kakadu Dunnart
Sminthopsis butleri Archer 1979	Butler's Dunnart
Sminthopsis crassicaudata (Gould 1844)	Fat-tailed Dunnart
Sminthopsis dolichura Kitchener, Stoddart & Henry 1984	Little Long-tailed Dunnart
Sminthopsis douglasi Archer 1979	Julia Creek Dunnart
Sminthopsis gilberti Kitchener, Stoddart & Henry 1984	Gilbert's Dunnart
Sminthopsis granulipes Troughton 1932	White-tailed Dunnart
Sminthopsis griseoventor Kitchener, Stoddart & Henry 1984	Grey-bellied Dunnart
Sminthopsis hirtipes Thomas 1898	Hairy-footed Dunnart
Sminthopsis leucopus (Gray 1842)	White-footed Dunnart
Sminthopsis longicaudata Spencer 1909	Long-tailed Dunnart
Sminthopsis macroura (Gould 1845)	Stripe-faced Dunnart
Sminthopsis murina (Waterhouse 1838)	Common Dunnart
Sminthopsis ooldea Troughton 1965	Ooldea Dunnart
Sminthopsis psammophila Spencer 1895	Sandhill Dunnart
Sminthopsis virginae (Tarragon 1847)	Red-cheeked Dunnart
Sminthopsis youngsoni McKenzie & Archer 1982	Lesser Hairy-footed Dunnart

Family Myrmecobiidae (Numbat)

Myrmecobius fasciatus Waterhouse 1836 — Numbat

Order Peramelemorphia (bandicoots and bilbies)

Family Peroryctidae (rainforest bandicoots)

Echymipera rufescens australis Tate 1948 — Rufous Spiny Bandicoot

Family Peramelidae (bandicoots)

Chaeropus ecaudatus (Ogilby 1838) — Pig-footed Bandicoot
Isoodon auratus (Ramsay 1887) — Golden Bandicoot
Isoodon macrourus (Gould 1842) — Northern Brown Bandicoot
Isoodon obesulus (Shaw 1797) — Southern Brown Bandicoot
Macrotis lagotis (Reid 1837) — Bilby
Macrotis leucura (Thomas 1887) — Lesser Bilby
Perameles bougainville Quoy & Gaimard 1824 — Western Barred Bandicoot
Perameles eremiana Spencer 1897 — Desert Bandicoot
Perameles gunnii Gray 1838 — Eastern Barred Bandicoot
Perameles nasuta Geoffroy 1804 — Long-nosed Bandicoot

Order Notoryctemorphia (marsupial moles)

Family Notoryctidae (marsupial moles)

Notoryctes caurinus Thomas 1920 — Northern Marsupial Mole
Notoryctes typhlops (Stirling 1889) — Southern Marsupial Mole

Order Diprotodontia (koala, wombats, possums, macropods)

Family Vombatidae (wombats)

Lasiorhinus krefftii (Owen 1872) — Northern Hairy-nosed Wombat
Lasiorhinus latifrons (Owen 1845) — Southern Hairy-nosed Wombat
Vombatus ursinus (Shaw 1800) — Common Wombat

Family Phascolarctidae (Koala)

Phascolarctos cinereus (Goldfuss 1817) — Koala

Family Phalangeridae (brushtail possums, cuscuses, Scaly-tailed Possum)

Phalanger intercastellanus Thomas 1895 — Southern Common Cuscus
Spilocuscus maculatus (Gould 1850) — Common Spotted Cuscus
Trichosurus caninus (Ogilby 1836) — Mountain Brushtail Possum
Trichosurus vulpecula (Kerr 1792) — Common Brushtail Possum
Wyulda squamicaudata Alexander 1919 — Scaly-tailed Possum

Family Burramyidae (pygmy possums)

Burramys parvus Broom 1896 — Mountain Pygmy Possum
Cercartetus caudatus (Mjöberg 1916) — Long-tailed Pygmy Possum
Cercartetus concinnus (Gould 1845) — Western Pygmy Possum
Cercartetus lepidus (Thomas 1888) — Little Pygmy Possum
Cercartetus nanus (Desmarest 1818) — Eastern Pygmy Possum

Family Tarsipedidae (Honey Possum)

Tarsipes rostratus Gervais & Verreaux 1842 — Honey Possum

Family Petauridae (Striped Possum, Leadbeater's Possum, wrist-winged gliders)

Dactylopsila trivirgata Thomas 1908 — Striped Possum
Gymnobelideus leadbeateri McCoy 1867 — Leadbeater's Possum
Petaurus australis Shaw 1791 — Yellow-bellied Glider
Petaurus breviceps Waterhouse 1839 — Sugar Glider
Petaurus gracilis (De Vis 1883) — Mahogany Glider
Petaurus norfolcensis (Kerr 1792) — Squirrel Glider

Family Pseudocheiridae (ringtail possums, Greater Glider)

Hemibelideus lemuroides (Collett 1884) — Lemuroid Ringtail Possum
Petauroides volans (Kerr 1792) — Greater Glider
Petropseudes dahli (Collett 1895) — Rock Ringtail Possum
Pseudocheirus occidentalis (Thomas 1888) — Western Ringtail Possum

Pseudocheirus peregrinus (Boddaert 1785) — Common Ringtail Possum
Pseudochirops archeri (Collett 1884) — Green Ringtail Possum
Pseudochirulus cinereus (Tate 1945) — Daintree River Ringtail Possum
Pseudochirulus herbetensis (Collett 1884) — Herbert River Ringtail Possum

Family Acrobatidae (Feathertail Glider)

Acrobates pygmaeus (Shaw 1794) — Feathertail Glider

Family Hypsiprymnodontidae (Musky Rat Kangaroo)

Hypsiprymnodon moschatus Ramsay 1876 — Musky Rat-kangaroo

Family Potoroidae (potoroos and bettongs)

Aepyprymnus rufescens (Gray 1837) — Rufous Bettong
Bettongia gaimardi (Desmarest 1822) — Southern Bettong
Bettongia lesueur (Quoy & Gaimard 1824) — Burrowing Bettong
Bettongia penicillata Gray 1837 — Woylie
Bettongia tropica Wakefield 1967 — Northern Bettong
Caloprymnus campestris (Gould 1843) — Desert Rat-kangaroo
Potorous gilbertii (Gould 1841) — Gilbert's Potoroo
Potorous longipes Seebeck & Johnston 1980 — Long-footed Potoroo
Potorous platyops (Gould 1844) — Broad-faced Potoroo
Potorous tridactylus (Kerr 1792) — Long-nosed Potoroo

Family Macropodidae (kangaroos, wallabies, tree kangaroos)

Dendrolagus bennettianus De Vis 1887 — Bennett's Tree Kangaroo
Dendrolagus lumholtzi Collett 1884 — Lumholtz's Tree Kangaroo
Lagorchestes asomatus Finlayson 1943 — Central Hare Wallaby
Lagorchestes conspicillatus Gould 1842 — Spectacled Hare Wallaby
Lagorchestes hirsutus Gould 1844 — Mala
Lagorchestes leporides (Gould 1841) — Eastern Hare Wallaby
Lagostrophus fasciatus (Peron & Lesueur 1807) — Banded Hare Wallaby
Macropus agilis (Gould 1842) — Agile Wallaby
Macropus antilopinus (Gould 1842) — Antilopine Wallaroo
Macropus bernardus Rothschild 1904 — Black Wallaroo
Macropus dorsalis (Gray 1837) — Black-striped Wallaby
Macropus eugenii (Desmarest 1817) — Tammar Wallaby
Macropus fuliginosus (Desmarest 1817) — Western Grey Kangaroo
Macropus giganteus Shaw 1790 — Eastern Grey Kangaroo
Macropus greyi Waterhouse 1845 — Toolache Wallaby
Macropus irma (Jourdan 1837) — Western Brush Wallaby
Macropus parma Waterhouse 1845 — Parma Wallaby
Macropus parryi Bennett 1835 — Whiptail Wallaby
Macropus robustus Gould 1841 — Euro, Wallaroo
Macropus rufogriseus (Desmarest 1817) — Red-necked Wallaby
Macropus rufus (Desmarest 1822) — Red Kangaroo
Onychogalea fraenata (Gould 1841) — Bridled Nailtail Wallaby
Onychogalea lunata (Gould 1841) — Crescent Nailtail Wallaby
Onychogalea unguifera (Gould 1841) — Northern Nailtail Wallaby
Petrogale assimilis Ramsay 1877 — Allied Rock Wallaby
Petrogale brachyotis (Gould 1841) — Short-eared Rock Wallaby
Petrogale burbidgei Kitchener & Sanson 1978 — Monjon
Petrogale coenensis Eldridge & Close 1992 — Cape York Rock Wallaby
Petrogale concinna Gould 1842 — Narbarlek
Petrogale godmani Thomas 1923 — Godman's Rock Wallaby
Petrogale herberti Thomas 1926 — Herbert's Rock Wallaby
Petrogale inornata Gould 1842 — Unadorned Rock Wallaby
Petrogale lateralis Gould 1842 — Black-flanked Rock Wallaby
Petrogale purpureicollis Le Souef 1924 — Purple-necked Rock Wallaby
Petrogale mareeba Eldridge & Close 1992 — Mareeba Rock Wallaby
Petrogale penicillata (Gray 1825) — Brush-tailed Rock Wallaby

Petrogale persephone Maynes 1982	Proserpine Rock Wallaby
Petrogale rothschildi Thomas 1904	Rothschild's Rock Wallaby
Petrogale sharmani Eldridge & Close 1992	Sharman's Rock Wallaby
Petrogale xanthopus Gray 1855	Yellow-footed Rock Wallaby
Setonix brachyurus (Quoy & Gaimard 1830)	Quokka
Thylogale billardierii (Desmarest 1822)	Rufous-bellied Pademelon
Thylogale stigmatica (Gould 1860)	Red-legged Pademelon
Thylogale thetis (Lesson 1827)	Red-necked Pademelon
Wallabia bicolor (Desmarest 1804)	Black Wallaby

Subclass Eutheria (eutherian or placental mammals)

Order Chiroptera (bats)

Family Pteropodidae (fruit bats)	
Dobsonia moluccensis (Quoy & Gaimard 1830)	Bare-backed Fruit Bat
Macroglossus minimus (Geoffroy 1810)	Least Blossom Bat
Nyctimene cephalotes Pallas 1767	Torresian Tube-nosed Bat
Nyctimene robinsoni Thomas 1904	Eastern Tube-nosed Bat
Pteropus alecto Temminck 1837	Black Flying-fox
Pteropus banakrisi Richards & Hall 2001	Torresian Flying-fox
Pteropus brunneus Dobson 1878	Dusky Flying-fox
Pteropus conspicillatus Gould 1850	Spectacled Flying-fox
Pteropus macrotis Peters 1867	Large-eared Flying-fox
Pteropus poliocephalus Temminck 1835	Grey-headed Flying-fox
Pteropus scapulatus Peters 1862	Little Red Flying-fox
Syconycteris australis (Peters 1867)	Eastern Blossom Bat
Family Megadermatidae (Ghost Bat)	
Macroderma gigas (Dobson 1880)	Ghost Bat
Family Rhinolophidae (horseshoe bats)	
Rhinolophus megaphyllus Gray 1834	Eastern Horseshoe Bat
Rhinolophus philippinensis Waterhouse 1843	Large-eared Horseshoe Bat
Family Hipposideridae (leaf-nosed bats)	
Hipposideros ater Templeton 1848	Dusky Leaf-nosed Bat
Hipposideros cervinus (Gould 1854)	Fawn Leaf-nosed Bat
Hipposideros diadema Troughton 1937	Diadem Leaf-nosed Bat
Hipposideros semoni Matschie 1903	Semon's Leaf-nosed Bat
Hipposideros stenotis Thomas 1913	Northern Leaf-nosed Bat
Rhinonicteris aurantius (Gray 1845)	Orange Leaf-nosed Bat
Family Emballonuridae (sheathtail bats)	
Saccolaimus flaviventris (Peters 1867)	Yellow-bellied Sheathtail Bat
Saccolaimus mixtus Troughton 1925	Papuan Sheathtail Bat
Saccolaimus saccolaimus (Temminck 1838)	Bare-rumped Sheathtail Bat
Taphozous australis Gould 1854	Coastal Sheathtail Bat
Taphozous georgianus Thomas 1915	Common Sheathtail Bat
Taphozous hilli Kitchener 1980	Hill's Sheathtail Bat
Taphozous kapalgensis McKean & Friend 1979	Arnhem Sheathtail Bat
Taphozous troughtoni Tate 1952	Troughton's Sheathtail Bat
Family Molossidae (freetail bats)	
Chaerephon jobensis (Miller 1902)	Northern Freetail Bat
Mormopterus beccarii Peters 1881	Beccari's Freetail Bat
Mormopterus norfolkensis (Gray 1839)	East Coast Freetail Bat
Mormopterus sp.	Little Northern Freetail Bat
Mormopterus sp.	Mangrove Freetail Bat
Mormopterus sp.	Eastern Freetail Bat
Mormopterus sp.	Inland Freetail Bat
Mormopterus sp.	Southern Freetail Bat

Mormopterus sp.	Western Freetail Bat
Mormopterus sp.	Hairy-nosed Freetail Bat
Tadarida australis (Gray 1838)	White-striped Freetail Bat
Family Vespertilionidae (ordinary bats)	
Chalinolobus dwyeri Ryan 1966	Large-eared Pied Bat
Chalinolobus gouldii (Gray 1841)	Gould's Wattled Bat
Chalinolobus morio (Gray 1841)	Chocolate Wattled Bat
Chalinolobus nigrogriseus (Gould 1856)	Hoary Wattled Bat
Chalinolobus picatus (Gould 1852)	Little Pied Bat
Falsistrellus mackenziei Kitchener, Caputi & Jones 1986	Western False Pipistrelle
Falsistrellus tasmaniensis (Gould 1858)	Eastern False Pipistrelle
Kerivoula papuensis Dobson 1878	Golden-tipped Bat
Miniopterus australis (Tomes 1858)	Little Bent-wing Bat
Miniopterus schreibersii (Kuhl 1817)	Common Bent-wing Bat
Murina florium Thomas 1908	Flute-nosed Bat
Myotis adversus (Morsfield 1824)	Large-footed Myotis
Nyctophilus arnhemensis Johnson 1959	Arnhem Long-eared Bat
Nyctophilus bifax Thomas 1915	Northern Long-eared Bat
Nyctophilus geoffroyi Leach 1821	Lesser Long-eared Bat
Nyctophilus gouldi Tomes 1858	Gould's Long-eared Bat
Nyctophilus howensis McKean 1973	Lord Howe Long-eared Bat
Nyctophilus timoriensis (Geoffroy 1806)	Greater Long-eared Bat
Nyctophilus walkeri Thomas 1892	Pygmy Long-eared Bat
Pipistrellus adamsi Kitchener, Caputi & Jones 1986	Cape York Pipistrelle
Pipistrellus westralis Koopman 1984	Mangrove Pipistrelle
Scoteanax rueppellii (Peters 1866)	Greater Broad-nosed Bat
Scotorepens balstoni Thomas 1906)	Inland Broad-nosed Bat
Scotorepens greyii (Gray 1843)	Little Broad-nosed Bat
Scotorepens orion (Troughton 1937)	Eastern Broad-nosed Bat
Scotorepens sanborni (Troughton 1937)	Northern Broad-nosed Bat
Scotorepens sp.	n-e. NSW species
Vespadelus baverstocki Kitchener, Jones & Caputi 1987	Inland Forest Bat
Vespadelus caurinus (Thomas 1914)	Northern Cave Bat
Vespadelus darlingtoni (Allan 1933)	Large Forest Bat
Vespadelus douglasorum (Kitchener 1976)	Kimberley Cave Bat
Vespadelus finlaysoni Kitchener, Jones & Caputi 1987	Inland Cave Bat
Vespadelus pumilus (Gray 1841)	Eastern Forest Bat
Vespadelus regulus (Thomas 1906)	Southern Forest Bat
Vespadelus troughtoni Kitchener, Jones & Caputi 1987	Eastern Cave Bat
Vespadelus vulturnus (Thomas 1914)	Little Forest Bat

Order Rodentia (rodents)

Family Muridae (rats and mice)	
Conilurus albipes (Lichenstein 1829)	White-footed Rabbit Rat
Conilurus penicillatus (Gould 1842)	Brush-tailed Rabbit Rat
Hydromys chrysogaster Geoffroy 1804	Water Rat
Leggadina forresti (Thomas 1906)	Desert Short-tailed Mouse
Leggadina lakedownensis Watts 1976	Tropical Short-tailed Mouse
Leporillus apicalis (Gould 1853)	Lesser Stick-nest Rat
Leporillus conditor (Stuart 1848)	Greater Stick-nest Rat
Mastacomys fuscus Thomas 1882	Broad-toothed Rat
Melomys burtoni (Ramsay 1887)	Grassland Melomys
Melomys capensis Tate 1951	Cape York Melomys
Melomys cervinipes (Gould 1852)	Fawn-footed Melomys
Melomys rubicola Thomas 1924	Bramble Cay Melomys
Mesembriomys gouldi (Gray 1843)	Black-footed Tree Rat

Mesembriomys macrurus (Peters 1876)	Golden-backed Tree Rat
Mus musculus Linnaeus 1758	House Mouse
Notomys alexis Thomas 1922	Spinifex Hopping Mouse
Notomys amplus Brazenor 1936	Short-tailed Hopping Mouse
Notomys aquilo Thomas 1921	Northern Hopping Mouse
Notomys cervinus (Gould 1853)	Fawn Hopping Mouse
Notomys fuscus (Wood Jones 1925)	Dusky Hopping Mouse
Notomys longicaudatus (Gould 1844)	Long-tailed Hopping Mouse
Notomys macrotis Thomas 1921	Big-eared Hopping Mouse
Notomys mitchelli (Ogilby 1838)	Mitchell's Hopping Mouse
Notomys mordax Thomas 1922	Darling Downs Hopping Mouse
Notomys sp.	Broad-cheeked Hopping Mouse
Pogonomys mollipilosus Peters & Doria 1881	Prehensile-tailed Rat
Pseudomys albocinereus (Gould 1845)	Ash-grey Mouse
Pseudomys apodemoides Finlayson 1932	Silky Mouse
Pseudomys australis Gray 1832	Plains Mouse
Pseudomys bolami Troughton 1932	Bolam's Mouse
Pseudomys calabyi Kitchener & Humphries 1987	Calaby's Pebble-mound Mouse
Pseudomys chapmani Kitchener 1980	Pilbara Pebble-mound Mouse
Pseudomys delicatulus (Gould 1842)	Delicate Mouse
Pseudomys desertor Troughton 1932	Desert Mouse
Pseudomys fieldi (Waite 1896)	Djoongari
Pseudomys fumeus Brazenor 1934	Smoky Mouse
Pseudomys glaucus Thomas 1910	Blue-grey Mouse
Pseudomys gouldi (Waterhouse 1839)	Gould's Mouse
Pseudomys gracilicaudatus (Gould 1845)	Eastern Chestnut Mouse
Pseudomys hermannsburgensis (Waite 1896)	Sandy Inland Mouse
Pseudomys higginsi (Trouessart 1897)	Long-tailed Mouse
Pseudomys johnsoni Kitchener 1985	Central Pebble-mound Mouse
Pseudomys laborifex Kitchener & Humphries 1986	Kimberley Mouse
Pseudomys nanus (Gould 1858)	Western Chestnut Mouse
Pseudomys novaehollandiae (Waterhouse 1843)	New Holland Mouse
Pseudomys occidentalis Tate 1951	Western Mouse
Pseudomys oralis Thomas 1921	Hastings River Mouse
Pseudomys pilligaensis Fox & Briscoe 1980	Pilliga Mouse
Pseudomys patrius (Thomas & Dollman 1909)	Eastern Pebble-mound Mouse
Pseudomys shortridgei (Thomas 1907)	Heath Mouse
Rattus colletti (Thomas 1904)	Dusky Rat
Rattus exulans (Peale 1848)	Pacific Rat
Rattus fuscipes (Waterhouse 1839)	Bush Rat
Rattus leucopus (Gray 1867)	Cape York Rat
Rattus lutreolus (Gray 1841)	Swamp Rat
Rattus norvegicus (Berkenhout 1796)	Brown Rat
Rattus rattus (Linnaeus 1758)	Black Rat
Rattus sordidus (Gould 1858)	Canefield Rat
Rattus tunneyi (Thomas 1904)	Pale Field Rat
Rattus villosissimus (Waite 1898)	Long-haired Rat
Uromys caudimaculatus (Krefft 1867)	Giant White-tailed Rat
Uromys hadrourus (Winter 1984)	Masked White-tailed Rat
Xeromys myoides Thomas 1889	Water Mouse
Zyzomys argurus (Thomas 1889)	Common Rock Rat
Zyzomys maini Kitchener 1989	Arnhem Land Rock Rat
Zyzomys palatalis Kitchener 1989	Carpentarian Rock Rat
Zyzomys pedunculatus (Waite 1896)	Central Rock Rat
Zyzomys woodwardi (Thomas 1909)	Kimberley Rock Rat

Family Sciuridae (squirrels)	
Funambulus pennanti Wroughton 1905	Five-lined Palm Squirrel
Order Sirenia (dugong and manatees)	
Family Dugongidae (dugong)	
Dugong dugon (Müller 1796)	Dugong
Order Carnivora (carnivorous eutherian mammals)	
Family Canidae (dogs and foxes)	
Canis lupus dingo Linnaeus 1758	Dingo
Vulpes vulpes Linnaeus 1758	Red Fox
Family Felidae (cats)	
Felis catus Linnaeus 1758	House Cat
Family Otariidae (eared seals)	
Arctocephalus forsteri (Lesson 1828)	New Zealand Fur Seal
Arctocephalus gazella (Peters 1875)	Antarctic Fur Seal
Arctocephalus pusillus doriferus (Scheber 1775)	Australian Fur Seal
Arctocephalus tropicalis (Gray 1872)	Subantarctic Fur Seal
Neophoca cinerea (Peron 1816)	Australian Sea-lion
Family Phocidae ('true' seals)	
Hydrurga leptonyx (de Blainville 1820)	Leopard Seal
Leptonychotes weddellii (Lesson 1826)	Weddell Seal
Lobodon carcinophagus (Hombron & Jacquinot 1842)	Crab-eater Seal
Mirounga leonina Gray 1827	Southern Elephant Seal
Ommatophoca rossii Gray 1844	Ross Seal
Order Perissodactyla (odd-toed ungulates)	
Family Equidae (horses)	
Equus asinus Linnaeus 1758	Donkey
Equus caballus Linnaeus 1758	Horse
Order Artiodactyla (even-toed ungulates)	
Family Bovidae (horned ruminants)	
Bubalus bubalus Linnaeus 1758	Swamp Buffalo
Capra hircus Linnaeus 1758	Goat
Bos javanicus D'Alton 1823	Bali Banteng
Family Camelidae (camels)	
Camelus dromedarius Linnaeus 1758	One-humped Camel
Family Cervidae (deer)	
Cervus axis (Erxleben 1777)	Chital
Cervus dama Linnaeus 1758	Fallow Deer
Cervus elaphus Linnaeus 1758	Red Deer
Cervus porcinus (Zimmerman 1780)	Hog Deer
Cervus timorensis de Blainville 1822	Rusa
Cervus unicolor Kerr 1792	Sambar
Family Suidae (pigs)	
Sus scrofa Linnaeus 1758	Pig
Order Lagomorpha (rabbits and hares)	
Family Leporidae (rabbits and hares)	
Lepus capensis Linnaeus 1758	Brown Hare
Oryctolagus cuniculus Linnaeus 1758	European Rabbit
Order Cetacea (whales, dolphins, porpoises)	
Family Delphinidae (dolphins, pilot whales and Killer Whale)	
Delphinus delphis Linnaeus 1758	Common Dolphin
Feresa attenuata Gray 1875	Pygmy Killer Whale

Globicephala macrorhynchus Gray 1846	Short-flippered Pilot Whale
Globicephala melas (Traill 1809)	Long-flippered Pilot Whale
Grampus griseus (Cuvier 1812)	Risso's Dolphin
Lagenodelphis hosei Fraser 1956	Fraser's Dolphin
Lagenorhynchus obscurus Gray 1828)	Dusky Dolphin
Lissodelphis peroni Lacérpède 1804	Southern Rightwhale Dolphin
Orcaella brevirostris Gray 1866	Irrawaddy Dolphin
Orcinus orca Linnaeus 1758	Killer Whale
Peponocephala electra Nishiwaki & Norris 1966	Melon-headed Whale
Pseudorca crassidens Owen 1846	False Killer Whale
Stenella attenuata Gray 1846	Pantropical Spotted Dolphin
Stenella coeruleoalba (Meyen 1833)	Striped Dolphin
Stenella longirostris (Gray 1828)	Spinner Dolphin
Steno bredanensis (Lesson 1823)	Rough-toothed Dolphin
Sousa chinensis (Osbeck 1757)	Indo-Pacific Humpback Dolphin
Tursiops aduncus (Ehrenberg 1832)	Long-beaked Bottlenose Dolphin
Tursiops truncatus Montagu 1821	Bottlenose Dolphin
Family Phocoenidae (porpoises)	
Australophocoena dioptrica (Lahille 1912)	Spectacled Porpoise
Family Ziphiidae (beaked whales)	
Berardius arnuxii Duvernoy 1851	Arnoux's Beaked Whale
Hyperoodon planifrons Flower 1882	Southern Bottlenose Whale
Indopacetus pacificus (Longman 1926)	Tropical Bottlenose Whale
Mesoplodon bowdoini Andrews 1908	Andrews' Beaked Whale
Mesoplodon densirostris (de Blainville 1817)	Blainville's Beaked Whale
Mesoplodon ginkgodens Nishiwaki & Kamiga 1958	Gingko-toothed Beaked Whale
Mesoplodon grayi (Van Haast 1876)	Gray's Beaked Whale
Mesoplodon hectori (Gray 1871)	Hector's Beaked Whale
Mesoplodon layardii (Gray 1865)	Strap-toothed Beaked Whale
Mesoplodon mirus True 1913	True's Beaked Whale
Tasmacetus shepherdi Oliver 1937	Shepherd's Beaked Whale
Ziphius cavirostris Cuvier 1823	Cuvier's Beaked Whale
Family Physeteridae (Sperm Whale)	
Physeter macrocephalus Linnaeus 1758	Sperm Whale
Family Kogiidae (pygmy sperm whales)	
Kogia breviceps (de Blainville 1838)	Pygmy Sperm Whale
Kogia simus (Owen 1866)	Dwarf Sperm Whale
Family Balaenidae (right whales)	
Eubalaena australis (Desmoulins 1822)	Southern Right Whale
Family Neobalaenidae (Pygmy Right Whale)	
Caperea marginata (Gray 1846)	Pygmy Right Whale
Family Balaenopteridae (rorquals)	
Balaenoptera acutorostrata Lacérpède 1804	Dwarf Minke Whale
Balaenoptera bonaerensis Burmeister 1867	Antarctic Minke Whale
Balaenoptera borealis Lesson 1828	Sei Whale
Balaenoptera edeni Anderson 1878	Bryde's Whale
Balaenoptera musculus Linnaeus 1758	Blue Whale
Balaenoptera physalus Linnaeus 1758	Fin Whale
Megaptera novaeangliae (Borowski 1781)	Humpback Whale

Identification keys to Australian mammals

Notes on using the keys

Keys to orders and genera of small mammals

These keys allow the identification to genus level of handheld specimens of those small Australian mammals that may prove difficult to identify using only the species accounts and illustrations. An identification key is a step-by-step process of elimination based on choices between two or more characteristics (or couplets). When the correct choices or decisions are made at each step a correct decision on the identity of the specimen will be quickly achieved. To progress through the key, decide which of the descriptions at each couplet best fits the specimen in question, and proceed to the couplet number indicated. Continue this process until a conclusion (genus) is reached. Once a decision on genus is reached, consult the species accounts, distribution maps and illustrations to determine the species. If none of the species descriptions match the specimen, return to the couplets in the key where difficulty was experienced in making a choice, and try the other alternatives.

Before the genus can be determined it is necessary to know the order to which the specimen in question belongs. Thus, the first key (Key A) distinguishes between orders of small terrestrial Australian mammals. It is assumed that readers can readily recognise members of the order Chiroptera (bats) and should proceed directly to the key to families of bats (Key F).

Key to species of marine mammals

The key to marine mammals is designed to allow the identification of stranded animals that can be closely examined. It includes all seals, whales, dolphins and porpoises recorded from the seas around Australia.

Index to keys

A Orders of small terrestrial and arboreal mammals page 24
B Genera of rodents page 24
C Genera of small polyprotodont marsupials (<2 kg) page 26
D Genera of small possums and gliders (<0.5 kg) page 27
E Families of Macropoidea and genera of rat kangaroos page 27
F Families of bats page 28
- F1 Genera of fruit bats page 29
- F2 Genera of vespertilionid bats page 29
- F3 Genera of freetail bats page 30
- F4 Genera of sheathtail bats page 31
- F5 Genera of bats with obvious noseleaves page 31

G Marine mammals page 31

Identification keys

Key A
Orders of small terrestrial and arboreal mammals

1 **a** One pair of large upper incisors and one pair of large lower incisors; females with paired teats in rows on lower abdomen sometimes extending to pectoral region
rodents—families Muridae and Sciuridae (Key B)

b Four, rarely five, pairs of small incisors in upper jaw and three pairs in lower jaw; females with teats in semi-circle in lower abdomen (with or without pouch covering)
polyprotodont marsupials—families Dasyuridae, Peramelidae, Peroryctidae and Thylacomyidae (Key C)

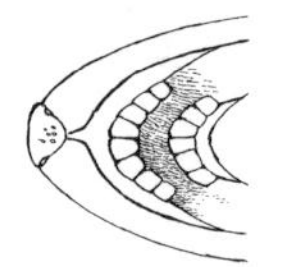

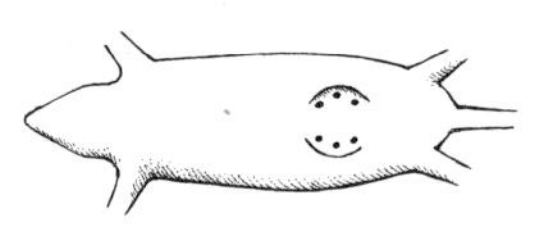

c Only one pair of lower incisors which are large, elongated and pointing forward; three or less pairs of upper incisors; females with pouch covering 4 teats
diprotodont marsupials—families Burramyidae, Petauridae, Tarsipedidae, and Acrobatidae (Key D), or Potoroidae and Hypsiprymnodontidae (Key E)

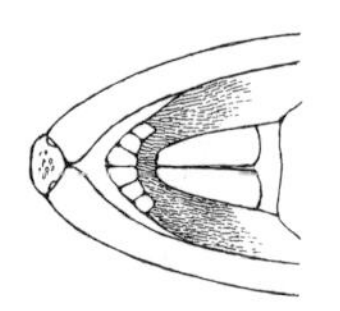

Key B
Genera of rodents

1 **a** Tail bushy for entire length — *Funambulus* (Plate 81)
b Tail not bushy, long hairs if present restricted to terminal tuft — 2

2 **a** Hindfoot partially webbed between the toes; claws as broad as tips of toes; tail completely covered with short dense fur — *Hydromys* (Plate 67)
b not as above — 3

3 **a** Tail clearly >head–body length, prehensile; upperside of tip flattened and lacking scales — *Pogonomys* (Plate 81)
b not as above — 4

4 **a** Only 2 obvious pads on sole of hindfoot, the 2 central interdigital pads (sometimes 2 small outer interdigital pads also present); hindfoot long and narrow; 5th toe short, barely reaches to base of 4th toe; tail very long and tufted — *Notomys* (Plates 68–69)
b 6 pads on sole of hindfoot; 5th toe reaches well past base of 4th toe — 5

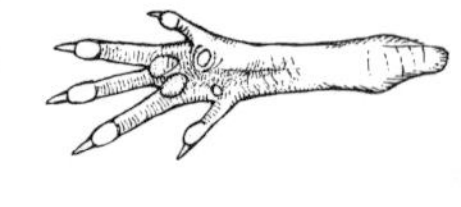

5 **a** Upperparts uniformly mid-grey; underparts white with sharp demarcation along throat and flanks; hairs of underparts entirely white; ears short (<15 mm) — *Xeromys* (Plate 67)

b Upperparts not uniformly grey; if hairs of underparts white, they are grey at base 6

6 **a** Tail fat at base with terminal tuft or crest of longer hairs; convex muzzle; large bulbous eyes *Zyzomys* (Plates 70–71)
b Tail not obviously fat at base 7

7 **a** Tail with obvious tuft of longer hairs towards tip 8
b Tail without obvious tuft of longer hairs 10

8 **a** Tail distinctly bicoloured, brown above and white below *Conilurus* (*C. albipes*) (Plate 67)
b Tail similar colour on upper and underside but colour may vary along length 9

9 **a** hindfoot <46 mm; tail about = head–body length *Conilurus* (*C. penicillatus*) (Plate 67)
b hindfoot >48 mm; tail clearly >head–body length *Mesembriomys* (Plate 66)

10 **a** Back of upper incisors with occlusal notch; tail about = head–body length; distinctive mousy odour *Mus* (Plate 71)
b Backs of upper incisors smoothly curved 11

11 **a** Tail scales not overlapping or in distinct rings; tail naked, with only sparse minute hairs 12
b Tail scales slightly overlapping and in clear rings, often hairy 13

12 **a** Large rats; hindfoot >35 mm; tail usually partially pinkish white, specially towards tip *Uromys* (Plate 66)
b Hind foot <30 mm, tail brown *Melomys* (Plate 65)

13 **a** Hindfoot <20 mm 14
b Hindfoot >20 mm 15

14 **a** Hindfoot broad, length about 4 times width at base of first digit; tail uniformly coloured 15
b Hindfoot narrow, length about 5 times width at base of first digit; tail often darker on upper side 16

15 **a** Hindfoot with elongate posthallucal pad; tail length variable *Rattus* (Plates 78–80)
b Hindfoot with nearly round posthallucal pad; tail about 3/4 head–body length, compact hunched posture *Mastacomys* (Plate 78)

16 **a** Tail clearly < head–body length, thick with little taper; muzzle broad and blunt *Leggadina* (Plate 71)
b Tail usually ⩾ head–body length, tapers to fine tip 17

17 **a** Hindfoot >40 mm; ears >27 mm — *Leporillus* (Plate 81)
b Hindfoot <38 mm; ears <27 mm — *Pseudomys* (Plates 72–77)

Key C
Genera of small polyprotodont marsupials (<2 kg)

1 **a** Pelage with distinct white spots, weight >100 g — *Dasyurus* (Plate 3)
b Pelage not spotted, or if so weight <30 g — 2

2 **a** Four separate toes on hindfeet (1st (inner) toe absent) — 3
b Five separate toes on hindfeet (1st (inner) toe present but may be reduced) — 4
c Second and 3rd toes joined giving the impression of a single digit with 2 claws (syndactyly) — 12

a b c

3 **a** Limbs long and delicate; tail about 150% of head–body length — *Antechinomys* (Plate 8)
b Limbs robust; tail < head–body length — *Dasyuroides* (Plate 5)

4 **a** Tail with distinct dense black brush for outer $^{1}/_{2}$ to $^{2}/_{3}$ — 5
b Tail not as above, slight terminal crest may be present — 6

5 **a** Tail with long black brush all around — *Phascogale* (Plate 4)
b Tail with long black fur on dorsal surface only — *Dasycercus* (Plate 5)

6 **a** Hindfeet relatively long and narrow (<3 mm); no post-interdigital pads and usually no enlarged granules on outer half of sole; may have dark midline or patch on forehead and crown; tail may have slight crest or tuft — *Sminthopsis* (Plates 12–16)
b Hindfeet relatively broad (>3 mm); post-interdigital pads present on outside half of sole; tail without brush or crest — 7

a b

7 **a** Very small—adult weight usually <12 g; if >12 g, then head flattened and triangular in shape and pads on forefeet without clearly defined striations — 8
b Adult weight greater than 12 g — 9

8 **a** Straight and flat posterior edge to supratragus of ear — *Planigale* (Plate 11)
b Posterior edge to supratragus curled — *Ningaui* (Plate 10)

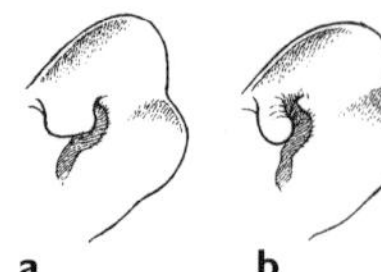

9 **a** Tail clearly shorter than head–body, thick at base and tapering evenly to a point; dorsal pelage uniformly coloured — 10
b Tail thin or if thick, reddish orange patches behind and below ears — 11

10 **a** Dorsal colour rich brown heavily grizzled with cream, strong contrast with creamy underparts — *Parantechinus* (Plate 8)
b Pelage uniformly russet or copper coloured above and below — *Dasykaluta* (Plate 8)

11 **a** Reddish or orange patches behind and below very large ears; tail may be swollen at base — *Pseudantechinus* (Plates 9–10)
b No obvious orange patches behind ears; tail never swollen — *Antechinus* (Plates 6–7)

12 **a** Tail with crest of long fur on upper surface — 13
b Tail with uniformly short hair — 14

13 **a** Forefoot with 5 digits — *Macrotis* (Plate 19)
b Forefoot with 4 digits — *Chaeropus* (Plate 19)

14 **a** Ears clearly longer than wide, and pointed — Perameles (Plate 18)
b Ears not obviously longer than wide, and rounded — 15

15 **a** Four pairs of upper incisors, fur highly spiny — *Echymipera* (Plate 19)
b Five pairs of upper incisors, fur coarse — Isoodon (Plate 17)

Key D
Genera of small possums and gliders (<0.5 kg)

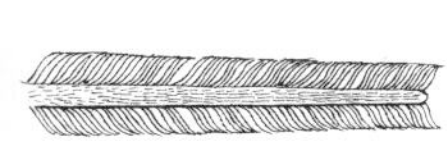

1 **a** Tail with fringe of hairs on either side, resembling a feather; gliding membrane from wrist to ankle — *Acrobates* (Plate 24)
b Tail sparsely furred, not resembling feather and no gliding membrane — 2
c Tail densely furred for entire length — 4

2 **a** Snout elongated; 3 longitudinal dorsal stripes — *Tarsipes* (Plate 24)
b Snout short, at most 1 dorsal stripe — 3

3 **a** Dense body fur extends about 2 cm along base of tail; rest of tail very sparsely furred; posterior premolars very large and blade-like with grooved sides — *Burramys* (Plate 24)

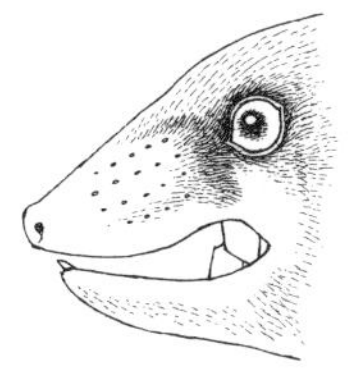

b Tail sparsely furred from base; tail may be thickened at base; posterior premolar blade-like; digits with small claws and large apical pads, claws extend only slightly (if at all) beyond tip of digit — *Cercartetus* (Plate 23)

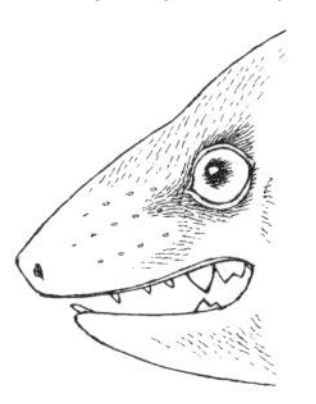

4 **a** Gliding membrane from outside of fifth digit to ankle, tail tapered towards tip — *Petaurus* (Plate 25)
b No gliding membrane, tail wider at tip than at base — *Gymnobelidius* (Plate 25)

Key E
Families of Macropoidea and genera of rat kangaroos (families Potoroidae and Hypsiprymnodontidae)

1 **a** All claws on forefeet approximately equal in length; tail used to support body in slow locomotion — kangaroos and wallabies (family Macropodidae, Plates 32–46)

b Claw on middle digit of forefoot longer than claws on outer digits; tail partially prehensile, used to carry nesting material not to support body
rat kangaroos (families Potoroidae and Hypsiprymnodontidae) 2

2 **a** Hindfoot with 5 toes though 2nd and 3rd toes combined in a sheath *Hypsiprymnodon* (Plate 32)
b Hindfoot with 4 toes, 2nd and 3rd combined in a sheath 3

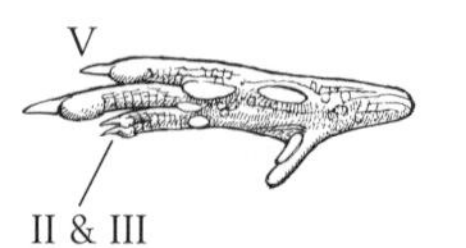

a

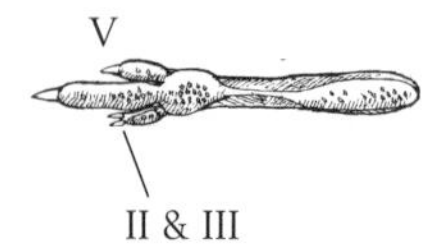

b

3 **a** Central part of rhinarium sparsely furred *Aepyprymnus* (Plate 29)
b Rhinarium entirely naked 4

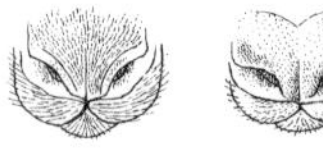

a b

4 **a** Tail clearly longer than head–body; hindfoot 10–12 cm *Caloprymnus* (Plate 30)
b Tail shorter than or not obviously longer than head–body 5

5 **a** Head broad; muzzle short; eyes large; tail similar length to head–body; dorsal fur colour varies on head, flanks and base of tail *Bettongia* (Plates 29–30)
b Head tapering to a narrow muzzle; profile flat on top; eyes small; tail 75% of head–body length; dorsal fur colour uniformly dark grey brown *Potorous* (Plate 31)

Key F
Families of bats

1 **a** Claws on thumb and 2nd finger (except *Dobsonia*); tail membrane absent or not joined between legs **Pteropodidae (Key F1, plates 47–49)**
b No claw on 2nd finger; tail membrane joined between legs 2

2 **a** Tail fully enclosed by tail membrane; or no tail but full tail membrane 3
b Tail extends beyond tail membrane; or end of tail projects through upper surface of membrane and is enclosed in membrane sheath 4

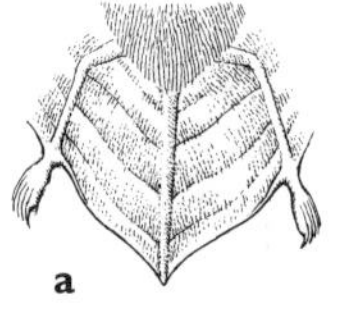

a

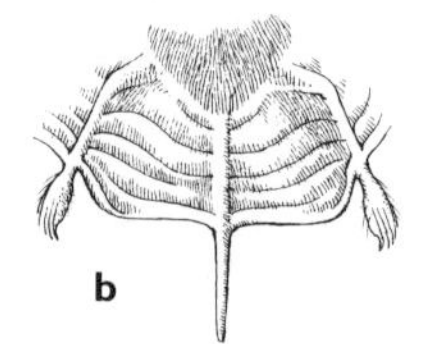

b

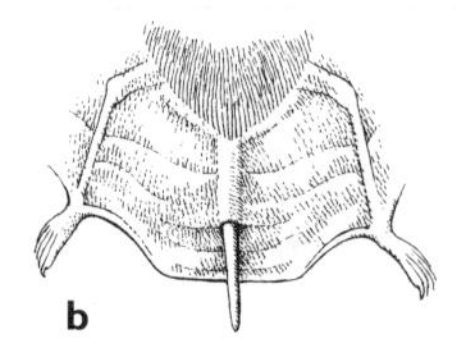

b

3 **a** Large complex noseleaf present; tragus absent or, if present, forked 5
b Noseleaf absent; tragus present **Vespertillionidae (Key F2, plates 56–64)**

4 **a** Tail clearly extends beyond tail membrane **Molossidae (Key F3, plates 54–55)**
b End of tail projects through upper surface of membrane and is enclosed in a sheath **Emballonuridae (Key F4, plates 52–53)**

5 **a** Tail absent; but complete tail membrane present, tragus forked **Megadermatidae (Key F5, plate 50)**
b Tail present and fully enclosed by tail membrane 6

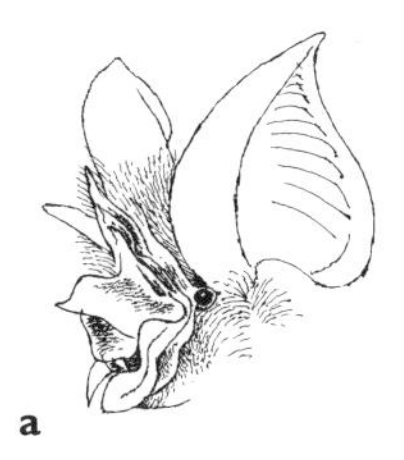

a

6 **a** Noseleaf large, elongate, lower leaf distinctly horseshoe-shaped and covering upper lip; large prominent forward projection from centre of noseleaf
Rhinolophidae (Key F5, plate 50)

b Noseleaf squarish or oval, without distinct central forward projection
Hipposideridae (Key F5, plates 50–51)

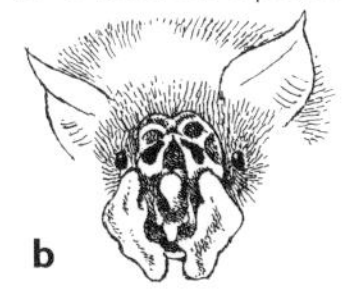

b

Key F1
Genera of fruit bats (family Pteropodidae)

1 **a** large size; forearm >100 mm — 2
b small size; forearm <70 mm — 3

2 **a** Wing membranes joined at midline of back; no claw on 2nd digit; short tail
Dobsonia (Plate 48)

b Wing membranes attached along sides of body; claw on 2nd digit; no tail
Pteropus (Plates 48–49)

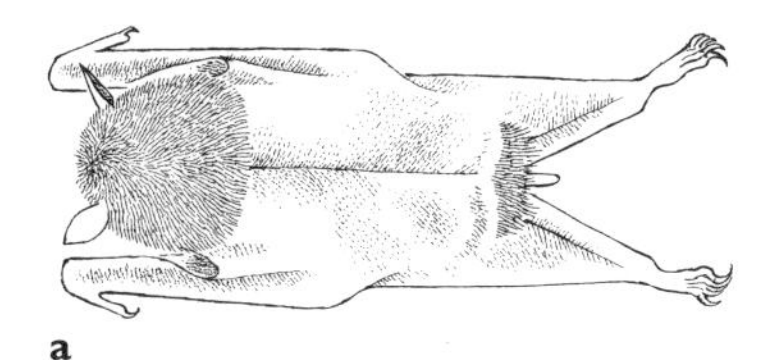

a

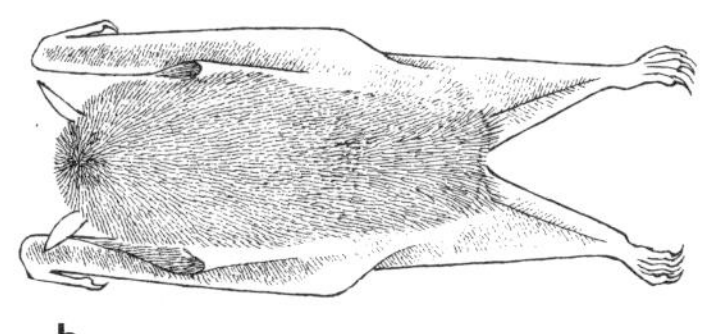

b

3 **a** forearm >55 mm; obvious tubular nostrils; yellow/green spots on wing membranes — *Nyctimene* (Plate 47)
b forearm <45 mm; small size; nostrils not tubular — 4

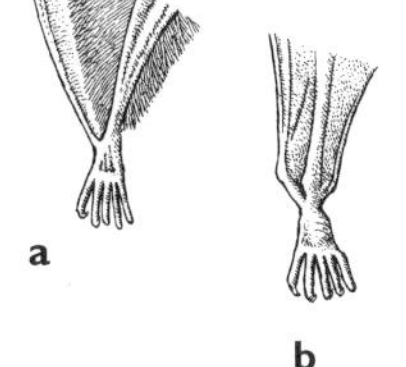

a b

4 **a** No tail or tail membrane; fringe of hairs along inside of legs
Syconycteris (Plate 47)

b No tail; tail membrane reduced to a narrow flap along inside of legs
Macroglossus (Plate 47)

Key F2
Genera of vespertilionid bats (family Vespertilionidae)

1 **a** Ears long and joined across forehead by flap of skin; simple noseleaf present in form of post-nasal ridge and nasal exfoliations — *Nyctophilus* (Plates 63–64)
b Ears not joined; no adornments to nostrils — 2

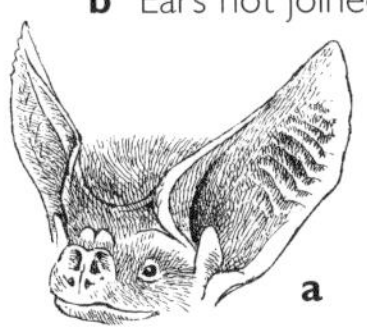

a

2 **a** terminal phalanx of 3rd finger about 3 times as long as sub-terminal phalanx
Miniopterus (Plate 60)

b terminal phalanx of 3rd finger about equal in length to sub-terminal phalanx — 3

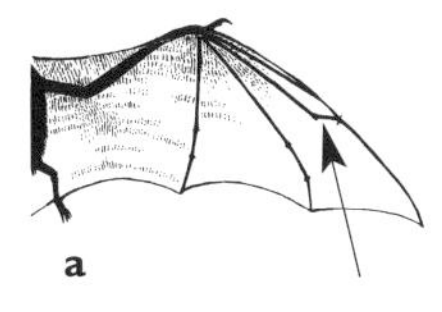

a

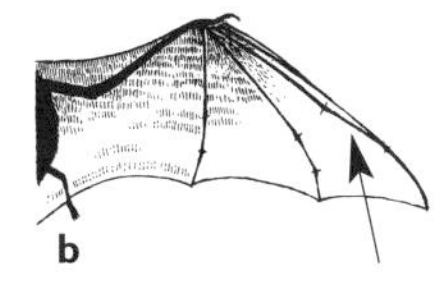

b

3 **a** Nostrils tubular and extend sideways; much of tail membrane furred *Murina* (Plate 57)
b not as above 4

4 **a** Fur curly, dark brown with golden tips; fur extends onto forearms, legs and tail *Kerivoula* (Plate 57)
b not as above 5

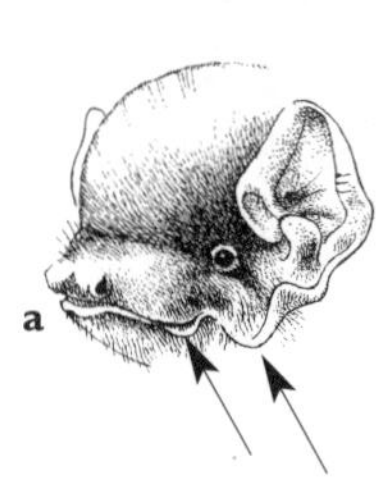

5 **a** Fleshy lobes at base of ear and on lower lip near corner of mouth; ear relatively short and rounded *Chalinolobus* (Plate 56)
b No lobe on lower lip or at corner of mouth; ear longer, more pointed 6

6 **a** Forearm >45 mm 7
b Forearm <45 mm 8

7 **a** Two pairs of upper incisors; outer pair minute; gap between upper incisors and canines; ears long, overlap when pressed together over head *Falsistrellus* (Plate 61)
b One pair of upper incisors; no gap between upper incisors and canines; ears barely touch when pressed together over head *Scoteanax* (Plate 61)

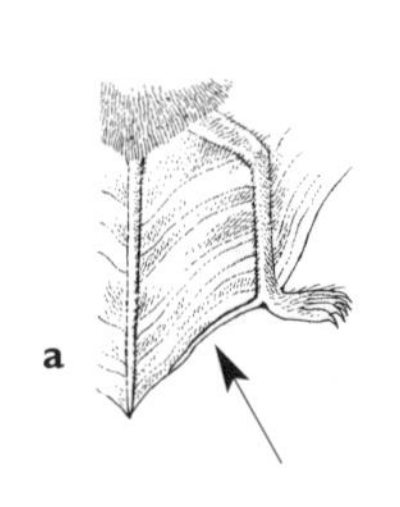

8 **a** Calcar long, extending $\frac{3}{4}$ of distance from ankle to tail tip; tragus long, slender, straight; foot large (8–11 mm, >$\frac{1}{2}$ length of tibia) *Myotis* (Plate 60)
b not as above 9

9 **a** One pair of upper incisors; muzzle broad, naked with swollen glandular areas *Scotorepens* (Plate 62)
b Two pairs of upper incisors (outer pair may be hidden behind inner pair); muzzle not broad; lacking swollen glandular areas 10

10 **a** Inner upper incisors with distinct cleft, giving 2 lobes easily seen from front on; outer upper incisors minute; only one pair of upper premolars *Vespadelus* (Plates 58–59)
b Inner upper incisors cleft but one lobe hidden behind other when viewed front on; outer upper incisor obvious, only slightly smaller than inner pair; 2 pairs of upper premolars, forward premolar only half height of rear one *Pipistrellus* (Plate 57)

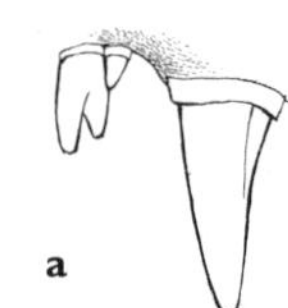
a

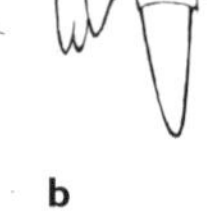
b

Key F3
Genera of freetail bats (family Molossidae)

1 **a** forearm >45 mm 2
b forearm <45 mm *Mormopterus* (Plates 54–55)

2 **a** Ears joined by a band of skin across forehead; no throat pouch *Chaerephon* (Plate 54)
b Throat pouch present; ears not joined along forehead; white stripes along flanks *Tadarida* (Plate 54)

Key F4
Genera of sheathtail bats (family Emballonuridae)

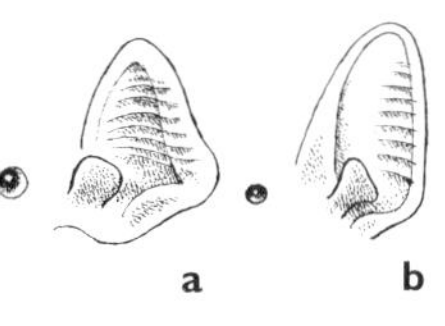

1 **a** Tragus broad, flat and uniform thickness; foot with small lateral granule on inner top surface near ankle — *Taphozous* (Plate 53)

b Tragus broad, flat and thicker towards top; foot without small lateral granule — *Saccolaimus* (Plate 52)

Key F5
Genera of bats with obvious noseleaves (families Megadermatidae, Rhinolophidae and Hipposideridae)

1 **a** No tail but full tail membrane; ears joined for lower half of their inner margins, forearm length >90 mm — *Macroderma* (Plate 50)

b Tail present and fully enclosed within tail membrane; forearm length <85 mm — **2**

2 **a** Noseleaf elongated; lower leaf distinctly horseshoe shaped and covering upper lip; upper leaf triangular and pointed; large prominent forward projection (the stella or lancet) from centre of noseleaf — *Rhinolophus* (Plate 50)

b Noseleaf squarish or oval without large central projection — **3**

3 **a** Noseleaf complex; lower leaf broad with distinct notch at bottom with subsidiary leaflet behind it; no large projecting upper leaf — *Rhinonicteris* (Plate 50)

b Noseleaf rounded or squarish with projecting upper leaf with or without 2 small club-shaped knobs — *Hipposideros* (Plates 51–52)

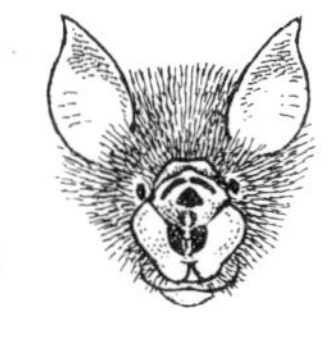

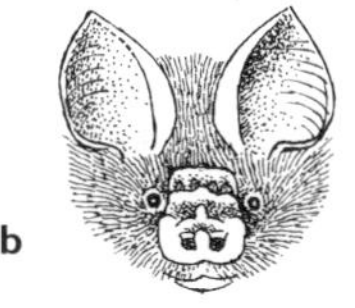

Key G
Marine mammals

Contributed by Graham J.B. Ross

Unfortunately, identifying marine mammals in the field to species level is not always a simple matter. Useful distinguishing characters, such as colour patterns and body form, change with age in many species, and their usefulness may be further compromised by damage during stranding or subsequent decomposition, especially for cetaceans. The identity of several species, particularly beaked whales, needs to be confirmed from skull characters; for example, the external features of the Tropical Bottlenose Whale are known only from two stranded juveniles and from photographs of animals at sea identified on characters apparent in those two juveniles. Furthermore, the true distributions of many species (as opposed to the distributions suggested by stranding records) are poorly known, so that keys for identification need to be broad enough to cover species that could be 'vagrants' to a region where they have not yet been recorded.

This key to Australian marine mammals incorporates characters that are least affected by age and decomposition, with the emphasis on those requiring little specialist knowledge and minimal inspection of the stranded animal. Such restrictions do limit our current ability to distinguish juveniles and adult females of at least two species of beaked whales in the field: Strap-toothed Beaked Whale and Shepherd's Beaked Whale. Beaked whales are usually identified from the form and position of the one or two pairs of lower teeth, which generally erupt only in adult males. This key uses shape and proportions of gape and melon; growth changes in these characters are most evident in calves and juveniles less than 2.5–3.0 m long. The identification of stranded beaked whales can usually be confirmed only if the skull is collected and examined by an expert. Identification of seal pups may also require a skull examination for a positive identification.

The cetacean section of the key, unlike the species accounts, includes three species that reasonably could be expected to occur in Australian waters: the Finless Porpoise, *Neophocaena phocaenoides*, which may reach northern Australia; the Peruvian Beaked Whale, *Mesoplodon peruvianus*, now recorded from New Zealand waters; and the Hourglass Dolphin, *Lagenorhynchus cruciger*, which occurs in the Southern Ocean.

Characters used in the key focus on the head region, supplemented by some features of the tail, fin and flippers. Parentheses around characters in key couplets indicate that they may be limited in their use; the relevant feature may be missing (for example, baleen may fall out during decomposition). Tooth counts may also be misleading (teeth may not be fully erupted in newborn animals, juveniles or females) but can generally be determined, even in newborn animals, from gum-covered bumps along each lower jaw; counts indicate the number of teeth on one side only. Colour patterns have not been used as characters, except when the effect of colour change after stranding is likely to be minimal. The range of total length (length at birth to maximum length) is given for each species, but the extent of overlap greatly diminishes the value of total length for identification purposes.

A **a** Hindlimbs absent; tail expanded into flukes — **B**
b Hindlimbs present, webbed; tail a short nub between hindlimb — **Pinnipedia 33**

B **a** Line of mouth directed downwards; muzzle and face bristly — **Dugong (Plate 104)**
b Line of the mouth (gape) horizontal or directed forwards and downwards; bristles absent on face — **Cetacea (whales and dolphins) 1**

1 **a** Two blowholes; (row of baleen plates on each upper jaw) — **Baleen Whales 2**
b Single blowhole; (erupted or non-erupted teeth present) — **Toothed Whales 7**

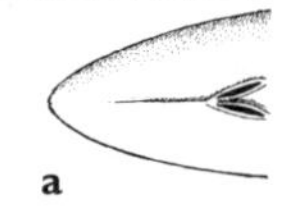

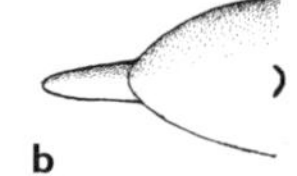

2 **a** Upper jaw and line of gape arched in profile; no throat grooves; baleen elongate, narrow — **Right Whales 3**
b Upper jaw and line of gape straight in profile; numerous longitudinal throat grooves; baleen broad — **Rorquals 4**

3 **a** Callosities on head; no dorsal fin; (baleen black); (6–17 m) — **Southern Right Whale (Plate 97)**
b No callosities on head; dorsal fin present; (baleen yellowish with dark outer edge); (ca 1.6–6.5 m) — **Pygmy Right Whale (Plate 98)**

4 **a** Large round knobs on rostrum and lower jaws; flipper very long (25–33% of tl), with knobbly leading edge and rounded tip; (4–13.5 m) — **Humpback Whale (Plate 97)**

b Rostrum and lower jaws smooth; flipper short (<14% of tl), with smooth leading edge and pointed tip 5

5 **a** Single dorsal ridge along midline of rostrum to tip 6

b Additional ridge on each side of mid-dorsal ridge from blowholes to rostral tip; throat grooves reach navel; (baleen black with coarse creamy inner fringe); (3.4–16 m) **Bryde's Whale (Plate 99)**

6 **a** In dorsal view rostrum broad and rounded at tip, and weak mid-dorsal ridge posteriorly on rostrum; throat grooves reach navel; (baleen black and symmetrical in colour); (6–30 m) **Blue Whale (Plate 99)**[1]

b In dorsal view rostrum tapers to acute tip, and well-formed mid-dorsal ridge on rostrum; throat grooves reach navel; (baleen grey with yellow streaks); (6.4–25 m) **Fin Whale (Plate 99)**

c In dorsal view rostrum tapers to acute tip, and prominent mid-dorsal ridge on rostrum; throat grooves do not reach navel; (baleen greyish-black, fringed with white); (4.5–17 m) **Sei Whale (Plate 98)**

d In dorsal view rostrum triangular with acute tip, with a prominent mid-dorsal ridge; throat grooves do not reach navel; (prominent white patch on flipper linked to whitish area on adjacent body) (baleen off-white, outer 30% of the width of larger plates black); (2.8–8 m) **Dwarf Minke Whale (Plate 98)**

e In dorsal view rostrum triangular with acute tip, with a prominent mid-dorsal ridge; throat grooves do not reach navel; (uniform grey flipper or two-tone grey patch on flipper) (baleen off-white, narrow black outer edge covering 10% of plate width on some larger plates); (2.8–9.4 m) **Antarctic Minke Whale (Plate 98)**

7 **a** Mouth narrow, set under the head; erupted teeth in lower jaw only; notch in centre of tail **Sperm Whales 8**

b Mouth opens anteriorly; pair of throat grooves forming a V; no notch in tail; (one or two large teeth in each lower jaw, usually erupting only in adult males) **Beaked Whales 9**

c Mouth opens anteriorly; notch in tail; three to 60 teeth in each lower jaw **Dolphins and Porpoises 18**

8 **a** S-shaped blowhole at top left front corner of the massive head; fin a series of humps; flipper rounded; 18–28 teeth in each jaw; (4–18 m) **Sperm Whale (Plate 97)**

---

1 As its name suggests, the Pygmy Blue Whale, *Balaenoptera musculus brevicauda*, is shorter overall (maximum 24 m) and the post-anal part of the body is slightly shorter (distance from anus to notch of flukes 22–31% of total length) than in ordinary blue whales (maximum length 30 m; anus to notch of flukes generally more than 31% of total length). Overlaps in measurements indicate caution in distinguishing the two forms.

b Blowhole more than 10% body length from snout tip; fin tiny, less than 5% of body length in height; 11–16 teeth in each lower jaw; (1.2–3.4 m)
Pygmy Sperm Whale (Plate 96)

c Blowhole less than 10% body length from snout tip; fin small, more than 5% of body length in height; 7–11, rarely 12 teeth in each lower jaw; (1.0–2.7 m)
Dwarf Sperm Whale (Plate 96)

9 **a** Crescentic form of blowhole is directed forwards; melon moderate in size, rising vertically from medium-length rostrum; fin a small, rounded nub; (two pairs of erupted teeth at tip of lower jaw in all adults); (ca 3–10 m)
Arnoux's Beaked Whale (Plate 103)

b Crescentic form of blowhole is directed backwards; well-developed, triangular to falcate fin — 10

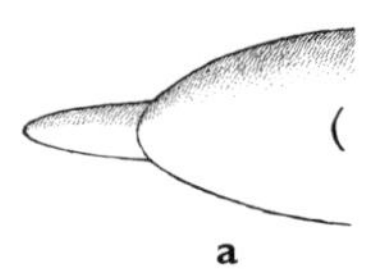

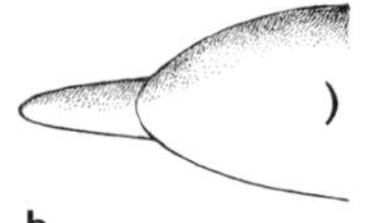

10 **a** In profile, rostrum short, snout tip to anterior point of melon less than half length of gape — 11

b In profile, rostrum moderate to long, snout tip to anterior point of melon half length of gape or more — 12

11 **a** In profile, melon rises steeply from medium-length rostrum, its apex defined by short crease, and overhangs rostrum in whales more than 3.5 m long; (pair of cylindrical teeth erupt at rostral tip in adult male); (ca 2.9–7.8 m)
Southern Bottlenose Whale (Plate 103)

b In profile, rounded melon, slightly swollen dorsally, rises steeply from the rostrum, curving evenly towards the blowhole; (one pair of teeth at tip of mandibles, presumed to erupt in adult males); (ca 2.9–7 m)
Tropical Bottlenose Whale (Plate 103)

c In profile, melon rises evenly, with no anterior crease, at less than 45° from the stubby rostrum, becoming bulbous anterior to the blowhole; (one pair of flattened columnar teeth erupt at the tip of the mandibles in adult males); (ca 2.2–5.4 m)
True's Beaked Whale (Plate 101)

d In profile, moderately swollen melon, rising at about 45° from or very close to rostral tip; no crease anterior to melon; (pair of cylindrical teeth erupt at rostral tip in adult male); (ca 2.6–7 m)
Goose-beaked Whale (Plate 102)

12 **a** In profile, gape is more or less straight, and may rise slightly, evenly or abruptly, at about its midlength — 13

b In profile, gape is straight for less than half the gape length before rising markedly or strongly, and then curving down in the direction of the eye — 16

13 **a** In profile, melon rounded and full, arising steeply from moderate to long rostrum — 14

b In profile, melon small to medium sized, moderately swollen and arises gradually from short to long rostrum — 15

14 **a** (One pair of long, strap-like teeth erupt in adult male anterior to the melon); (ca 2.2–6.3 m)
Strap-toothed Beaked Whale (Plate 101)

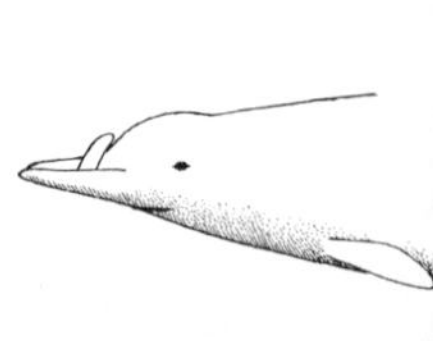

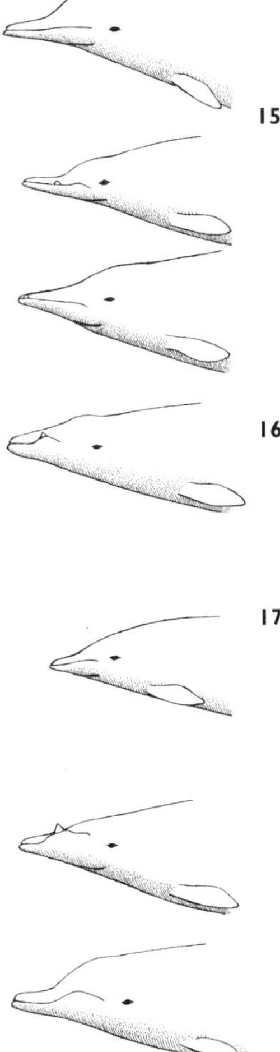

b (One pair of enlarged erupted teeth at tip of lower jaws in adult males; 17–29 smaller erupted teeth posteriorly in adults of both sexes); (ca 3–7 m)
Shepherd's Beaked Whale (Plate 102)

15 **a** In profile, rostrum long and slender, and the small, moderately swollen melon rises evenly from rostrum and curves steeply towards the blowhole; (one pair of flattened triangular teeth erupt at midlength of gape in adult males); (ca 1.8–5.6 m) **Gray's Beaked Whale (Plate 102)**

b In profile, rostrum medium in length tapering to the tip, and the small, evenly curved melon rises slightly above the line between rostral tip and blowhole; (one pair of flattened triangular teeth erupt ca 3 cm posterior to rostral tip in adult males); (ca 1.6–5 m) **Hector's Beaked Whale (Plate 101)**

16 **a** In profile, length of gape approximately 50% of length rostrum tip to eye, arched moderately, or strongly in adult males; (single tooth at mid-length of mandible, erupted in adult male); (ca 1.9–4.5 m) **Andrew's Beaked Whale (Plate 100)**

b In profile, gape more than 50% of length rostrum tip to eye 17

17 **a** In profile, melon small, rising from rostrum in an even curve; posterior corner of gape below or level with eye; (pair of flattened columnar teeth erupt on raised pulpit 33–50% of length of gape from rostral tip in adult males); (up to 3.7 m)
Peruvian Beaked Whale, *Mesoplodon peruvianus*

b In profile, melon small, rising from rostrum in an even curve, distinctly flattened anteriorly; gape rises sharply at about midlength nearly to level of rostrum before curving down to end above the level of the eye; (pair of massive tusk-like teeth rise up to 40 mm above the level of the rostrum at midlength of the gape in adult males); (1.9–4.8 m) **Blainville's beaked Whale (Plate 100)**

c In profile, melon medium to large, full, rising steeply from rostrum and descending slightly to the blowhole; gape rises smoothly at about midlength in females, sharply to the tooth pulpit in adult males, and descends to end at the level of the eye; (pair of large flattened leaf-shaped teeth at midlength of gape which erupt a few millimetres in adult males); (up to ca 5 m)
Ginkgo-toothed Beaked Whale (Plate 100)

18 **a** Dorsal fin absent 19
b Dorsal fin present 20

19 **a** Blunt, rounded head; small tubercles on back in place of fin; 13–22 teeth in each jaw, some spatulate; (0.55–1.9 m)
Finless Porpoise, *Neophocaena phocaenoides*
b Distinct broad beak in front of sloping melon; 44–49 conical teeth in each jaw; (ca 1.0–2.9 m) **Southern Right Whale Dolphin (Plate 92)**

20 **a** In profile, head round or blunt, melon generally large, and beak absent or very short 21
b In profile, distinct beak at all ages anterior to small to medium-sized melon 28

21 **a** Fin about one-third body length from head 22
b Fin at about midlength of body 23

22 **a** Flipper length 14–19% of body length; 7–10 teeth in each jaw
Short-flippered Pilot Whale (Plate 95)

b Flipper length 18–27% of body length; 9–12 teeth in each jaw
Long-flippered Pilot Whale (Plate 95)

23 **a** Short beak present at all ages 24
b Beak absent at all ages 25

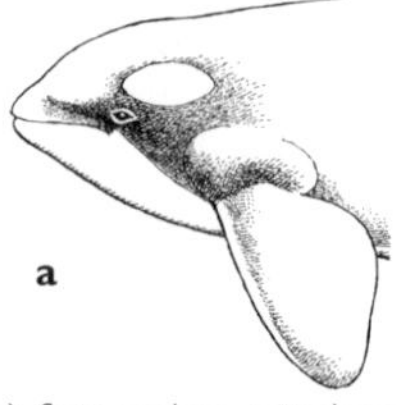

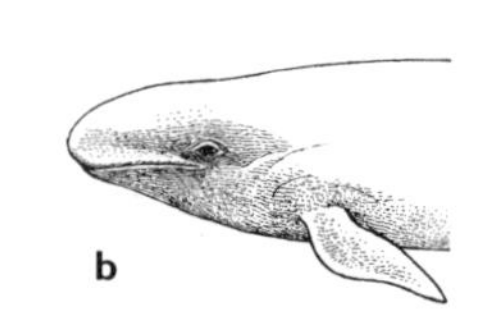

24 **a** 10–12 large teeth (less than 15 mm in diameter), flattened on anterior and posterior surfaces, in each jaw **Killer Whale (Plate 94)**
b 21–25 small teeth (more than 10 mm in diameter), in each jaw
Melon-headed Whale (Plate 95)

25 **a** 2–7 teeth in anterior of each lower jaw only; shallow longitudinal groove on anterior face of melon **Risso's Dolphin (Plate 92)**
b 8 or more teeth in lower jaw; anterior face of melon smoothly rounded 26

26 **a** Leading edge of flipper S-shaped; 8–11 large teeth (more than 10 mm in diameter) in each jaw **False Killer Whale (Plate 96)**
b Leading edge of flipper entirely convex; teeth less than 10 mm in diameter 27

27 **a** 8–13 pointed teeth in anterior two-thirds of each jaw; shallow groove on midline of belly **Pygmy Killer Whale (Plate 95)**
b 12–20 pointed teeth over whole length of each jaw; shallow groove on midline of belly **Irrawaddy Dolphin (Plate 91)**
c 16–23 spatulate teeth over whole length of each jaw; no groove on belly
Spectacled Porpoise (Plate 92)

c

28 **a** Melon merges smoothly and indistinctly into rostrum 29
b Apex of melon distinct at its junction with the rostrum 30

29 **a** Rounded, full melon; triangular fin on broad base; 29–34 teeth in each jaw
Indo-pacific Humpback Dolphin (Plate 91)
b Sloping melon (head conical in profile); falcate dorsal fin; 20–27 teeth in each jaw
Rough-toothed Dolphin (Plate 91)

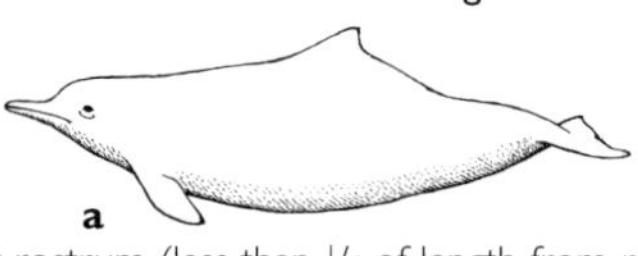

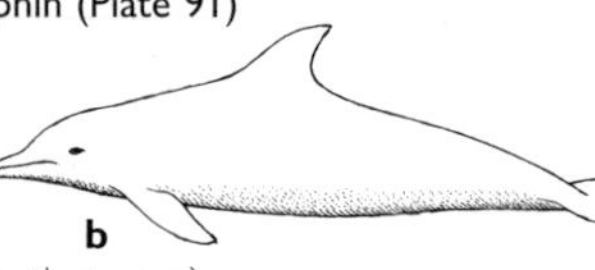

30 **a** Short to very short rostrum (less than 1/4 of length from rostrum tip to eye) 31
b Short to medium-length rostrum (1/4 to 1/3 of length from rostrum tip to eye); 19–29 teeth in each jaw **Bottlenose Dolphin (Plate 94)**[2]

long-beaked form

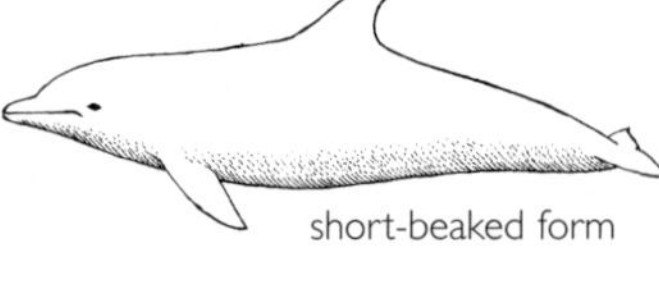
short-beaked form

2 The taxonomic status of the *aduncus* form of *Tursiops truncatus* requires further resolution; recent morphological and molecular studies indicate that it is a separate species; adult *aduncus* in northern Australia have numerous dark spots and blotches on the chest and belly.

c Medium-length to long rostrum (more than 1/3 of length from rostrum tip to eye) 32

31 **a** Rostrum 10% or less of length from tip of rostrum to eye; grey stripe from flipper to near posterior end of gape; 34–44 teeth in each jaw; (broad dark stripe from gape through eye to vent in animals longer than 2.2 m)
Fraser's Dolphin (Plate 93)

b Rostrum about 20% of length from tip of rostrum to eye; black stripe from flipper to corner of gape; 24–36 teeth in each jaw; (chevron pattern on posterior flank) Dusky Dolphin (Plate 92)

c Rostrum about 20% of length from tip of rostrum to eye; broad black stripe from flipper to eye; about 28 teeth in each jaw; (two linked white patches forming "hourglass" on side) Hourglass Dolphin *Lagenorhynchus cruciger*

32 **a** Dark stripe from flipper to eye, and eye to anus; 39–52 teeth in each jaw
Striped Dolphin (Plate 93)

b Dark stripe from flipper to eye; 45–59 teeth in each jaw
Spinner Dolphin (Plate 93)

c Dark stripe from flipper to gape; 34–48 teeth in each jaw
Pantropical Spotted Dolphin (Plate 93)

d Dark stripe from flipper to mid-length of gape; 47–60 teeth in each jaw
Common Dolphin (Plate 94)[3]

33 **a** No external ears. Phocidae, True Seals 34

b Small external ears present. Otariidae, Fur Seals and Sea-lions 37

---

3 The taxonomic status of Common Dolphins needs further resolution. Characters of *Delphinus delphis* and *D. capensis* currently recognised elsewhere are present in Australian specimens, but animals with intermediate skeletal features occur; *Delphinus tropicalis* may also be present in northern waters.

34 **a** Snout short, less than its breadth at base; diameter of canine tooth less than 3 times that of post-canine teeth **35**

b Length of snout greater than breadth at base; diameter of canine tooth less than 3 times that of post-canine teeth **36**

c Length of snout about equals breadth at base, or snout extended to form a short trunk; dark grey or dark brown fading to greyish brown; females 1.2–2.6 m, males up to 5.0 m total length; diameter of large canine tooth more than 3 times that of post-canine teeth **Southern Elephant Seal (Plate 89)** **c**

35 **a** Greyish-brown to black dorsally, streaked on the flanks and greyish white below; 1.5–3.3 m total length; outer upper incisor twice size of inner incisor; post-canines simple with one cusp **Weddell Seal (Plate 90)** **a**

b Silver-grey dorsally, silver-white ventrally, and throat streaked with grey; 1.0–2.4 m total length; incisors similar in size; primary cusp of post-canines flanked by small indistinct secondary cusps **Ross Seal (Plate 90)** **b**

36 **a** Silvery brownish grey dorsally, paler ventrally, with variable brown markings on the shoulders and flanks; a neck region not readily apparent; 1.5–2.6 m total length; primary cusp of post-canine teeth flanked by two anterior and two or three posterior secondary cusps **Crabeater Seal (Plate 90)** **a**

b Dark grey on the back, paler underneath, with distinctive dark spots on neck, shoulder and flanks; distinct neck region emphasised by the large head; 1.6–3.0 m total length; primary cusp of post-canine teeth flanked each side by a strong secondary cusp **Leopard Seal (Plate 89)** **b**

37 **a** Tip of snout squarish or blunt; neck of large males white, smaller animals silvery grey/fawn dorsally; females 0.7–1.8 m, males to 2.5 m nose to tail; 5 upper post canines **Australian Sea-lion (Plate 89)** **a**

b Tip of snout pointed; postnatal animals dark brown dorsally; 6 upper post-canines **Fur Seals 38**

38 **a** Pale creamy-white facial colour extends behind and above eye; upperparts dark greyish brown, underparts creamy, especially in adults; snout short; females 0.6–1.4 m, males up to 2.0 m total length; erectile fur crest on forehead of subadult and adult males; single cusp of post-canines may bear a minute anterior secondary cusp, posterior post-canine small **Sub-antarctic Fur Seal (Plate 88)** **a**

b Adult males with grizzled grey-brown fur, females grey-brown above, white to grey below; in lateral view, snout short, forehead rises steeply relative to snout (adult males) or very little (females and subadults); 0.65–1.5 m in adult females, adult males to 2 m total length; small erectile fur crest may be present on some subadult and adult animals; post-canines bear no accessory cusps, and two most posterior teeth are minute **Antarctic Fur Seal (Plate 90)** **b**

c Adult males greyish black overall, females brownish-grey, paler below; no crest on forehead; in lateral view, snout elongate and robust, forehead rises little relative to snout; 0.6–1.7 m in females, to 2.3 m in males total length; post-canine teeth robust with developed secondary cusps **Australian Fur Seal (Plate 88)** **c**

d Adults dark grey-brown dorsally, paler below; no crest on forehead; in lateral view, snout elongate and distinctly pointed, forehead rises steeply relative to snout; 0.55–1.3 m in females, to 2.2 m in adult males total length; post-canine teeth moderately robust, slightly larger anteriorly, with or without small secondary cusps **New Zealand Fur Seal (Plate 88)** **d**

tropical rainforest

wet sclerophyll forest

Examples of broad habitat types utilised by Australian mammals *(photographs by author).*

dry sclerophyll forest

tropical savannah

mallee

arid shrubland

gibber plain

coastal heathland

mangrove

tussock grassland

hummock grassland

rock scree

Abbreviations

ACT	Australian Capital Territory
alt.	altitude
arch.	archipelago
Aust.	Australia
C	canine tooth
cm	centimetre
e.	east, eastern
el	ear length from base of ear notch to distal apex
est.	estimated
fa	forearm length, the length of the ulna
g	gram
GDR	Great Dividing Range
hb	length of head and body combined—from tip of snout to anus with head pointed forward
hf	hindfoot length from heel to tip of longest toe, excluding claw
I	incisor tooth
I. Is	island, islands
in	number of pairs of inguinal teats
intr.	introduced
kg	kilogram
km	kilometre
m	metre
M	molar tooth
min.	minute
mm	millimetre
Mt	Mount, mountain
n.	north, northern
nm	nautical mile
NP	national park
NR	nature reserve
NSW	New South Wales
NT	Northern Territory
P	premolar tooth
pe	number of pairs of pectoral teats
Pen.	peninsula
pl.	plural
Qld	Queensland
R.	river
s.	south, southern
SA	South Australia
sh	shoulder height
st	standing height, from ground to crown of head
Stn	station (ranch), as in cattle station
t	tail length, from anus to distal tip
Tas	Tasmania
th	number of pairs of thoracic teats
tl	total length, from tip of snout to end of tail
Vic	Victoria
w.	west, western
WA	Western Australia
wt	body weight
<	less than
>	greater than

Key to distribution maps

terrestrial

current distribution

former distribution

intermittent occupation

? possible occurrence

marine

regular occurrence

irregular occurrence

? possible occurrence

Short-beaked Echidna *Tachyglossus aculeatus*

Spiny Anteater

hb 230–350 mm; **t** 85–95 mm; **wt** 2–7 kg
Robust ground-dweller with strong, sharp spines covering top of head, back and tail; snout tubular and naked with tiny mouth and nostrils at tip; bulbous forehead; powerful digger with short legs and long claws; foreclaws spade-like; hindclaws directed backward, second hindclaw long, used for grooming. Length and density of fur varies geographically—sparser and shorter in n. and central Aust, in Tas can almost obscure the spines. Fur varies from dark brown to straw coloured; naked skin blackish; spines straw-coloured with black tips. Characteristic diggings and cylindrical, shiny scats indicate presence. **Distribution, habitat and status** Throughout, including Tas, King, Flinders and Kangaroo Is; in almost all terrestrial habitats except intensively managed farmland. Common to sparse. **Behaviour** Moves with slow, determined, rolling gait; when disturbed burrows vertically down with surprising ease until only spiny back is exposed; solitary within large home range (40–70 ha); active day or night to avoid extremes of heat or cold; shelters in logs, crevices, burrows or piles of litter. A toothless, highly specialised feeder on ants, termites and other soil invertebrates, particularly beetle larvae. Food exposed by powerful digging and tearing into soil or rotten wood with forelimbs, then licked up with long, sticky tongue. Mating period (June–Sept) is the only time when aggregations may be found—usually a group of males attending a receptive female. The single egg and, later, the juvenile is carried in pouch on belly; later left in nest in burrow, weaned when about 8 months old.

scats

Platypus *Ornithorhynchus anatinus*

hb 310–400 mm; **t** 100–150 mm; **wt** 700–2400 g
Amphibious. Streamlined body, short limbs, webbed feet, naked leathery snout resembling duck's bill, and broad flat tail. Small eyes and ear openings located in a furrow on each side of crown. Upperparts uniform mid brown, paler below; snout blue-grey, soft, pliable, highly sensitive. Frontal shield extends back over fur of forehead and throat. Webbing of forefeet extends beyond the toes and can be folded under the palm to expose the foreclaws when walking or burrowing. Adult males have a sharp spur about 12 mm long on each hind ankle through which poison can be injected. **Distribution, habitat and status** Locally common and widespread throughout e. coast from Cooktown area, Qld, to sw. Vic, but absent from lower reaches of Murray–Darling system; also Tas; King, and Kangaroo (intr.) Is. Inhabits freshwater streams, ranging from alpine creeks to tropical lowland rivers; also lakes, shallow reservoirs and farm dams. Prefers areas with steep, vegetated banks in which to burrow; entrances concealed under overhangs or vegetation. Locally common. Has declined in Murray–Darling system. **Behaviour** Shy and difficult to observe. When on the surface only tip of snout, top of head, back and tail are visible. Mostly crepuscular and nocturnal but may be diurnal. Largely solitary but groups inhabit high quality habitat. Feeds on benthic invertebrates; food is stored in cheek pouches until it can be masticated while floating on the surface. Mating occurs in Aug–Oct, being earlier in the n. Two eggs laid into nest in long, branched nursery burrow; hatch after 2 weeks; young left in burrow and fed by mother for 4–5 months. Outside breeding season shelters in den in a short burrow in bank. **Similar species** Water Rat and Brown Rat are the only other Aust mammals that frequently swim in freshwater. Platypus is distinguished by snout and tail shape, lack of external ears, smooth swimming style and low silhouette.

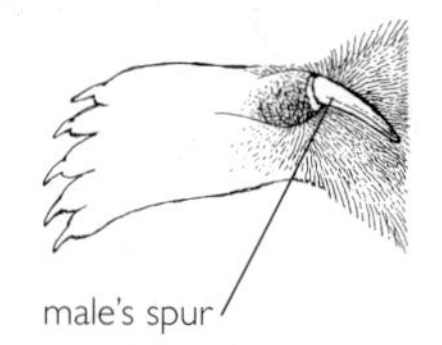

male's spur

PLATE 1

Thylacine *Thylacinus cyanocephalus*

Tasmanian Tiger

hb 100–130 cm; **t** 50–68 cm; **sh** to 60 cm; **hf** 14–17 cm; **el** 8 cm; **wt** to about 35 kg

Largest living marsupial carnivore at time of European settlement. Appearance superficially wolf-like, especially head and forelimbs, but hindlimbs and broad-based tail typically marsupial. Muzzle long and narrow. Tail broad at base, tapering rapidly, short-haired and about half hb length; ears short, erect, furred. 5 foretoes and 4 hindtoes. Fur coarse, short; sandy-brown above, creamy below with whitish patches on cheeks, above eyes and at base of ears; parallel dark brown bars across back from behind shoulders to base of tail and extending onto thighs, longest and broadest over rump. **Voice** Terrier-like yapping when hunting cooperatively; guttural, coughing bark when disturbed. **Distribution, habitat and status** During historical times known only from Tas, where it occupied sclerophyll forest, woodland and rock outcrops. Never common, now presumed extinct. **Behaviour** Pursuit carnivore, capable of bursts of speed, sometimes hunted cooperatively. Nocturnal and crepuscular, rested during day in lair amongst rocks, logs or dense vegetation; main prey kangaroos and wallabies but probably not fussy, taking whatever vertebrates were available. Births occurred in all seasons, mostly winter and spring; 2 or 3 young reared in backward-opening pouch with 4 teats; after leaving pouch young left in lair until weaned then accompanied mother till independent. **Similar species** Domestic Dog lacks bands across back, has obvious point where tail joins rump; mangy Red Fox can appear similar but is smaller, has lighter build, narrow pointed muzzle, more bounding gait. Both Dogs and Red Fox have 3 not 4 upper incisors and 4 not 3 upper and lower premolars.

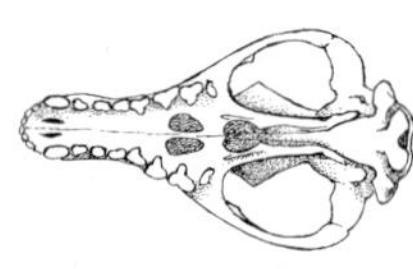

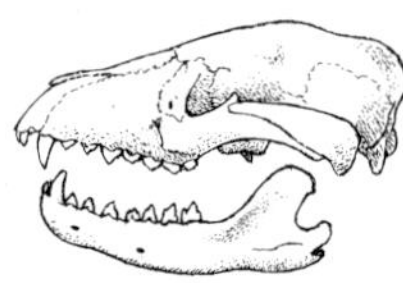

3 upper and 3 lower premolars

Tasmanian Devil *Sarcophilus harrisi*

hb 550–650 mm; **t** 240–260 mm; **hf** 80–95 mm; **el** 60–73; **wt** 7–9 kg

Largest of the living marsupial carnivores; resembles a small stocky Dog with a short muzzle and short limbs; forelegs longer than hindlegs. Males larger with a broader head and neck. Ears short and rounded; 5 foretoes, 4 hindtoes; long, sharp claws. Fur dense and black, usually with irregular white patches on chest and sometimes on shoulders or rump; fur sparse on muzzle, ears and around eyes, exposing pink skin. **Voice** Varied low growls, snorts, whines and loud gurgling screeches. **Distribution, habitat and status** Widespread and common in Tas, absent from Bass Strait Is. Found in most habitats including outer suburban areas; most common in dry sclerophyll forest and coastal woodland with patches of open grassland. **Behaviour** Nocturnal and crepuscular; scavenger and predator; home ranges of 8–20 km^2 overlap extensively and several animals may gather at a carcass, leading to squabbles, growls and screeches, and ritualised baring of teeth; feeds on a variety of live prey and carcasses, ranging from large invertebrates to large mammals (kills mammals up to small wallaby size); with a steady, loping gait patrols beaches for beachcast animals and roads for traffic kills. Capable tree climber. Mating occurs in March, births in April, and up to 4 young are carried in the backward opening pouch until Aug; when fully furred, young are left in a den and begin to explore outside during Nov; reach independence in Jan at about 40 weeks old.

Thylacine

Tasmanian Devil

PLATE 2

Northern Quoll *Dasyurus hallucatus*

Northern Native Cat

hb 200–310 mm; **t** 180–340 mm; **wt** 300–1000 g

Size of a large kitten; grey-brown to brown above with large white spots on body but not tail, which is sparsely furred; underparts cream-grey. Muzzle finely pointed; eyes and ears large, prominent. Distinctive pungent odour. **Voice** Hissing and sniffing sounds. **Distribution, habitat and status** Formerly across n. Aust from North-west Cape, WA to se. Qld but has contracted seriously. Now restricted to 6 main areas—Hamersley Range and Kimberley, WA; n. and w. Top End, NT; n. Cape York, Atherton–Cairns area, Carnarvon Range–Bowen area, Qld. Most abundant in rocky eucalypt woodland but occurs in range of vegetation types, mostly within 200 km of coast. **Behaviour** Mostly nocturnal and crepuscular; aggressive carnivore; preys on small mammals, reptiles, arthropods, fruit. Dens in tree hollows and rock crevices. Can become confiding around people. Births occur in mid dry season (June–Sept); all males die after mating.

Spot-tailed Quoll *Dasyurus maculatus*

Spotted-tailed Quoll, Tiger Quoll, Tiger Cat

hb 350–750 mm; **t** 350–550 mm; **wt** males to 7 kg, mostly 3–4, females to 4 kg, mostly 1.5–2

Long-bodied, long-tailed and short-legged; thick, coarse fur, rufous or dark brown above, with bold white spots on body, legs and tail; underparts pale yellowish grey. Race *gracilis* from ne. Qld is smaller, more delicately built. **Voice** Deep hissing and screeches. **Distribution, habitat and status** Formerly widespread in se. Aust, now sparse from Fraser I in Qld to sw. Vic; widespread in Tas. Found from sea-level to sub-alps in many habitats—rainforest, wet and dry sclerophyll forest, coastal heath and scrub, sometimes Red Gum forest along inland rivers. Race *gracilis* isolated in ne. Qld upland rainforest from Mt Finnigan to s. Atherton Tableland, is endangered. Elsewhere vulnerable. **Behaviour** Solitary, mostly nocturnal, partly arboreal. Dens in tree hollow, hollow log, rock crevice. Eats small to medium-sized mammals and birds (up to possums, bandicoots, rosellas) and large arthropods. Mating occurs in late autumn or early winter (April Aug). **Similar species** Only Aust mammal with spotted tail. Muzzle less sharply pointed and ears shorter than Eastern or Northern Quoll; Western Quoll has long blackish hairs on outer half of tail, contrasting with rest of upperparts.

V

I

Western Quoll *Dasyurus geoffroyi*

Chudich, Western Native Cat

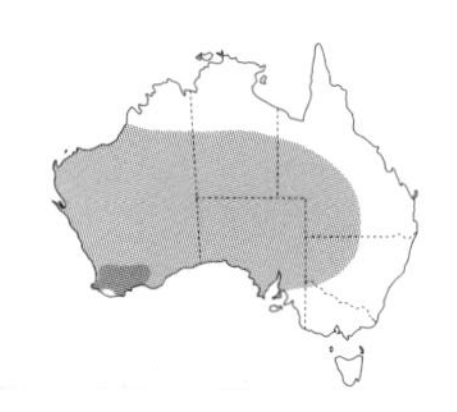

hb 260–400 mm; **t** 210–350 mm; **wt** males to 2 kg, females to 1 kg

Medium-sized quoll. Upperparts rufous grey with white spots on body and legs but not tail; underparts cream; tail sparsely furred rufous with blackish outer half; hindfoot pads not striated. **Distribution, habitat and status** Formerly present across semi-arid southern Aust from sw. WA to w. Qld and NSW, now confined to wet and dry sclerophyll and mallee remnants in sw. WA. Vulnerable. **Behaviour** Solitary, mostly nocturnal, terrestrial but can climb trees; dens in burrows or hollow logs. Generalist predator, eats small vertebrates, freshwater crays, large arthropods, carrion. Mating occurs late April – early July. **Similar species** Spot-tailed Quoll is larger, has white spots on uniformly brown tail. Eastern Quoll has finely pointed muzzle, more delicate build, lacks black tail tip.

Eastern Quoll *Dasyurus viverrinus*

Native Cat, Eastern Native Cat

hb 280–400 mm; **t** 190–280 mm; **wt** 0.7–2.0 kg

Delicately built; fine pointed muzzle and long ears; graceful, agile movements; does not climb, lacks 'big' toe. Fur colour either fawn-rufous with greyish back and nape, or black; body with sparse white spots; no spots on tail, which often has a whitish tip. Underparts creamy white. **Distribution, habitat and status** Formerly widespread in e. NSW, much of Vic and se. SA, probably extinct on mainland but common and secure in Tas. Inhabits range of open forest, scrubland and heath, especially where interspersed with grassy clearings. Dens in burrow, hollow log or rock crevice. **Behaviour** Mostly solitary, largely nocturnal, terrestrial. Opportunistic omnivore eating mostly arthropods and fruit, some small vertebrates, grass, carrion. Mates in late May – early June. **Similar species** Black morph unmistakable; rufous morph distinguished by delicate build, finely pointed muzzle, long ears, lack of spots on tail, lack of 1st toe.

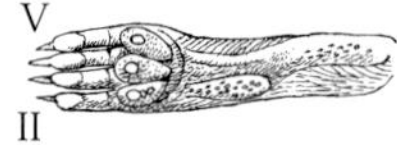

Northern Quoll
Spot-tailed Quoll
Western Quoll
Eastern Quoll

PLATE 3

Red-tailed Phascogale *Phascogale calura*

Red-tailed Wambenger

hb 95–125 mm; **t** 120–145 mm; **wt** 38–70 g

Smaller, browner, paler version of Brush-tailed Phascogale. Brown-grey above with rufous tinge, particularly on flanks, towards rump and behind ears; upperside of inner half of tail rusty red; outer half a brush of long blackish hairs, partly erectile. Clear demarcation between grey-rufous upperparts and cream to whitish underparts. Ears long, thin, reddish. Pale cream eye-rings. **Distribution, habitat and status** Formerly widespread in woodland habitats in inland s. and central Aust. Now restricted to remnants of mature Wandoo or Rock Oak woodland in s. WA wheatbelt where annual rainfall is 300–600 mm. Endangered. **Behaviour** As for Brush-tailed Phascogale. Mating occurs in a 3 week period in July after which all males die; after a gestation of 28–30 days females give birth to up to 13 young but only 8 teats are available; young are independent at about 4 months of age. **Similar species** Brush-tailed Phascogale is almost twice as long and 3–4 times heavier, blue-grey without rufous tinge or contrasting underparts, has more pronounced tail brush.

Brush-tailed Phascogale *Phascogale tapoatafa*

Tuan (Vic), Common Wambenger (WA)

hb 150–260 mm; **t** 165–235 mm; **wt** 110–310 g

Rat-sized, sharp-snouted hunter with a diagnostic black 'bottle-brush' tail. Dorsal fur uniform mid-grey, cream to white underside; distal 2/3 of tail a conspicuous black 'bottle-brush' with hairs up to 55 mm long, capable of being 'erected' when excited. Eyes large and bulging; ears long, thin, naked, grey-pink. Race *pirata* (WA) is paler and smaller (70% of weight of Vic specimens). **Distribution, habitat and status** Sparsely distributed outside the semi-arid zone in dry sclerophyll forest and monsoonal forest and woodland. Generally rare and threatened by habitat fragmentation, most common in sw. WA. **Behaviour** Nocturnal, mostly solitary, shy, rarely seen; avoids spotlight beam by moving behind branch. Agile, rapid, jerky movements; forages on trees, especially rough-barked eucalypts and dead branches; spirals upwards, jumps, clings underneath branches, runs downwards head-first. Probes for invertebrates amongst bark and rotting wood, tearing off bark with teeth. When alarmed, taps forefeet on substrate. Nests of bark, feathers and fur built in hollow branches and stumps. Eats mostly arthropods; also small vertebrates and nectar. Mating occurs during 3 weeks between mid-May and early July; all males then die; females may live to 3 years old. Gestation lasts 30 days; litter size 7–8 in e. Aust, 6 in sw. WA. **Similar species** See under Red-tailed Phascogale.

Numbat *Myrmecobius fasciatus*

Banded Anteater

hb 200–275 mm; **t** 165–210 mm; **wt** 300–715 g

Small, brightly coloured, diurnal terrestrial marsupial with a finely pointed muzzle and a bushy tail. General body colour grizzled yellowish-rust grading to brick red on back and blackish on rump; back and hindquarters crossed by 5–7 prominent white bands; conspicuous black stripe from side of muzzle through eye to base of ear, framed above and below by white patches; belly sandy grey. Head flat, ears long, narrow. **Distribution, habitat and status** Once occurred across much of arid and semi-arid southern Aust; now restricted to a few remnant forests of Wandoo, Powderbark Wandoo or Jarrah in sw. WA, e.g. Dryandra and Perup Forests. Vulnerable. **Behaviour** The only marsupial adapted to a diet of termites—like the Echidna, lacks effective teeth and has a long, cylindrical, sticky tongue that is used to lick up termites exposed by digging and tearing with strong foreclaws. Forages alone during daylight and shelters in nest of leaves and bark in hollow log or shallow burrow. Mating occurs from Feb–April; 2–4 young born 14 days later, weaned at about 7 months and dispersed by Dec.

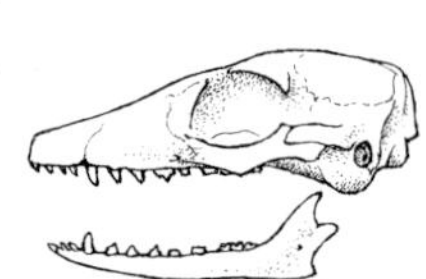

Red-tailed Phascogale

Brush-tailed Phascogale

Numbat

PLATE 4

Kowari *Dasyuroides byrnei*

hb 138–180 mm; **t** 110–135 mm; **wt** 70–140 g
Stocky carnivore with diagnostic tail: whitish, short-haired basal half contrasts strongly with dense black terminal brush. Upperparts olive-grey, sometimes tinged rufous; underparts whitish, sharply demarcated. Face fox-like with a long pointed muzzle, large upright ears, large dark eyes framed by conspicuous pale eye-rings. Feet noticeably pale, 5 foretoes, 4 hindtoes. **Distribution, habitat and status** Confined to scattered populations in sparsely vegetated, stony desert country of Lake Eyre drainage basin. Declining, possibly extinct, w. of Lake Eyre. Vulnerable. **Behaviour** Mostly solitary within home range of several km^2. Shelters in burrow during day; mostly nocturnal, may bask in morning sun. Eats large arthropods and small vertebrates. Breeds mostly May–Dec following rain. **Similar species** See under Mulgara.

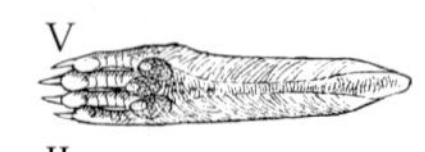

Mulgara *Dasycercus cristicauda*

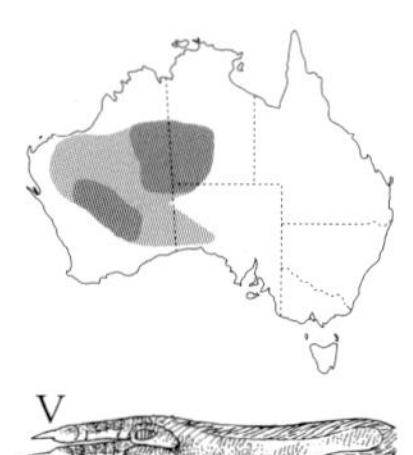

hb 125–220 mm; **t** 75–120 mm; **wt** 70–170 g
Stocky, short-limbed, short-tailed; head broad with pointed muzzle and short, rounded ears. Fur pale sandy-brown or rufous above, greyish-white below. Tail just over 1/2 hb; thick (sometimes carrot-shaped); bright rufous at base, tapering to tuft of longer black hairs. **Distribution, habitat and status** Widespread but patchy in sandy regions of arid central Aust and WA. Inhabits hummock grass plains, sand ridges, mulga shrubland on loamy sand. Populations fluctuate greatly with quality of seasons. Has declined in s. and e. of range. Vulnerable. **Behaviour** Sociable; constructs burrows on dune swales or lower slopes; mostly nocturnal but diurnal at times, suns in burrow entrance on warm winter mornings. Eats wide variety of invertebrates and small vertebrates, particularly large insects, spiders, scorpions, small rodents. Can live without free water. Mating occurs mid May to mid June; the 5–8 young are weaned at 3–4 months of age, in Oct–Nov. **Similar species** Ampurta is probably indistinguishable in the field. Kowari is paler and has distinctive tail colour and brush.

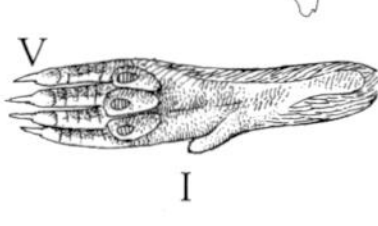

Ampurta *Dasycercus hillieri*

Until recently considered a subspecies of the Mulgara; the two are probably indistinguishable in the field. **Distribution, habitat and status** Core distribution is the central Simpson Desert with outlying populations in other parts of ne. SA and sw. Qld. Mostly found on dunes with canegrass. Endangered. **Behaviour** Presumably similar to Mulgara. Diet mostly large arthropods with some small vertebrates.

Southern Marsupial Mole *Notoryctes typhlops*
Northern Marsupial Mole *Notoryctes caurinus*

hb 120–160 mm; **t** 20–25 mm; **wt** 40–70 g
These two species little known, and probably indistinguishable in the field, so treated together here. Unmistakable. Extreme specialisation for burrowing through sand, including: mole-like tubular body with cone-shaped head and short, strong limbs; no functional eyes; ears reduced to a simple opening; leathery shield over muzzle; small mouth; 3rd and 4th foretoes with large shovel-like claws for digging. Fur dense and fine, uniformly pale golden white. **Distribution, habitat and range** Sparsely distributed across much of arid Aust, in sandy desert country, including dunefields and river flats. Northern species found in nw. WA. Endangered. **Behaviour** Solitary; lives mostly underground, said to come to the surface after heavy rain; tunnels through soft sand just below surface but also constructs deep burrows. Eats mostly insect larvae and pupae. Breeding system unknown.

Kowari

Mulgara

Southern Marsupial Mole

PLATE 5

Yellow-footed Antechinus *Antechinus flavipes*

Mardo (WA)

hb 90–160 mm; **t** 65–140 mm; **wt** 20–75 g; **teats** 8–14

Fur colour varies geographically, but always distinguished by change from grizzled slate-grey on head and shoulders to yellowish or russet flanks, rump, belly, legs and feet; chin and throat palest. Animals from ne. Qld (subspecies *rubeculus*) larger, more reddish; those from sw. WA (subspecies *leucogaster*) smaller, duller, darker with less contrast between head and rump, and whitish underparts. Tail well-furred, warm brown becoming grey or blackish towards tip where the hairs distinctly longer. Prominent pale yellow-grey eye-rings. **Distribution, habitat and status** The most widespread antechinus, subspecies *flavipes* extends from Eungella Qld s. in broad band to Mt Lofty Ranges SA (s. of about Port Stephens NSW to Grampians in w. Vic occurs only on inland side of GDR); subspecies *leucogaster* in sw. WA; subspecies *rubeculus* in ne. Qld from Mt Spec to Cooktown. In s. (Vic, SA and WA) mostly in dry sclerophyll forest and heathy woodland, some semi-arid shrubland; in n. also coastal heath, swampland and damp woodland; subspecies *rubeculus* inhabits tropical vine forest. Locally common. **Behaviour** More diurnal than other antechinus; forages with rapid jerky movements, on ground, in trees and among rocks; eats mostly invertebrates, also small vertebrates, eggs, nectar; builds nest of dry eucalypt leaves in tree cavity, amongst rocks or in building. Mates during 2-week period in Aug (s. Aust), Oct (se. Qld) or June–July (subspecies *rubeculus*); all males then die. **Similar species** Contrasting dorsal colours and yellowish fur on feet diagnostic.

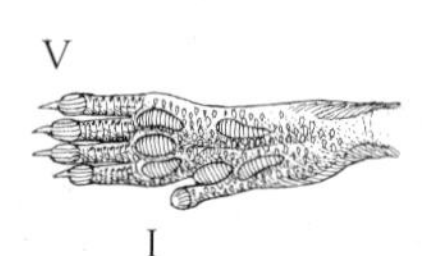

Atherton Antechinus *Antechinus godmani*

hb 90–155 mm; **t** 88–145 mm; **wt** 55–125 g; **teats** 6

Body colour dull brown, warmer on rump and sides of face, which can be bright ginger; obvious black guard hairs; underparts greyish fawn. Eyes small. Tail very sparsely furred except underside, which has thin crest of fur. **Distribution, habitat and status** Confined to rainforest above 600 m with annual rainfall >2700 mm, in the 130 km strip between Mt Bellenden Ker and Cardwell, ne. Qld. Rare and little-known. **Behaviour** Nocturnal or crepuscular; forages for invertebrates on forest floor; builds nest of leaves in tree hollow or in litter of epiphyte. Mates in July or Aug. All males die after mating. **Similar species** Distinguished from Yellow-footed Antechinus by heavier build, lack of eye-rings, lack of basal ear patches, and brown not yellow feet; from Rusty Antechinus (see page 252) by larger size, ginger hues and near-naked tail.

Cinnamon Antechinus *Antechinus leo*

hb 110–160 mm; **t** 85–140 mm; **wt** 34–120 g; **teats** 10

Upperparts rufous-grey flecked with yellow and cinnamon; blackish stripe along top of muzzle; sides of head, flanks, rump and legs rich cinnamon; underparts warm cinnamon to pale yellow; feet warm pale brown. Tail sparsely furred, bicoloured—brown above, darker and longer haired towards tip, pale cinnamon below. **Distribution, habitat and status** Only mammal endemic to Cape York Pen.; restricted to semi-deciduous rainforest in McIlraith and Iron Ranges, a north–south distribution of only 150 km. Locally common. **Behaviour** Mostly arboreal and nocturnal; shelters in nest in tree hollow; eats invertebrates gleaned from tree trunks, epiphytes, rotting logs. Mates in mid Sept and all males are dead by mid Oct. **Similar species** Only antechinus within distribution. Distinguished from Chestnut Dunnart by much larger size; from Red-cheeked Dunnart by cinnamon wash over most of body, not just face.

Fawn Antechinus *Antechinus bellus*

hb 112–145 mm; **t** 95–125 mm; **wt** 25–65 g; **teats** 10

Chunky build. Uniformly pale to medium grey upperparts, sometimes with pinkish tinge; pale eye-rings; cream to pale grey underparts; whitish feet and throat (sometimes stained yellow in males). Tail uniformly pale brown, slightly darker above. **Distribution, habitat and status** Only antechinus in NT; in tall open forest in monsoonal tropics n. of 14°S. Locally common but very patchy. **Behaviour** Arboreal, terrestrial. Shelters in tree hollows and hollow logs. Mates during a 2-week period in Aug, then all males die. Births occur in late Sept and early Oct. Litters of up to 10 weaned in Jan. **Similar species** Sandstone False Antechinus is darker, has noticeably longer ears with red patches behind. Red-cheeked Dunnart has grizzled brown fur with orange patches on sides of head.

Yellow-footed Antechinus

Atherton Antechinus

Cinnamon Antechinus

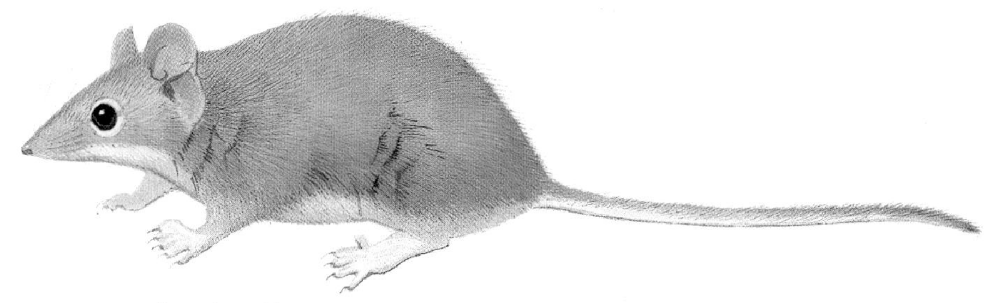
Fawn Antechinus

PLATE 6

Agile Antechinus *Antechinus agilis*

hb 80–116 mm; **t** 75–102 mm; **wt** 16–44 g; **teats** 6–10
Very similar to Brown Antechinus but slightly smaller, fur more greyish. **Distribution, habitat and status** Along the GDR and adjacent coastal plains through s., central and e. Vic to se. NSW, n. to Kioloa on the coast and Orange on the western slopes. Most types of wet or moist forest, heath and woodland are utilised, from sea-level to above 1800 m. Replaced by Yellow-footed Antechinus in drier box and ironbark forest on the inland slopes but can overlap narrowly. Common and secure. **Behaviour** As for Brown Antechinus. **Similar species** See under Brown Antechinus.

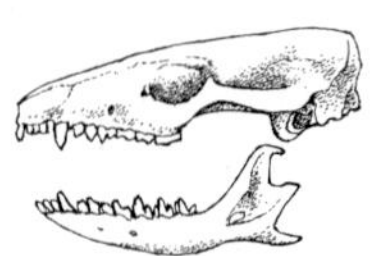

Brown Antechinus *Antechinus stuartii*

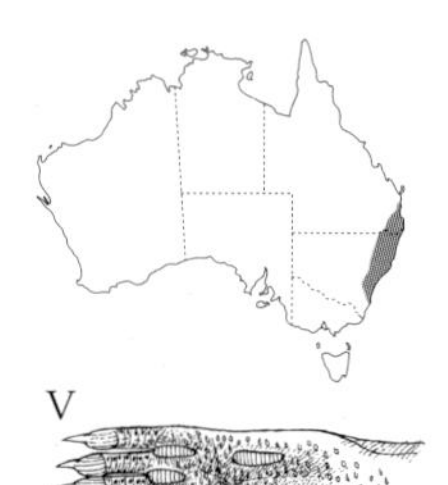

hb 93–130 mm; **t** 92–120 mm; **wt** 18–60 g; **teats** 6–10
Uniform light chocolate brown above, flecked with buff; paler brown below and on tail; upper surface of feet pale brown not greyish or yellowish; no pale eye-rings or coloured patches behind ear; t = hb. **Distribution, habitat and status** Along GDR and coastal plains from Kioloa, NSW, to se. Qld in wide range of moist habitats—rainforest, sclerophyll forest, woodland, heath—from sea level to sub-alps. In se. Qld apparently restricted to sub-tropical vine forest, both lowland and upland. Common. **Behaviour** Mostly nocturnal; climbs readily with rapid staccato movements; gleans invertebrates from litter and fissures in tree trunks. Large communal nests of dry leaves are built in tree hollows. Mating highly synchronised to a 2-week period, which varies from mid Aug in s. to late Sept in n., followed by death of all males. **Similar species** Agile Antechinus distinguishable only with experience—generally smaller, greyish-brown rather than light chocolate-brown. Yellow-footed Antechinus has contrasting dorsal colour pattern. Dusky Antechinus is larger, darker, with longer muzzle, black granules on sole of hindfoot, long pale claws. See Subtropical Antechinus (p. 252).

Dusky Antechinus *Antechinus swainsonii*

Swainson's Antechinus

hb 90–185 mm; **t** 75–120 mm; **wt** 38–170 g; **teats** 6–10
Large antechinus, thickset, particularly in hindquarters; long, slender muzzle; t 80% of hb. Eyes small, ears short, do not project far above fur. Claws long and pale. Black granules on soles. Upperparts uniformly dark brown to blackish, underparts pale grey; tail lightly furred, brown, paler below. Largest and darkest at higher altitudes in wet forest and heath; smaller, more grey-brown and grizzled on coast in heath and heathy sclerophyll forest. **Distribution, habitat and status** Patchy distribution along GDR and coastal plains from se. Qld to sw. Vic and Tas, but not Bass Strait Is. Inhabits dense wet vegetation, from coastal heath to wet sclerophyll forest, rainforest and subalpine heath. Uncommon to locally common. **Behaviour** Entirely terrestrial, partly diurnal, nocturnal; digs for invertebrates in topsoil and litter, may also eat small vertebrates such as skinks. Females build spherical nest of dry leaves in shallow burrow. Mating occurs in a 2-week period in Aug (Sept at high altitudes), followed by death of all males. **Similar species** See under Swamp Antechinus, Brown Antechinus and Subtropical Antechinus (p. 252).

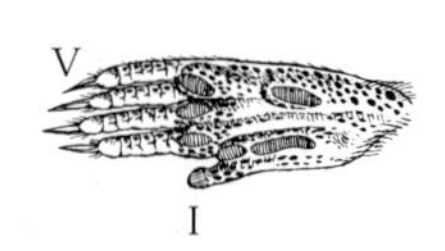

Swamp Antechinus *Antechinus minimus*

hb 95–140 mm; **t** 65–100 mm; **wt** 28–100 g; **teats** 6 in Tasmania, 8 on mainland
Similar in size, shape and appearance to Dusky Antechinus but paler and more yellow-ginger in colour. Upperparts grizzled brown, with distinct yellowish or rufous wash, particularly on rump, flanks and hindlegs; underparts paler, grey-yellow or buff. Pale eye-rings; long, strong foreclaws. t 70% of hb. **Distribution, habitat and status** The southernmost antechinus, restricted to Tas, Bass Strait Is, and coastal mainland between Corner Inlet, Vic and Robe, SA. Occupies dense wet heath and heathy woodland, sedgeland and dense tussock grassland. In Tas also found in adjacent wet forest gullies and up to 1000 m, but on mainland rarely above 200 m. Locally common. **Behaviour** Entirely terrestrial, digs for invertebrates in topsoil and litter. Builds nest of dry grass in shallow burrow. Mating highly synchronised within a population, occurs in May, June or July, followed by complete die-off of males. **Similar species** Dusky Antechinus has uniformly darker fur above, is less grizzled and usually lacks yellow or rufous wash on rump, lacks eye-ring; stronger contrast in colour between upper and underparts. Agile Antechinus is uniformly grey-brown, t about = hb.

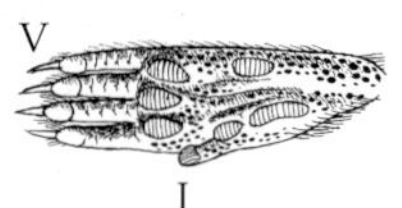

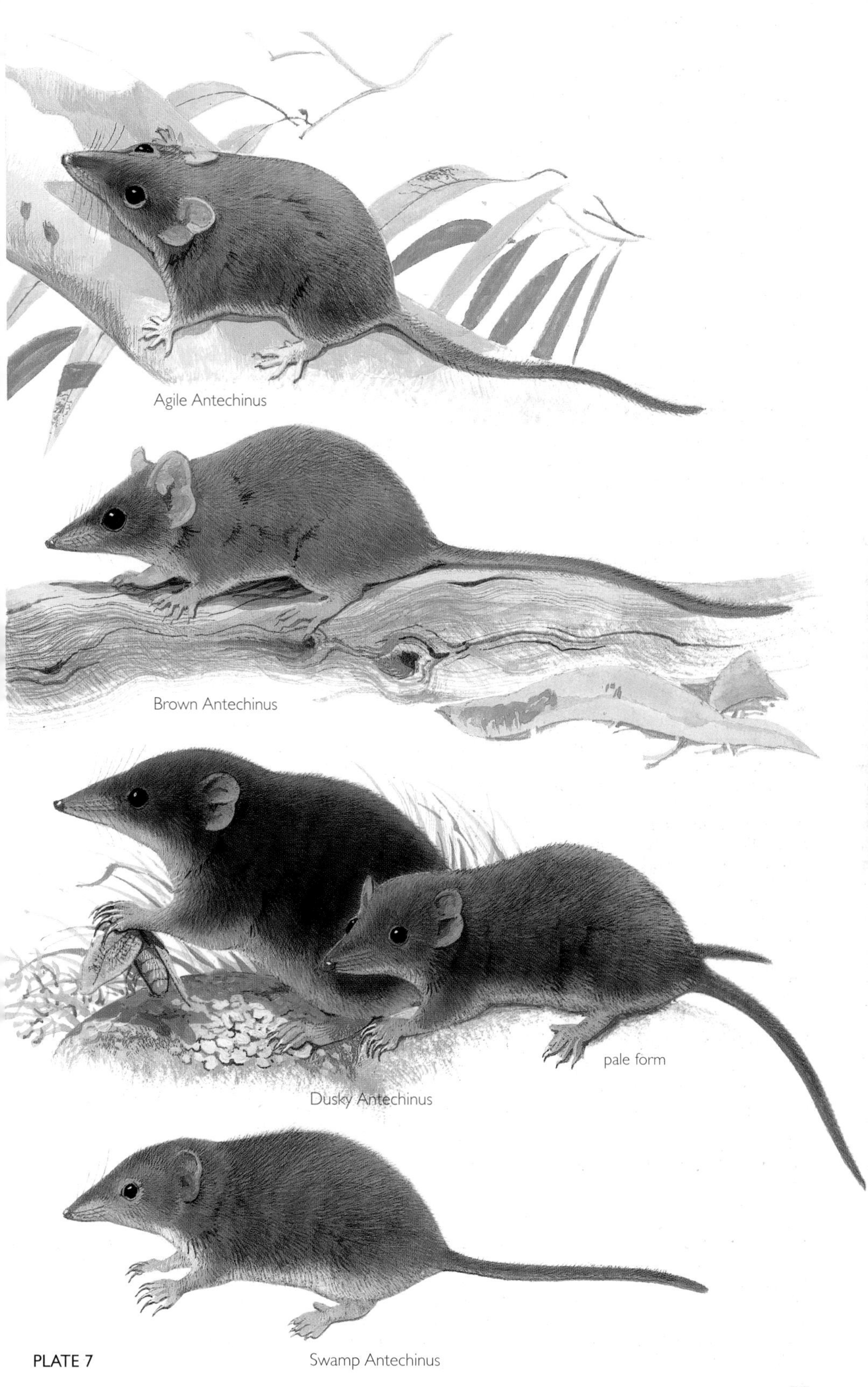
Agile Antechinus
Brown Antechinus
pale form
Dusky Antechinus
Swamp Antechinus

PLATE 7

Dibbler *Parantechinus apicalis*

Southern Dibbler

hb 140–145 mm; **t** 90–115 mm; **wt** 40–100 g

Upperparts rich brown, strongly flecked with cream; flanks and outer surface of limbs washed cinnamon, grading to yellowish-grey underparts. Distinct white eye-rings; upper surfaces of feet rufous grey. Tail well-furred with obvious taper, similar colour to body. Ears short and well-furred. **Distribution, habitat and status** Confined to old-growth mallee heath in coastal sw. WA between Fitzgerald R. NP and Torndirrup NP, plus Boulenger and Whitlock Is off Jurien Bay where it occurs in low heath. Endangered. **Behaviour** Solitary, mostly nocturnal; forages for invertebrates and small vertebrates in leaf litter, also climbs shrubs for insects and nectar. Mates in March–April. **Similar species** Tapering tail, bold eye-rings and rich brown, pale flecked upperparts diagnostic.

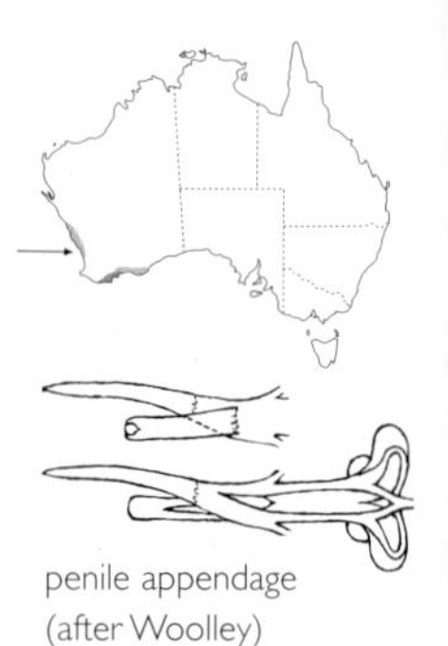

penile appendage
(after Woolley)

Kaluta *Dasykaluta rosamondae*

Little Red Kaluta, Little Red Antechinus, Spinifex Antechinus

hb 90–105 mm; **t** 55–70 mm; **hf** 15–18 mm; **el** 11–13 mm; **wt** 20–40 g

Small, robust, shaggy-furred, with short, thick, tapering tail. Uniform russet-brown or copper above, paler below; muzzle short—eyes closer to nostril than to ears; ears short and furred, barely protruding above crown; feet well-furred; t 60–66% of hb, often slightly swollen at base. **Distribution, habitat and status** Confined to subtropical arid hummock grassland in Pilbara and w. Little Sandy Desert regions of nw. WA. Locally common. **Behaviour** Mainly nocturnal, terrestrial, feeds on invertebrates. Mates in Sept and gives birth about 7 weeks later in Nov. Up to 8 young are reared and become independent at 14–16 weeks old. All males die after mating but some females may live to breed a second time. **Similar species** None; head profile, uniform dark colouration and shaggy fur are diagnostic.

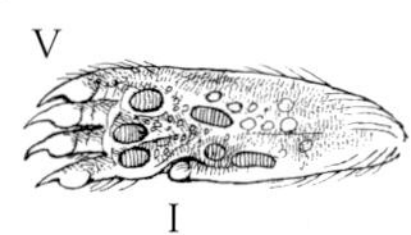

Kultarr *Antechinomys laniger*

hb 70–98 mm; **t** 100–145 mm; **hf** 27–30 mm; **wt** 20–30 g

Size of House Mouse but delicate, long-limbed, elegant; long, thin hindfeet with only 4 toes; interdigital pads fused into single tri-lobed pad granulated at front, hairy behind; tail thin, clearly > hb, outer half with dark blackish tuft; sharply pointed muzzle; ears very large, rounded; eyes large, protruding. Fur fawn-grey to sandy-brown above, darker mid-line on face and crown, dark eye-rings; underparts including lower limbs whitish; tail darker above. **Distribution, habitat and status** Generally uncommon, populations fluctuate seasonally. Patchily distributed across most of arid Aust in broad band from sw. Qld and w. NSW through n. SA and s. NT to central and s. WA. Occupies open, arid, tree-less plains and claypans, gibber, sandy desert with sparse grasses and small shrubs, Mitchell Grass downs. Isolated population in saltmarsh near mouth of Roper R., NT presumed extinct. **Behaviour** Usually solitary; forages at night with rapid, graceful, bounding gait for terrestrial invertebrates, e.g. spiders, crickets, cockroaches; builds nest of dry grass in soil cracks, burrows constructed by other animals, beneath rocks or dense vegetation. Mates in winter and spring; pouch young recorded Aug–Nov. **Similar species** Long limbs and tufted tail diagnostic. Dunnarts have shorter limbs and 5 hindtoes. Hopping Mice have 'Roman nose' not a finely pointed muzzle, lack dark eye-rings, have short forelimbs and a hopping not bounding gait.

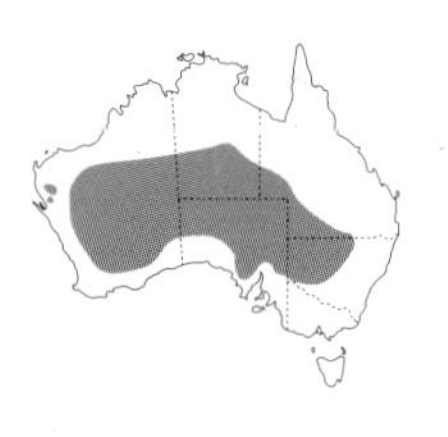

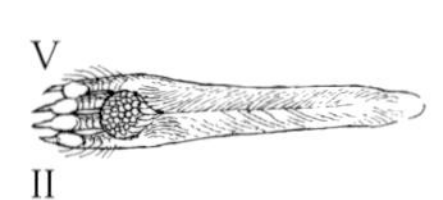

Dibbler

Kaluta

Kultarr

PLATE 8

Fat-tailed False Antechinus *Pseudantechinus macdonnellensis*

Fat-tailed Antechinus, Red-eared Antechinus

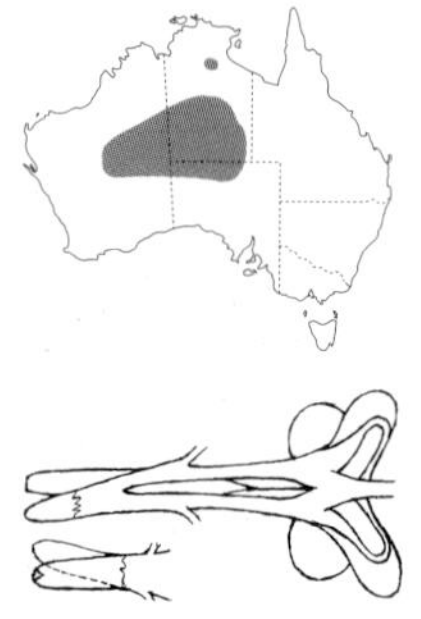

penile appendages (after Woolley)

hb 75–94 mm; **t** 72–90 mm; **hf** 12.5–15.0 mm; **el** 13–19 mm; **wt** 18–33 g
Sturdily built with a short, sparsely furred, carrot-shaped tail. Gingery brown above; whitish below; large, distinctive orange patches behind and below each ear, grading into whitish throat; upper surfaces of feet pale. Sharply pointed muzzle and bulging eyes. Males have distinct appendage to penis; females have 6 teats. Lacks P_3; P^3 greatly reduced. **Distribution, habitat and status** Patchy and uncommon on sparsely vegetated rocky slopes and adjacent plains across central and w. Aust, n. to Pine Creek NT, s. to Copper Hills SA. **Behaviour** Little studied. Mostly nocturnal but will sun; primarily insectivorous. Litters of up to 6 young born July–Sept in central Aust, later in the west. **Similar species** Woolley's False Antechinus very similar but slightly larger, usually has P_3, males lack penile appendage; Ningbing False Antechinus smaller, tail proportionately longer, females have only 4 teats. Tan False Antechinus (p. 252) is smaller and more reddish, has obviously bicoloured tail. Dunnarts may have swollen tails but are smaller, daintier, lack orange patches behind ears.

Woolley's False Antechinus *Pseudantechinus woolleyae*

3 lower premolars

hb 77–99 mm; **t** 75–97 mm; **hf** 13.0–15.5 (14.0) mm; **el** 18–20 mm; **wt** 18–43 g
The largest False Antechinus, t clearly < hb, moderately furred, often carrot-shaped. Upperparts rich brown with cinnamon patch behind ears, underparts pinkish buff. **Distribution, habitat and status** Sparsely distributed on rocky hillsides with acacia scrub and hummock or tussock grass in central-w. of WA: Pilbara, Ashburton and Murchison regions. **Behaviour** Little known. Births occur Sept–Oct; sexually mature at 10 months; breeding life 2 or more years. **Similar species** Tan False Antechinus (p. 252) is smaller, more reddish, lacks P_3, males have penile appendage. See under Fat-tailed False Antechinus and Ningbing False Antechinus.

Sandstone False Antechinus *Pseudantechinus bilarni*

Sandstone Antechinus

hb 80–96 mm; **t** 99–120 mm; **hf** 14.8–18.4 (16.6) mm; **wt** 21–30 g
Very long narrow muzzle. Greyish-buff above, cinnamon patches behind and below ears; pale grey or whitish below. t > hb (109–140%), thin, sparsely furred. Ears large, naked. **Distribution, habitat and status** Patchy and locally common in nw. Top End and w. coast of Gulf of Carpentaria. Inhabits boulder scree and rock slabs in a variety of vegetation; coastal rock platforms on Marchinbar I., NT. **Behaviour** Solitary, partly diurnal, insectivorous. Mates May–July, births occur mid Aug–Sept, sexually mature at 11 months old. Most males die after mating but about 25% of both sexes survive to breed in a second year. **Similar species** Carpentarian False Antechinus is smaller, t < hb. Fawn Antechinus lacks cinnamon patches behind smaller ears, t < hb. Red-cheeked Dunnart has orange face.

Carpentarian False Antechinus *Pseudantechinus mimulus*

hb 80–90 mm; **t** 68–70 mm; **hf** 13.3–13.5 mm; **wt** 15–18 g.
The smallest false antechinus. Tail 75–85% of hb, moderately well furred, can be carrot-shaped. Lacks P_3. Upperparts uniform buff-brown, distinct cinnamon patch behind ears; underparts pale grey. **Distribution, habitat and status** Known from near Mt Isa, Qld and from 3 Is in ne. NT: North, Central and South West Is, Sir Edward Pellew Group. Type specimen from Alexandria Stn, ne. NT. Occupies stony hillsides with shrubby open woodland and hummock grass understorey. Vulnerable. **Behaviour** Not described. **Similar species** Sandstone False Antechinus has t > hb. Fat-tailed False Antechinus similar, though larger, has greatly reduced P^3.

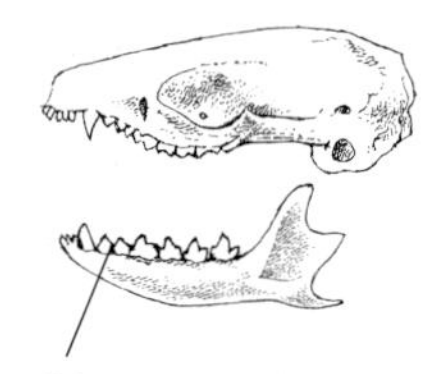

2 lower premolars

Fat-tailed False Antechinus

Woolley's False Antechinus

Sandstone False Antechinus

Carpentarian False Antechinus

PLATE 9

Ningbing False Antechinus *Pseudantechinus ningbing*

hb 73–93 mm; **t** 74–94 mm; **hf** 13.0–15.4 (14.3) mm; **wt** 15–33 g

Pale greyish-brown above, greyish-buff below; distinct cinnamon patches behind and below the large, lightly furred ears. Tail often carrot-shaped, similar length to hb, lightly furred. **Distribution, habitat and status** Locally common throughout Kimberley region, e. to Gregory NP, NT. Occupies rocky outcrops, both sandstone and limestone, in a range of vegetation types. **Behaviour** Little known; mates in June, gestation long (45–52 days), births July–Aug, weaning Oct–Nov. **Similar species** Similar in form to Fat-tailed and Woolley's False Antechinus but t = hb, females have 4 not 6 teats, P_3 absent, males have short and indistinct penile appendage.

Wongai Ningaui *Ningaui ridei*

Ride's Ningaui

hb 58–75 mm; **t** 60–70 mm; **wt** 6.5–10.5 g

Tiny carnivore with spiky, unkempt appearance; narrow muzzle, small, close-set eyes; ears barely protrude above fur of crown; thin tail similar in length and colour to hb. Fur grizzled gingery-brown to greyish-brown above with distinctive protruding black guard hairs; yellowish-grey flanks and whitish underparts. Sides of face and base of ears tinged ginger. **Distribution, habitat and status** Widespread in arid Aust from w. of Kalgoorlie across n. half of SA, sw. Qld and s. half of NT. Restricted to hummock grassland with scattered trees and shrubs growing on dunes and sand plains. Sparse, locally common. **Behaviour** Usually solitary; shelters in nest under spinifex clump, in hollow log or shallow tunnel; hunts at night for invertebrates (mostly <10 mm long, but also larger grasshoppers, spiders, cockroaches). Undergoes torpor during unfavourable conditions. Births occur Sept–Oct; litter size 5–7; young left in nest after about 6 weeks old and are independent at about 13 weeks. Two litters may be raised per year and few animals survive to breed in a second year. **Similar species** Dunnarts have sleeker fur, large eyes and ears and pale feet and tails; planigales have a noticeably flattened, triangular head and relatively broader hindfeet. Mallee Ningaui overlaps narrowly w. of Kalgoorlie, WA; distinguished by first toe not reaching beyond base of interdigital pads; reddish colour on face is restricted to a diffuse cinnamon crescent below and behind eye, and a patch below the ear.

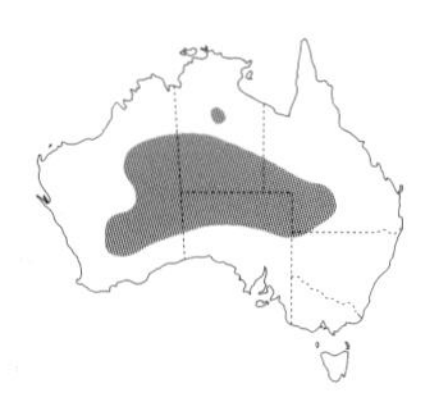

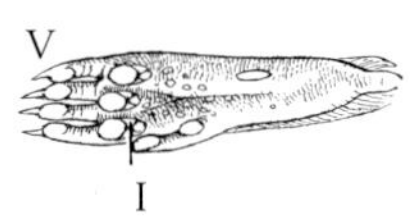

1st toe level with interdigital pads

Pilbara Ningaui *Ningaui timealeyi*

hb 45–58 mm; **t** 60–76 mm; **wt** 5.0–9.4 g

Similar to Wongai Ningaui but with rufous wash across face, ears, flanks. Tail > hb. Females have only 6 teats. **Distribution, habitat and status** Inhabits dense to mid-dense hummock grassland, usually along drainage line or runoff area and with upper stratum of open mallee or acacia scrub in the Pilbara region, WA. Can be locally common, e.g. in Hamersley Ranges. **Behaviour** Similar to Wongai Ningaui, partly arboreal. Breeding seasonal, depending on duration of rainy period. Mating begins in Sept and some females are still lactating in March. Litter size 4–6. **Similar species** See under Wongai Ningaui.

curled supratragus of ningauis

Mallee Ningaui *Ningaui yvonneae*

Southern Ningaui

hb 50–74 mm; **t** 57–70 mm; **hf** 12–14 mm; **el** 13–14 mm; **wt** 5–10 g

As for Wongai Ningaui but fur tawny-olive to greyish-olive grading to pale grey underparts, whitish chin; face and muzzle dark grey; diffuse cinnamon crescent below and behind eye; cinnamon patch below ear. Tail = hb. Always 7 teats arranged in 2 crescents of 3 with an anterior one completing the circles. **Distribution, habitat and status** Sparsely distributed across southern Australia from Lake Cronin, sw. WA to Big Desert, Sunset Country and Annuello, Vic, and Round Hill, NSW. Always associated with spinifex (*Triodia*), either as a component of the understorey in semi-arid heath and mallee scrub, or as the dominant species in hummock grassland. Can be locally common. **Behaviour** Mostly solitary, nocturnal; shelters beneath spinifex hummocks or other dense vegetation; undergoes torpor during unfavourable conditions. Preys on wide range of invertebrates. A single litter of 5–7 young is born in spring or early summer; young remain in the pouch for about 30 days and are independent at 70–80 days old. **Similar species** See under Wongai Ningaui, dunnarts and planigales.

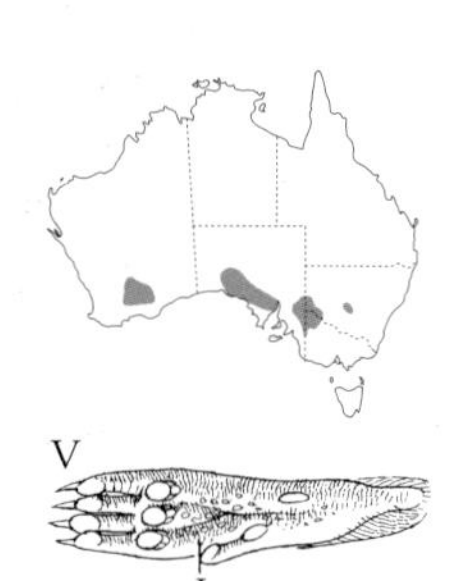

1st toe short of level of interdigital pads

Ningbing False Antechinus

Wongai Ningaui

Pilbara Ningaui

Mallee Ningaui

PLATE 10

Giles' Planigale *Planigale gilesi*

Paucident Planigale

hb 60–80 mm; **t** 55–70 mm; **wt** 6–15 g

A large planigale; upperparts uniform mid-grey with cinnamon wash; underparts whitish or buff, palest on throat and sides of neck; ears short, furred. Claws black. Tail usually thin but can become swollen in good seasons, t < hb. Only 2 premolars in each tooth row. **Distribution, habitat and status** Sparsely distributed in arid and semi-arid clay-soil areas from e. of Lake Eyre, SA to Moree, NSW and s. NT to Murray R. floodplain w. of Mildura, Vic. Inhabits dense vegetation such as lignum, canegrass and sedgeland on floodplains and dune swales, and grassy plains with cracking clay soils. **Behaviour** Forages rapidly through litter, low vegetation and soil crevices for invertebrates and small vertebrates. Mostly nocturnal, suns in winter, enters torpor during unfavourable conditions. Breeds mid July till Jan, peaking in Sept; some females raise 2 litters per season; young independent at 75 days old. **Similar species** See under Narrow-nosed Planigale. All other planigales have 3 premolars in each tooth row.

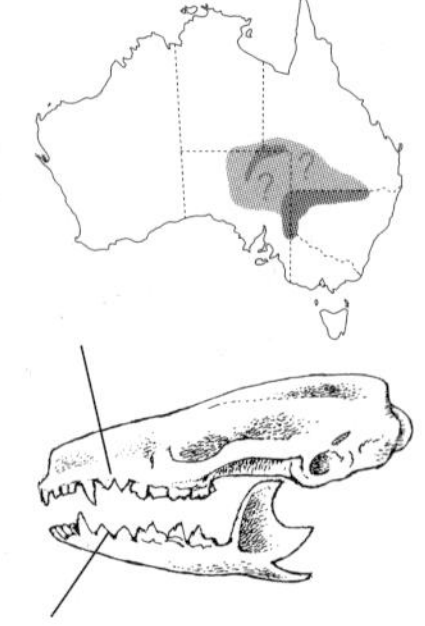

2 upper and 2 lower premolars

Common Planigale *Planigale maculata*

hb 70–95 mm; **t** 60–90 mm; **wt** 6–12 g

Large planigale, skull not extremely flattened; uniform rufous-brown above, sometimes with small dull spots, rarely with bold white spots; cinnamon wash to flanks and lower face; underparts pale yellowish grey, yellowest on chin and throat. Tail thin, usually < hb. **Distribution, habitat and status** Dense sedgeland and scrub along edges of floodplains in n. Kimberley of WA and n. NT; sand dunes in e. Arnhem Land and Groote Eylandt; sclerophyll and rainforest on e. coast from Cape York to Upper Hunter R., NSW. **Behaviour** Terrestrial, nocturnal, hunts amongst litter and crevices for invertebrates. In NT litters of 8–12 born throughout year; on e. coast litters of 5–10 mostly born Oct–Jan. Commonly 2 or more litters per year and males may breed until at least 24 months old. **Similar species** Long-tailed Planigale is much smaller, has markedly flattened head and tiny eyes. Giles' Planigale similar size but plainer grey, has only 2 upper and 2 lower premolars.

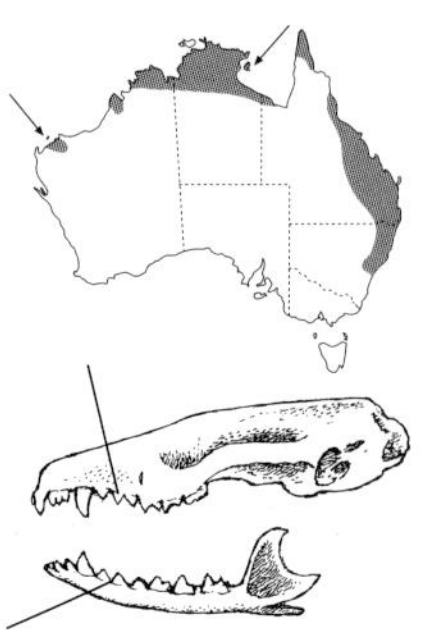

3 upper and 3 lower premolars

Narrow-nosed Planigale *Planigale tenuirostris*

hb 50–75 mm; **t** 50–65 mm; **wt** 4–9 g

Much smaller than House Mouse; flat triangular head; tail slightly < hb. Pale rufous-brown above, paler on shoulders; pale grey-buff below, whitish chin and lower face; claws caramel. **Distribution, habitat and status** Sparsely distributed through arid and semi-arid e. Aust from s. NT to w. slopes of NSW. Occupies more open habitats than Giles' Planigale—e.g. Mitchell Grass, saltbush and canegrass on alluvial plains, dry lake beds, drainage lines, also mallee with cracking clay soils, and gibber. **Behaviour** Active hunter of invertebrates in soil cracks and beneath litter; mostly nocturnal, terrestrial but does climb in low vegetation. Breeding period extends through late winter, spring and summer; 10–12 teats in temporary pouch. **Similar species** Giles' Planigale is larger, uniformly pale grey, has blackish claws and only 2 upper and 2 lower premolars. Ningauis have cone-shaped (not flattened) head and 'bristly' fur.

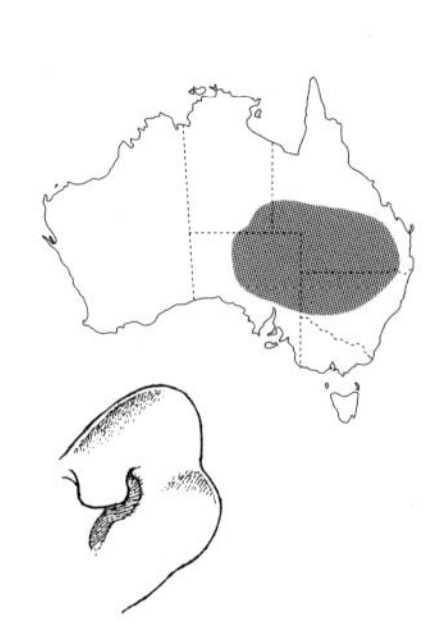

flat supratragus of planigales

Long-tailed Planigale *Planigale ingrami*

Northern Planigale

hb 55–65 mm; **t** 45–60 mm; **wt** 4–6 g

Tiny—smallest marsupial; remarkably flat, triangular skull—skull depth (4 mm) 20% of skull length. Tail thin, length similar to hb. Colour variable: upperparts pale grey-brown with buff tips; underparts paler with yellowish tinge, cheeks and chin whitish. Central Qld animals browner. Ears large, naked and translucent. **Distribution, habitat and status** Common in seasonally flooded grasslands and savannah woodlands with cracking clay soils, riparian areas and blacksoil plains across n. Aust from the North Pilbara, Great Sandy Desert and Kimberley in WA, Barkly Tablelands in NT, to Townsville, Qld. **Behaviour** Active nocturnal hunter of invertebrates and small vertebrates in soil crevices and leaf litter. In NT litters of 4–6 young born Feb to April; in ne. Qld litters of 4–12 born year round, but mostly Dec-March. Young carried in well-developed, rear-opening pouch till 6 weeks old then left in concealed nest; independent at about 12 weeks. **Similar species** See under Common Planigale.

Gile's Planigale

Common Planigale

Narrow-nosed Planigale

Long-tailed Planigale

PLATE 11

White-footed Dunnart *Sminthopsis leucopus*

hb 70–110 mm; **t** 70–90 mm; **wt**: 19–27 g. Tas animals slightly larger.
Mouse-sized; muzzle finely pointed. Upperparts greyish-buff, darker on top of muzzle, warmer buff on sides of head. Underparts pale grey. Tail thin, never swollen, slightly < hb, bicoloured in same shades as rest of pelage. Feet whitish. **Distribution, habitat and status** The most southerly dunnart. Patchily distributed in lowland heathy woodland and forest, coastal scrub, coastal dune grassland in se. NSW, s. Vic, Tas including Cape Barren, East Sister and West Sister Is of Furneaux Group. Outlying population, apparently of this species, occurs in upland rainforest in the Paluma area of ne. Qld. **Behaviour** Terrestrial, nocturnal, predator of arthropods and small skinks. Constructs bark nest beneath fallen timber or dense litter. Births in Aug or Sept; pouch life 7–8 weeks. **Similar species** Common Dunnart almost indistinguishable, usually lacks striations on foot pads, shorter muzzle and steeper forehead; occupies drier habitats. Fat-tailed Dunnart has fully fused interdigital pads; House Mouse lacks deep ear notch, has distinctive odour, sparsely furred flesh-coloured tail.

Common Dunnart *Sminthopsis murina*

Mouse Dunnart

hb 65–100 mm; **t** 68–90 mm; **wt** 12–28 g
Mouse-grey above, darker on head and neck with buff wash on face and flanks; whitish underparts and feet. Ears large, rounded; tail never swollen, length about = hb. Underside of hindfoot hairy at sides and heel, interdigital pads granulated, rarely striated. **Distribution, habitat and status** Widespread in variety of heathy dry sclerophyll forest and mallee heath across se. Aust from Sunshine Coast, Qld to Fleurieu Pen., SA but not along coast s. of Sydney. Locally common in se. Qld and ne. NSW, rare elsewhere. Isolated population of uncertain taxonomic status (*S. m. tatei*) on Atherton Tableland, ne. Qld. **Behaviour** Nocturnal, terrestrial carnivore, eats mostly insects and spiders. Breeds Aug-March; gestation 12 days; 8–10 young may be born, weaned at about 65 days old and mature at 150 days. **Similar species** See under White-footed Dunnart.

Fat-tailed Dunnart *Sminthopsis crassicaudata*

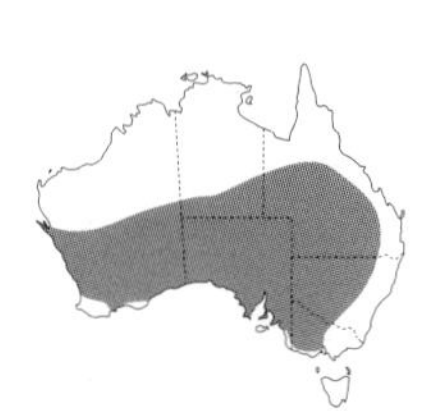

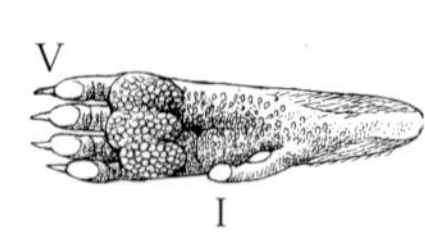

hb 60–90 mm; **t** 45–70 mm; **hf** 14–16 (15) mm; **wt** 10–20 g
Small, colourful dunnart with short, spindle-shaped, greyish tail (longer in n. of range), very large ears and a sharply pointed muzzle. Upperparts buff or fawn, warmer on face; darker patches around eyes and on crown and forehead; sometimes whitish crescents behind and below ears. Underparts and feet whitish, sharply demarcated from upperparts. Interdigital pads fused but still distinguishable as 3 pads. **Distribution, habitat and status** Common and widespread in great variety of open habitats in s. and central Aust: tussock and hummock grassland, gibber plain, saltbush and bluebush plains, claypans, rough pasture and the edges of stubble paddocks. **Behaviour** Nocturnal, terrestrial, forages for invertebrates in open spaces adjacent to cover. Shelters communally in nest of grass beneath log or rock; will become torpid during unfavourable conditions. Breeds May–June, gestation lasts 12 days and 8–10 young are born; weaned after about 70 days. **Similar species** See under Stripe-faced Dunnart. Julia Creek Dunnart is 3 to 4 times heavier and hf is > 20 mm. Hairy-footed and Lesser Hairy-footed Dunnarts have covering of fine hairs on soles of hindfeet.

Stripe-faced Dunnart *Sminthopsis macroura*

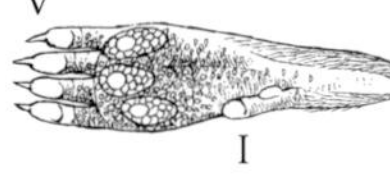

hb 75–98 mm; **t** 80–100 mm; **hf** 17–18 mm; **wt** 15–25 g
Pale grey-brown above with distinct dark head stripe to between ears. Cinnamon wash on face, flanks and base of tail. Sometimes whitish patches behind ears. Tail > hb, often swollen to carrot shape, hairs extend beyond tail tip. White underparts and feet. **Distribution, habitat and status** Scattered throughout much of arid zone, e. to near Coonabarabran, NSW. Locally common. Mostly on sandy substrates but also cracking clay and stony plains. Vegetation includes dune hummock grasslands, tussock grasslands, arid shrublands. **Behaviour** Broadly similar to Fat-tailed Dunnart, often eats termites. Females receptive July–Feb; gestation 11 days; litter size 6–8; pouch life 40 days and young weaned at about 70 days. Two litters per season not unusual. **Similar species** Fat-tailed Dunnart lacks obvious head stripe, has shorter, spindle-shaped tail with hairs never extending beyond tip. Julia Creek Dunnart is 2 to 3 times heavier, hf > 20 mm.

White-footed Dunnart

Common Dunnart

Fat-tailed Dunnart

Stripe-faced Dunnart

PLATE 12

Little Long-tailed Dunnart *Sminthopsis dolichura*

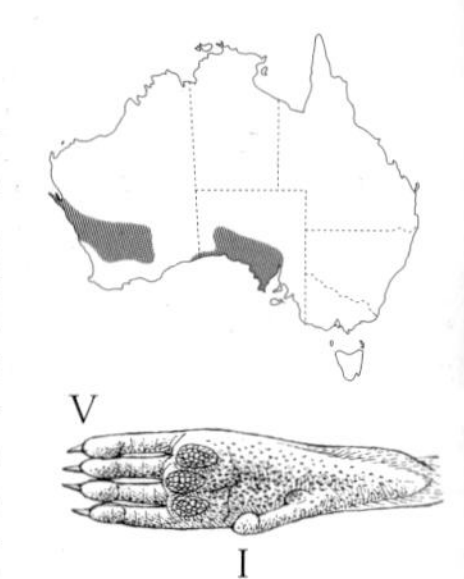

hb 65–95 mm; **t** 85–105 mm; **hf** 16–17 mm; **el** 17–19 mm; **wt** 10–20 g
Grizzled grey above; face, cheeks and small patches behind ears washed cinnamon; narrow black eye-rings; whitish underparts, including chin and feet. Tail 20% >hb, bicoloured in same shades as rest of pelage; never swollen. **Distribution, habitat and status** Two discrete populations: n. Goldfields and Geraldton hinterland, WA, and coastal SA w. of Port Augusta. In dry sclerophyll forest, semi-arid woodland, mallee and heath. Most common in early seral stages (3–4 years) following fire. **Behaviour** Nocturnal, terrestrial insectivore/carnivore; males in particular have large drifting home ranges—an adaptation to exploiting patchy and transitory habitats. Females receptive March–Aug; one litter of up to 8 young produced per year. **Similar species** Differs from other dunnarts within range by combination of predominantly grey upperparts and white underparts; long tail; lack of evenly granulated terminal pads on toes.

Gilbert's Dunnart *Sminthopsis gilberti*

hb 80–90 mm; **t** 75–90 mm; **hf** 18 mm; **el** 21 mm; **wt** 14–25 g
Sooty grey above, darker crown; whitish patches behind and below ears; whitish underparts, including chin and feet. Tail ≤ hb, bicoloured in same shades as rest of pelage; never swollen. **Distribution, habitat and status** Locally common in heath and heathy forest on near-coastal ranges and southern wheatbelt of WA. Isolated population on Roe Plain near WA–SA border. Overlaps with Little Long-tailed Dunnart in Jarrah forest near Collie but not with Grey-bellied Dunnart, which occurs more coastally. **Behaviour** Nocturnal, insectivore; may nest above ground in hollows or dense vegetation. Females receptive Sept–Dec, young weaned Jan–Feb. **Similar species** Little Long-tailed Dunnart has tail longer than hb; coloured patch behind ears brownish not white; ears and hindfeet shorter (19.3 vs 21; 16.7 vs 18 mm respectively). White-tailed Dunnart has thick, pinkish tail and is larger.

White-tailed Dunnart *Sminthopsis granulipes*

Ash-grey Dunnart

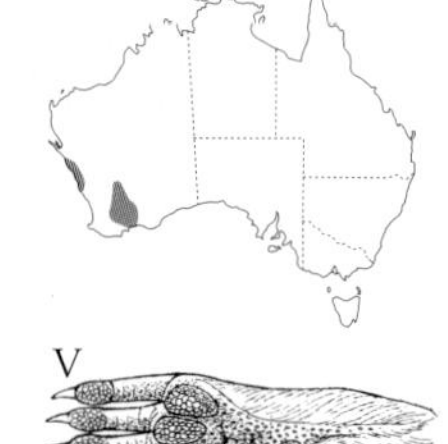

hb 70–100 mm; **t** 56–68 mm; **wt** 18–35 g
Upperparts uniform light fawn grading to white underparts; feet pinkish white. Tail diagnostic—t < hb, often swollen at base, pinkish-white with a narrow brown stripe along upper surface. Has fine even granulations over much of sole of hindfoot. **Distribution, habitat and status** Locally common in sparse to mid-dense low shrubland with or without emergent mallee eucalypts on sandy substrates in the western Goldfields Region, WA and coastal heath between Kalbarri and Jurien, WA. **Behaviour** Little known, presumably a nocturnal, terrestrial insectivore; births probably occur in winter and young are weaned by late Oct.

Grey-bellied Dunnart *Sminthopsis griseoventor*

hb 65–95 mm; **t** 65–98 mm; **hf** 16–17 mm; **el** 17–18 mm; **wt** 15–25 g
Fur long, soft; upperparts uniformly mid-dark grey, paler on flanks; face and cheeks pale olive-grey; no white patches behind ears; underparts grey not whitish; tail = hb, thin and sparsely haired giving a pinkish appearance. **Distribution, habitat and status** Locally common on coastal plains and adjacent lateritic ranges of sw. WA from Gairdner Range in n. to Cape Arid NP in e.; also Boulenger Is. Occupies wide range of habitats—heathy forest and woodland, banksia woodland, melaleuca swampland, dense mature heath. **Behaviour** Nocturnal, insectivore/carnivore; shelters in nest just below ground. One litter of up to 8 born in Aug (Boulenger Is.) or Oct (mainland). **Similar species** Combination of grey underparts, resulting in reduced contrast between upper and underparts, and pinkish tail is diagnostic.

Little Long-tailed Dunnart

Gilbert's Dunnart

White-tailed Dunnart

Grey-bellied Dunnart

PLATE 13

Chestnut Dunnart *Sminthopsis archeri*

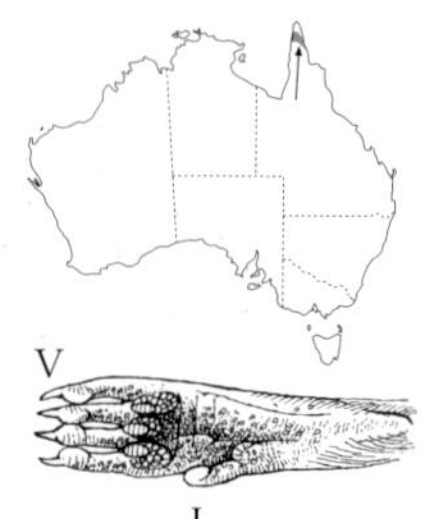

hb 85–105 mm; **t** 82–105 mm; **hf** 17–20 mm; **el** 17–19 mm; **wt** 15–20 g
Upperparts tan or greyish tan; darker around eye; buff wash on cheeks and behind ears. Pale grey underparts including chin and tops of feet. Tail = hb; thin and almost naked. Apical granules on interdigital pads large, oval and striated. Slight Roman-nose profile. **Distribution, habitat and status** Locally common in tall stringybark woodlands on red earth soils of n. Cape York. **Behaviour** Little known; breeds in dry season, July–Oct. **Similar species** Red-cheeked Dunnart has orange face; Butler's Dunnart lacks striations on apical granules of interdigital pads. Cinnamon Antechinus is at least twice the weight, has yellowish underparts.

Butler's Dunnart *Sminthopsis butleri*

hb 75–88 mm; **t** 72–90 mm; **hf** 16–19 mm; **el** 14–17 mm; **wt** 10–20 g
Buffy brown to mouse-grey above; buff wash on cheeks and flanks; whitish underparts and feet. Tail thin; sparsely furred; bicoloured in similar shades to rest of pelage; t = hb. Sole of hindfoot relatively hairy, interdigital pads fused at base with conspicuously enlarged, unstriated apical granule. **Distribution, habitat and status** Known only from a few specimens—near Kalumburu, n. Kimberley, WA, and Bathurst and Melville Is, NT. All from eucalypt or melaleuca forest on sandy substrates within 20 km of coast. Vulnerable. **Behaviour** Not known. **Similar species** Kakadu Dunnart has separate interdigital pads with striated apical granule, t > hb. Red-cheeked Dunnart has orange face; Stripe-faced Dunnart has bold stripe on top of muzzle and fat tail. Planigales have darker underparts.

Kakadu Dunnart *Sminthopsis bindi*

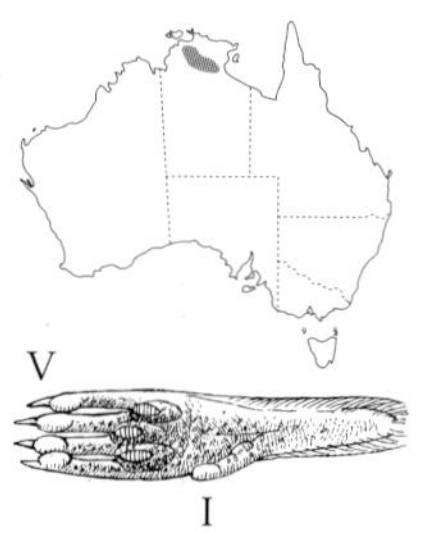

hb 50–85 mm; **t** 60–105 mm; **wt** 10–25 g
Delicately-built. Uniform pale gingery grey above, pale buff cheeks; white underparts and feet. Tail 20–25% > hb; thin, sparsely-furred, grey-brown. Hindfoot very narrow, heel covered with hair, interdigital pads separate, with striated apical granule. **Distribution, habitat and status** Locally common on gravelly hillsides carrying eucalypt woodland in the Top End of NT. **Behaviour** Little known; probably breeds during dry season and preys on a variety of arthropods. **Similar species** Butler's Dunnart has shorter tail, fused interdigital pads with smooth apical granules.

Julia Creek Dunnart *Sminthopsis douglasi*

hb 100–135 mm; **t** 110–130 mm; **hf** 22–24 mm; **wt** 40–70 g
Large. Buffy brown above with dark triangle from muzzle to between ears, buff or rufous wash from below eyes to behind ears. Underparts whitish or buff; feet pinkish white. Ears relatively short. Tail stout, t about = hb, uniformly drab olive. Hindfoot > 20 mm. **Distribution, habitat and status** Known only from an 8000 km^2 area of Mitchell Grass downs country between Julia Creek and Richmond, n. central Qld. Endangered. **Behaviour** Little known; probably shelters in cracks in the clay soils and may breed opportunistically after rain. **Similar species** Fat-tailed and Stripe-faced Dunnarts are smaller, have relatively longer ears, fatter tails and shorter hindfeet. Fat-tailed lacks bold head stripe and Stripe-faced lacks rufous on cheeks.

Chestnut Dunnart

Butler's Dunnart

Kakadu Dunnart

Julia Creek Dunnart

PLATE 14

Hairy-footed Dunnart *Sminthopsis hirtipes*

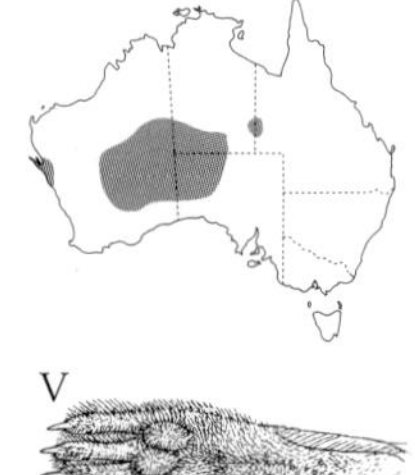

hb 72–85 mm; **t** 75–95 mm; **hf** 16–19 mm; **wt** 13–19 g
Distinguished by covering of fine silver hairs on soles of broad hindfeet, with a fringe of longer bristles around the sole. Interdigital pads fused, forming one large pad. Upperparts pale yellow-brown, variously streaked with black hairs, particularly on crown and back; darker eye-patch. Cheeks and underparts white. Tail slightly > hb; short-furred and pinkish white, can be thickened at base. **Distribution, habitat and status** Locally common in variety of arid and semi-arid woodland, heath and hummock grassland communities from w. Qld to the WA coast at Kalbarri. **Behaviour** Eats arthropods and small reptiles; shelters in burrows, mostly built by other animals such as spiders or bull-ants. Pouch young have been found in Oct and juveniles as late as April. **Similar species** Lesser Hairy-footed Dunnart is smaller and has granulated and far less hairy soles, and its interdigital pads are joined but still clearly 3 pads.

Lesser Hairy-footed Dunnart *Sminthopsis youngsoni*

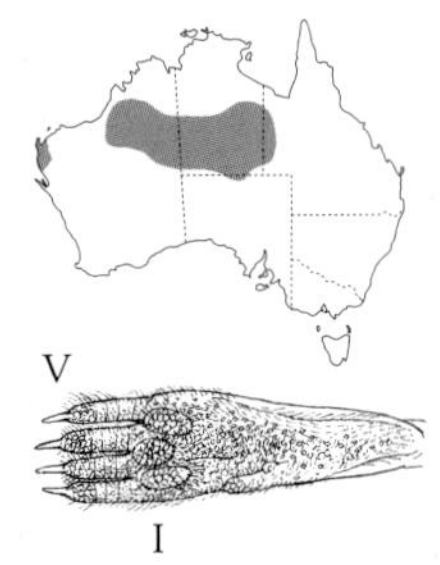

hb 66–70 mm; **t** 62–70 mm; **wt** 9–14 g
Upperparts yellow-brown, in patches finely streaked with black; face boldly patterned—dark patch around eye extends forward to muzzle, bordered above by whitish 'eyebrow', sharp contrast below between white cheek and dark eye, pale yellow on muzzle, darker patch on forehead. Underparts white, including flanks, cheeks back to behind ear, throat and feet. Tail = hb, pinkish white, basal half slightly swollen. Interdigital pads joined but not fully fused. **Distribution, habitat and status** Common where hummock grassland is present on sandy substrates in the arid interior of subtropical WA, NT and w. Qld. **Behaviour** Shelters in burrows, often those dug by dragon lizards; entirely insectivorous; 5–6 young born Sept–Nov, independent young captured Nov–Feb. **Similar species** See under Hairy-footed Dunnart.

Ooldea Dunnart *Sminthopsis ooldea*

Troughton's Dunnart

hb 55–80 mm; **t** 60–93 mm; **hf** 13–15 mm; **wt** 10–18 g
Greyish yellow above with darker patches before eye and on forehead and crown; pale grey face. Underparts including feet whitish. Tail > hb, rarely narrowly carrot-shaped, sparsely furred, pinkish. Outline of ear broadly triangular rather than oval. Interdigital pads hairless, fused at base. **Distribution, habitat and status** Probably common in variety of arid habitats—eucalypt and acacia woodland, mallee and hummock grassland between Tanami Desert, NT and Ooldea, SA. **Behaviour** Little-known; up to 8 young born Sept–Nov. **Similar species** Fat-tailed Dunnart has sandy coloured pelage rather than greyish, large oval ears, proportionately shorter tail and interdigital pads are fused for most of their length.

Long-tailed Dunnart *Sminthopsis longicaudata*

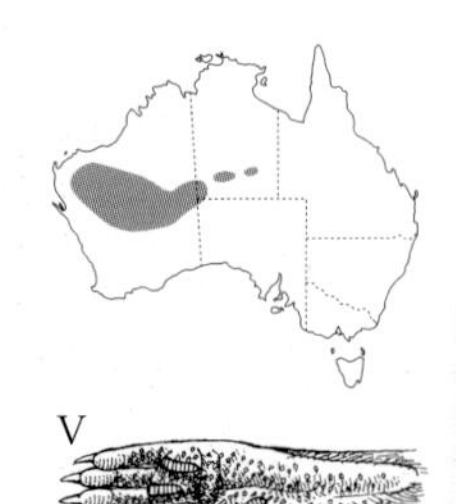

hb 80–96 mm; **t** 180–210 mm; **wt** 15–20 g
The only dunnart with a tail at least twice length of hb with terminal tuft of long hairs. Clearly striated foot pads. **Distribution, habitat and status** Known only from several widespread localities between Pilbara, ne. Goldfields and Gibson Desert, WA, and w. MacDonnell Ranges, NT. Found in rocky screes with hummock grasses and shrubs. **Behaviour** Bounds with great agility between rocks, using the long, mobile tail for balance. Mating occurs in Oct–Nov and up to 6 young are born.

Hairy-footed Dunnart
Lesser Hairy-footed Dunnart
Ooldea Dunnart
Long-tailed Dunnart

PLATE 15

Kangaroo Island Dunnart *Sminthopsis aitkeni*

Sooty Dunnart

hb 80–93 mm; **t** 90–105 mm; **hf** 17.5 mm; **el** 18 mm; **wt** 20–25g
The darkest dunnart; upperparts (including legs) sooty grey; paler on face; underparts light grey; chin and feet whitish. Tail thin, clearly bicoloured, length > hb. **Distribution, habitat and status** Known only from a few specimens from mallee heath on laterite soils on the w. half of Kangaroo I., SA. Kangaroo Island's only endemic mammal. Endangered. **Behaviour** Not described.

Sandhill Dunnart *Sminthopsis psammophila*

hb 85–95 mm; **t** 110–128 mm; **hf** 22–26 mm; **wt** 26–40 g
A large dunnart with diagnostic tail colour. Drab grey to buff above with blackish guard hairs. Dark eye-rings and darker triangle on crown and forehead; face and flanks buff. Underparts and feet whitish. Tail > hb, thin, pale grey above, darker below (opposite pattern to all other dunnarts), vertical crest of short, stiff blackish hairs on last quarter. **Distribution, habitat and status** Endangered. Known from 4 widely-spaced areas of arid southern Aust: near L. Amadeus, sw. NT; central Eyre Pen., SA; sw. edge of Great Victoria Desert and nearby Queen Victoria Spring, WA; Yellabinna Sand Dunes, central SA. Habitat is low parallel sand dunes carrying open woodland with diverse low shrub layer and hummock grass. **Behaviour** Little known. Insectivore, shelters in hummock grass. Females probably receptive in spring and early summer; a female with 5 pouch young captured in Dec.

tail pattern

Red-cheeked Dunnart *Sminthopsis virginiae*

hb 80–135 mm; **t** 87–135 mm; **wt** 18–75 g
A large, short-furred, distinctively coloured dunnart of the tropics. Pale grizzled grey above with a blackish stripe along top of muzzle; distinct rufous-orange wash over sides of face, beneath ears, down to forelegs. Underparts greyish cream. Tail similar in length to hb, usually thin and pale pink. **Distribution, habitat and status** Locally common in savannah grasslands, around swamps and soaks and margins of tropical forest in n. Kimberley, WA, Top End of NT, and coastal ne. Qld. **Behaviour** Little known. Females receptive Oct–March; up to 8 young born after a gestation of 15 days; weaned at 65–70 days and mature at 4–6 months. Several litters may be reared in succession. **Similar species** Sandstone False Antechinus lacks orange face and muzzle stripe. All eastern antechinuses are more uniformly coloured.

Kangaroo Island Dunnart

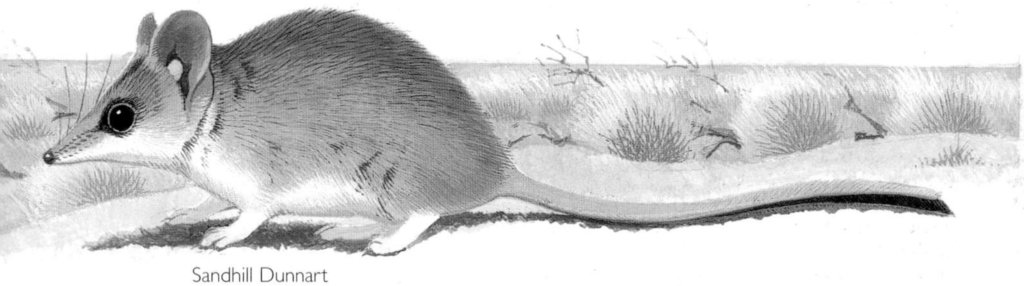
Sandhill Dunnart

Red-cheeked Dunnart

PLATE 16

Golden Bandicoot *Isoodon auratus*

hb 190–290 mm; **t** 85–120 mm; **wt** 250–650 g
The smallest and most colourful short-nosed bandicoot. Muzzle particularly long, flat and pointed. Golden brown above, heavily streaked with shiny black guard hairs; merging to russet on flanks and lower face; underparts, including throat, pale honey coloured; tail and upper surfaces of feet similar colour to upperparts. Eyes small and black; ears short and rounded; claws long. **Distribution, habitat and status** Formerly widespread in range of arid and semi-arid habitats through central and northern Aust. Now confined to hummock grass on sandstone, grassy woodlands and deciduous vine thickets in the Kimberley region, and on Augustus, Barrow and Middle Is off nw. WA where locally common, plus Marchinbar I., NT. Vulnerable. **Behaviour** Nocturnal, solitary, digs in topsoil for arthropods, tubers. Shelters in grass nest beneath dense cover. Births occur in all months, particularly after good rains. **Similar species** Northern Brown Bandicoot is much bigger and duller in colour.

Northern Brown Bandicoot *Isoodon macrourus*

Brindled Bandicoot

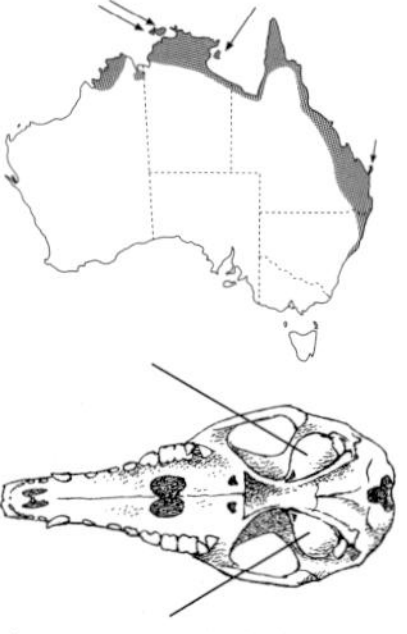

large tympanic bullae in *Isoodon*

hb 300–470 mm; **t** 80–210 mm; **wt** males 500–3000 g females 500–1500 g
The largest Aust bandicoot. Fur coarse, brindled brown, buff and black above, sometimes with rufous wash to back and flanks; underparts and forefeet cream; tail dark brown above, creamy yellow below. Ear short, barely extending above crown. **Distribution, habitat and status** Common in wet tropical and subtropical forest, woodland, scrub, grassland and gardens across n. and e. Aust from n. Kimberley in WA to Hawkesbury R., NSW. Extends inland in riverine vegetation to about the 650 mm isohyet. **Behaviour** Mostly nocturnal, solitary, territorial and aggressive. Omnivorous, digs in topsoil for arthropods, tubers, fruits and seeds, leaving characteristic conical pits in soil and lawns. Shelters in nest of vegetation beneath dense cover. Births occur in all months, but not winter in the s.; females can produce a litter of up to 4 young every 10 weeks. **Similar species** Southern Brown Bandicoot very similar but smaller, muzzle shorter and blunter. Potoroos are dark brown, have longer blackish tail, stance more kangaroo-like, fur soft, not coarse.

Southern Brown Bandicoot *Isoodon obesulus*

Short-nosed Bandicoot, Quenda (WA)

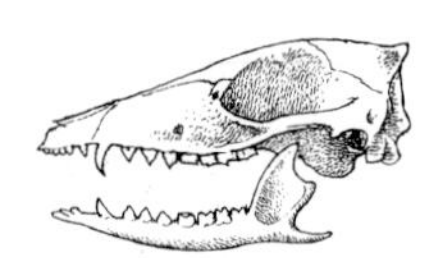

hb 280–355 mm; **t** 80–130 mm; **wt** males 500–1500 g females 400–1000 g
Brindled brown, buff and black above, less black on flanks and face; underparts and forefeet creamy white or pale yellow; tail brown above, creamy yellow below. Ear short, rounded, barely extends above crown. **Distribution, habitat and status** Declining in inland parts of range but locally common in heathy forest, heath and coastal scrub from Sydney, NSW s. and w. to Eyre Pen., SA, Tas including Flinders I., and sw. WA n. to Moore R. mouth. Subspecies *nauticus* confined to East and West Franklin Is, SA. Isolated subspecies *peninsulae* on e. Cape York may be a separate species. **Behaviour** Mostly nocturnal, solitary, territorial, aggressive and shy. Omnivorous, digs in topsoil for arthropods, tubers and fungi, leaving characteristic conical pits in soil and lawns. Shelters in nest of vegetation beneath dense cover. Births occur from late winter to late summer; 2 or 3 litters of 2–4 may be reared per year. **Similar species** See under Northern Brown Bandicoot.

Golden Bandicoot

Northern Brown Bandicoot

Southern Brown Bandicoot

Western Barred Bandicoot *Perameles bougainville*

Marl (WA)

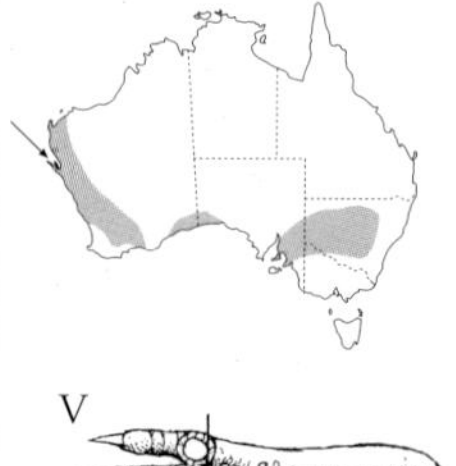

hb 170–235 mm; **t** 60–100 mm; **wt** 170–285 g

Small and delicate with a finely pointed muzzle. Upperparts mid-brown flecked with buff and rufous, darkest on rump, forehead and crown; dark eye-rings; flanks and lower face grey-buff, grading to cream underparts, forelegs and feet. One broad dark brown bar across rump and flanks anterior to hindlimbs; another 1 or 2 thinner bars across thighs from near base of tail. Tail about 1/3 of hb, narrow, creamy with thin brown line along upper surface. Ears brown. **Distribution, habitat and status** Formerly widespread in semi-arid shrublands across southern Australia; now confined to Bernier and Dorre Is in Shark Bay, WA, where abundant in dense shrubby vegetation behind dunes. Being re-introduced to nearby mainland. Endangered. **Behaviour** Nocturnal, solitary, primarily insectivorous (mostly beetles, grasshoppers, crickets), some plant material and small vertebrates also eaten. Shelters in well-concealed nest of dry vegetation in a scrape beneath bushes. Births occur April–Oct, litters usually comprise 2 young, sometimes 1 or 3. **Similar species** None within range.

V

I

insertion of toes II and III

Desert Bandicoot *Perameles eremiana*

hb 180–285 mm; **t** 77–135 mm; **wt** approx. 250 g

Similar to Western Barred Bandicoot but rufous-orange colouration on face, sides and rump; tail darker brown above, proportionally longer and tapered to a point. **Distribution, habitat and status** Possibly a form of Western Barred Bandicoot. Presumed extinct since the 1960s. Formerly common in central Aust and WA sandy deserts; last specimen obtained on Canning Stock Route in 1943. Occupied sand plain and sand dune country with hummock or tussock grassland. **Behaviour** Not recorded but probably similar to Western Barred Bandicoot.

Eastern Barred Bandicoot *Perameles gunnii*

hb 270–350 mm; **t** 70–110 mm; **wt** 500–1100 g, heavier in Tas

Upperparts grizzled buff-brown with black guard hairs, darkest from muzzle to rump and around eyes, grading to buff or rufous, streaked with black hairs on flanks and face below eyes; underparts and top of feet uniformly creamy white; 3–4 whitish bars across rump and thighs broken along the midline. Tail entirely pinkish white, < 1/3 hb. Ears long and pale pinkish brown. **Distribution, habitat and status** Locally common in Tas, endangered in Vic where confined to a few colonies derived from captive-bred animals and dependent on on-going control of foxes and cats. Occupies open grassland, including introduced pasture, with patches of dense vegetation for shelter. **Behaviour** Nocturnal, solitary, digs in topsoil for invertebrates and tubers. Births mostly July–Nov; litters of 1–5 (mostly 2–3) born after 13 day pregnancy and weaned at 3–5 months when mother may quickly give birth to another litter. **Similar species** See under Long-nosed Bandicoot.

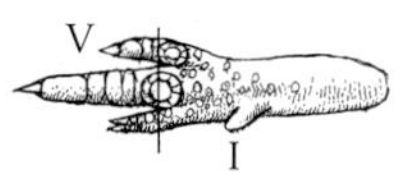

insertion of toes II and III

Long-nosed Bandicoot *Perameles nasuta*

small tympanic bullae in *Perameles*

hb 310–450 mm; **t** 120–150 mm; **wt** 850–1100 g

Upperparts shiny drab brown flecked with buff; grading to fawn on flanks and lower face. Underparts, including chin, throat and tops of feet, creamy white. Tail < half hb, thinly furred, pinkish brown above, cream below. Ears long and narrow, almost naked, pinkish brown. Strong horn-coloured claws. Young animals can have pale bars on rump. **Voice** Characteristic alarm call a loud, sneezed 'ke'. **Distribution, habitat and status** Locally common along e. coast and adjacent ranges from Ravenshoe, ne. Qld to Naringal, sw. Vic. Found in wet sclerophyll forest, scrub, rank grass and suburban gardens. **Behaviour** Nocturnal and crepuscular; solitary; digs in topsoil for invertebrates and tubers. Births occur throughout year in the n. but not in winter in s. Litters of 2–3 weaned at about 7 weeks old and sexually mature at 20 weeks. **Similar species** Brown Bandicoots have short ears, more yellowy buff colouration, including tail. Eastern Barred Bandicoot has bold barring on hindquarters and creamy white tail.

Western Barred Bandicoot

Desert Bandicoot

Eastern Barred Bandicoot

Long-nosed Bandicoot

PLATE 18

Bilby *Macrotis lagotis*

Greater Bilby, Rabbit-eared Bandicoot, Rabbit Bandicoot, Dalgyte

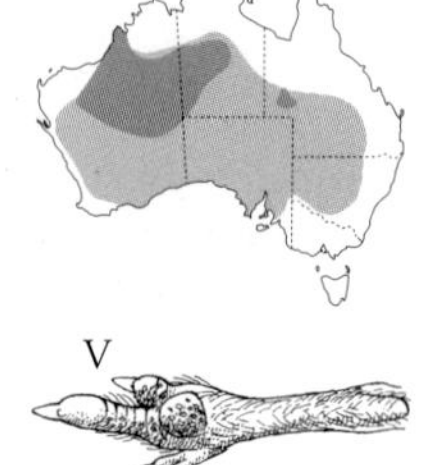

hb 300–550 mm; **t** 200–290 mm; **el** 80–95 mm; **wt** 800–2400 g

Distinctively coloured, long-tailed, long-eared, burrowing desert bandicoot. Fur long and silky; blue-grey tinged with rufous brown, particularly on flanks and base of tail. Underparts and feet whitish. Tail unique—long, tricoloured and crested—basal third similar to back, mid third black, terminal third white. Ears long, naked, mobile, pinkish. Muzzle long, pointed, pinkish. **Distribution, habitat and status** Formerly throughout arid and semi-arid Aust, e. to the Western Slopes of NSW. Now in scattered colonies in acacia shrubland and hummock grassland from Tanami Desert, NT w. to near Broome and s. to Warburton, WA. Also clay and stony downs in Channel Country of sw. Qld. Vulnerable. **Behaviour** Mostly solitary. Constructs long, deep burrow systems for daytime shelter. Emerges well after dark to forage in topsoil for arthropods, tubers and fungi, leaving numerous pits to 10 cm deep. Holds tail stiffly aloft while running with cantering gait. Births mostly March–May but can occur at any time if conditions good. Litters of 2 normal. **Similar species** Lesser Bilby smaller, duller, tail entirely whitish.

Lesser Bilby *Macrotis leucrura*

White-tailed Bilby

hb 200–270 mm; **t** 120–170 mm; **el** 63 mm; **wt** 300–435 g

Smaller and less colourful than Bilby. Grey-brown above, pale grey below. Tail long-haired and entirely white except for grey underneath inner third. **Distribution, habitat and status** Extinct. Known only from desert country in ne. SA and sw. NT, and Gibson and Great Sandy Deserts of WA. Occupied dunes and sandy plains with hummock grass or canegrass. **Behaviour** Poorly known. Nocturnal, sheltered during day in deep burrow in dune, with the entrance covered with sand when occupied. Apparently partially carnivorous. **Similar species** See under Bilby.

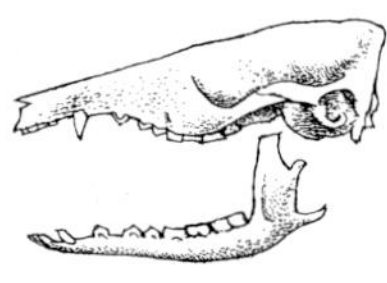

Pig-footed Bandicoot *Chaeropus ecaudatus*

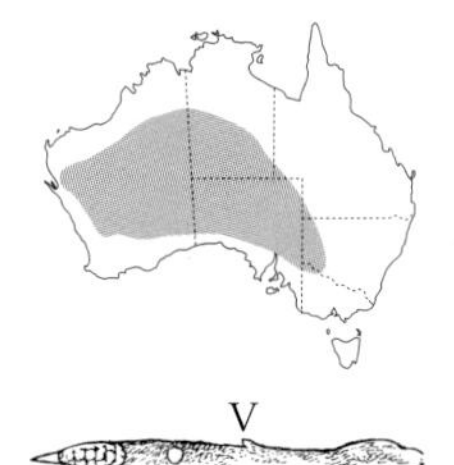

hb 230–260 mm; **t** 100–150 mm; **wt** approx. 200 g

Small, finely built bandicoot with unique arrangement of toes. Upperparts pale orange-brown grading to fawn beneath. Limbs, muzzle and ears long and slender. Tail half hb, slim with a low blackish crest merging to a paler end tuft. Forefeet with 2 functional toes (3rd, 4th) (1st and 5th absent, 2nd minute) resembling cloven hoof of a small deer; only 4th toe of hindfoot used for locomotion (1st absent, 2nd and 3rd fused, 5th minute). **Distribution, habitat and status** Extinct. Formerly widespread in arid and semi-arid Aust in light soil areas with grassy woodland, shrubland or hummock grass. Last specimen collected 1901, but Aboriginal people in the central deserts of WA knew it until perhaps the 1950s. **Behaviour** Poorly known. Nocturnal, sheltered during day in grassy nest beneath dense vegetation. Apparently partially vegetarian.

V
II & III

Rufous Spiny Bandicoot *Echymipera rufescens*

Rufescent Bandicoot

hb 300–400 mm; **t** 80–100 mm; **wt** 600–2000 g

A long-snouted, dark-coloured bandicoot from Cape York Pen. Upperparts coarse rufous brown, strongly flecked with buff and black spiny hairs; darkest on crown and rump. Underparts and feet cream. Tail short, almost hairless, dark brown above, buff below. **Distribution, habitat and status** Confined to vine forest, riverine forest and grassy woodlands of e. Cape York s. to McIlwraith Range. Locally common. **Behaviour** Poorly known. Nocturnal, omnivorous. **Similar species** Northern Brown Bandicoot has paler yellowish fur, shorter ears and paler tail. All other Aust bandicoots have 5 not 4 pairs of upper incisors.

Bilby

Lesser Bilby

Pig-footed Bandicoot

Rufous Spiny Bandicoot

PLATE 19

Koala *Phascolarctos cinereus*

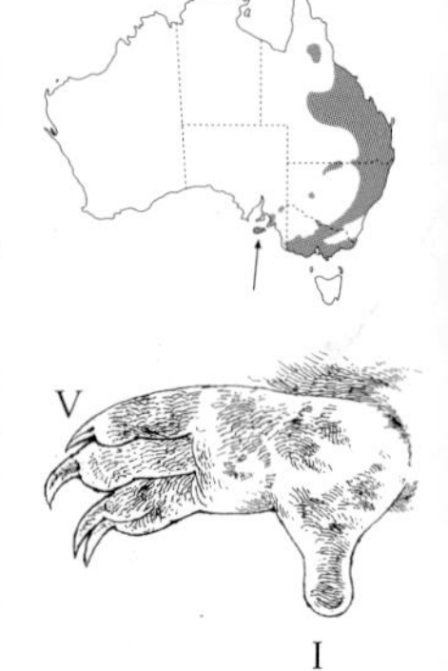

wt Vic males to 14 kg; Qld to 9 kg; Vic females to 10 kg; Qld to 7 kg
Unmistakable. Stocky; arboreal; no tail; large, round head; distinct naked, black rhinarium extending up to eye level; large, oval, furry ears at side of head; small eyes. Upperparts grey-brown; often irregular pale grey patches on rump. Underparts including collar to near base of ears whitish or pale yellow-grey. Southern animals larger, longer furred, often chocolate brown across shoulders and upper back, have long white ear tufts; northern animals smaller, shorter furred and greyer. Bright yellow eye-shine. **Voice** Males give loud snorting, bellowing territorial calls, particularly in spring and summer. **Distribution, habitat and status** Widespread in sclerophyll forest and woodland on foothills and plains on both sides of GDR from about Chillagoe in n. Qld to Mt Lofty Ranges in SA. Extends inland to Desert Uplands of central Qld and along riverine forests in central NSW and north-central Vic. Declining in some densely settled coastal areas, secure elsewhere. **Behaviour** Arboreal, agile climber, descends to ground when necessary; mostly solitary; diet almost entirely eucalypt leaves; inactive for 20 hrs in 24. Mates in spring and early summer; single young remains in pouch for about 6 months then carried by mother until independent at 12 months old when mother usually gives birth again.

Common Wombat *Vombatus ursinus*

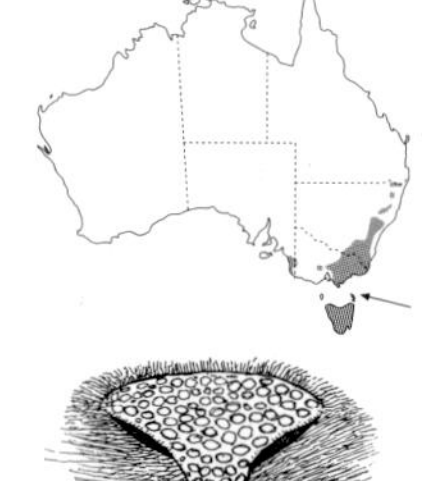

hb 700–1100 mm; **t** 25 mm; **wt** 20–35 kg, Flinders I. animals the smallest.
Large, thick-set grazer; limbs short, powerful; claws strong; head broad, muzzle short; ears short, rounded, barely protruding above crown; tail vestigial. Coarse fur usually uniformly grey-brown to blackish, can be patchy grey and buff or uniformly cream. **Distribution, habitat and status** From near Stanthorpe, se. Qld to se. SA and Tas, including Flinders Is. In Qld and n. NSW found only in wet forest above about 600 m. Further s. also inhabits drier forest, coastal scrub and heath from sea level to above snowline. Locally common, declining in w. Vic and SA. **Behaviour** Mainly crepuscular and nocturnal, solitary. Shelters in large burrows dug into sloping ground. Eats grass, sedges, tubers. Timing of breeding varies with latitude and altitude, probably related to growing season of major food plants. The single young leaves the pouch after 6–9 months and follows mother on foot until weaned at about 20 months old. Only 1 young produced every 2 years. **Similar species** Both hairy-nosed wombats have longer, pointed, almost naked ears, longer square-cut muzzle, and fine white hairs over rhinarium.

naked rhinarium

Northern Hairy-nosed Wombat *Lasiorhinus krefftii*

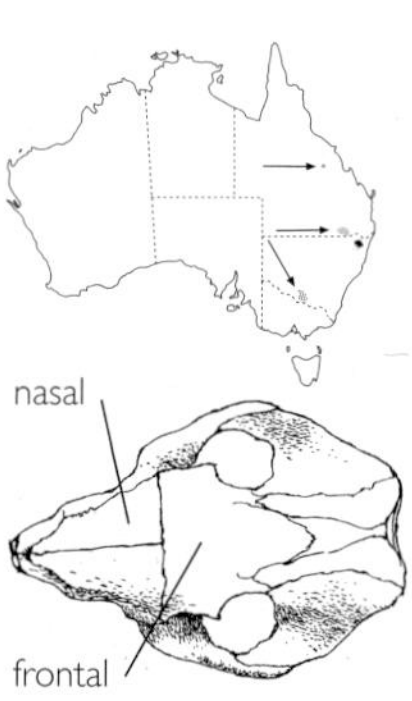

hb 900–1100 mm; **t** 25 mm; **wt** 26–35 kg
Large, rotund, heavily set, with pointed ears projecting well above crown and very square muzzle. Fur soft, shiny, uniformly mid-grey, sometimes with fawn or brownish wash. **Distribution, habitat and status** Critically endangered; reduced to about 70 animals in 300 ha of sandy grassy woodland in Epping Forest NP, nw. of Clermont, Qld. Formerly also in Riverina District of NSW and near St George, s. Qld. **Behaviour** Mainly crepuscular and nocturnal, solitary. Shelters in large burrows dug into sandy soil and often clustered. Grazes on perennial native grasses. Young born spring and summer, spend 10–11 months in pouch. Females can produce 2 young every 3 years in good conditions. **Similar species** None within range; see under Common Wombat. Differs from Southern Hairy-nosed Wombat by longer, very square muzzle, uniformly darker pelage, lack of whitish patches below eyes, and nasal bones shorter than frontal bones.

Southern Hairy-nosed Wombat *Lasiorhinus latifrons*

hb 775–935 mm; **t** 25 mm; **wt** 20–32 kg
As for Northern Hairy-nosed Wombat but pelage grizzled grey or tan, with whitish patches on face, particularly beneath eyes; paler grey underparts. **Distribution, habitat and status** Patchily distributed in semi-arid shrubland and mallee in s. Nullarbor Plain, Eyre and York Peninsulas (SA), Murray Mallee (SA) and far sw. NSW. May be vulnerable. **Behaviour** Mainly crepuscular and nocturnal, solitary. Shelters in large burrows clustered to form a warren with well-developed trails between the entrances and to preferred feeding sites. Most births occur Sept–Dec; the single young remains in the pouch for 6–9 months, weaned at about 12 months. Survival of young highly dependent on rainfall. **Similar species** See under Northern Hairy-nosed Wombat.

haired rhinarium

Koala
southern form
northern form
Common Wombat
Northern Hairy-nosed Wombat
Southern Hairy-nosed Wombat

PLATE 20

Southern Common Cuscus *Phalanger intercastellanus*

Grey Cuscus

hb 330–400 mm; **t** 280–350 mm; **wt** 1.5–2.2 kg

Grey-brown above, narrow dark brown midline from between eyes to mid-back; buff wash on face and at base of ears; underparts creamy, tinged yellow on chest and neck of males, abdomen sometimes grey-brown. Tail prehensile, naked and dark brown for outer $^2/_3$. Ears not hidden by fur, short, rounded, grey-pink. Eyeshine red. **Voice** Guttural screeches and grunts. **Distribution, habitat and status** Restricted to primary rainforest and adjacent acacia scrubs on the e. side of Cape York Pen. between McIlwraith Range and Iron Range. **Behaviour** Solitary, nocturnal, vegetarian. Shelters in hollow or den during daylight, moves slowly and deliberately through canopy in search of leaves, fruit and flowers. **Similar species** Differs from Common Spotted Cuscus by more prominent ears, lack of dorsal mottling, dark midline, dark brown rather than pink skin and lighter build. Distinguish from Daintree River Ringtail by naked prehensile tail.

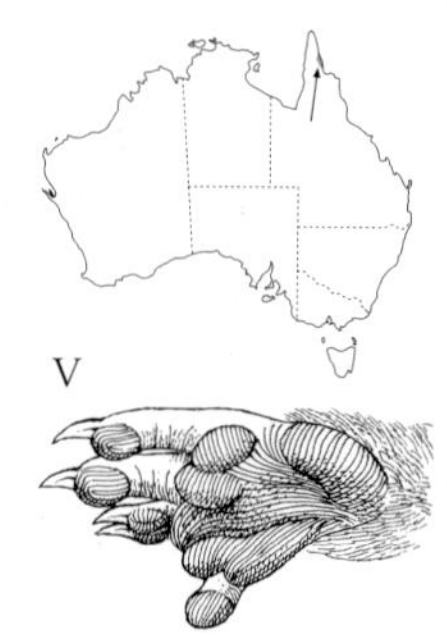

Common Spotted Cuscus *Spilocuscus maculatus*

hb 350–500 mm; **t** 310–430 mm; **wt** 1.5–4.5 kg

Face large, forward-looking, round; short ears barely visible through the dense fur; reddish eyes with vertical pupils. Upperparts usually grey with whitish cheeks; underparts whitish. Males have extensive whitish mottling on back and rump. Some animals all white, or strongly orange-tinged. Bare skin of muzzle and feet yellow-pink. Tail strongly prehensile, naked for outer $^2/_3$ and covered with rough bumps underneath. Eyeshine red. **Voice** Guttural screeches and grunts. **Distribution, habitat and status** Sparsely distributed on Cape York Pen. n. of Stewart R. and Coen–Archer R. system. Inhabits lowland rainforest to about 800 m, and adjacent eucalypt and paperbark open-forest, and mangroves. **Behaviour** Solitary, mostly nocturnal, omnivorous. Spends daylight sitting on branch or temporary platform of twigs, moves slowly and deliberately through canopy in search of leaves, fruit, flowers, insects and small vertebrates. **Similar species** See under Southern Common Cuscus.

C^1 close to I^3

Scaly-tailed Possum *Wyulda squamicaudata*

hb 300–390 mm; **t** 300 mm; **wt** 1.3–2.0 kg

Stocky body with short muzzle and ears. Upperparts uniformly pale grey flecked brown, grading to rufous on rump and base of tail; indistinct, narrow midline from between eyes to rump. Underparts cream. Prehensile tail thickly furred at base, then naked and covered with rasp-like scales, pigmented dark grey with whitish outer third. **Distribution, habitat and status** Patchily distributed in coastal nw. Kimberley (e.g. Mitchell Plateau, Kalumburu) in low open woodland, riparian forest and vine thickets where tumbled boulders provide shelter. **Behaviour** Solitary, nocturnal, shy, agile climber in trees and rocks, hangs from prehensile tail to reach food; forages in trees for leaves, flowers and fruit; spends daylight in rock crevices. Breeds during dry season (Mar–Aug); the single young leaves the pouch at about 25 weeks old, weaned at about 32 weeks. **Similar species** Common Brushtail Possum has long ears, 'foxy' face and furred tail. Rock Ringtail has half of tail furred, steep forehead, short pointed muzzle, white patches around eyes.

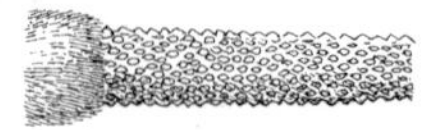

tail detail

Southern Common Cuscus

Common Spotted Cuscus

Scaly-tailed Possum

PLATE 21

Mountain Brushtail Possum *Trichosurus caninus*

Bobuck

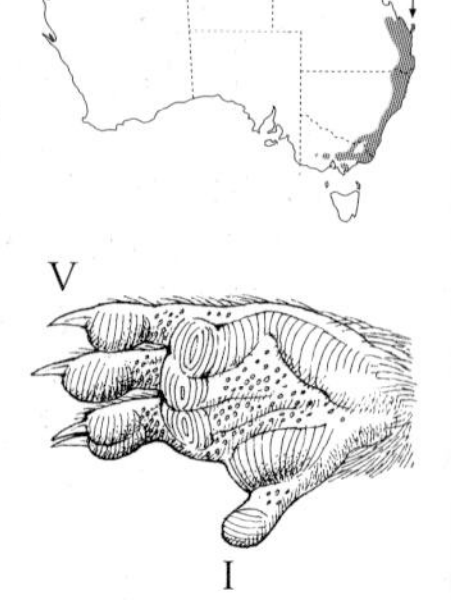

hb 400–550 mm; **t** 320–400 mm; **el** 40–50 mm; **wt** 2.5–4.5 kg
A large heavy-set possum of wet forest; upperparts usually uniformly dark grey flecked with buff, but can be blackish, dark brown or reddish; underparts cream or yellow/orange; tail thick and blackish. In ne. NSW some individuals entirely black. Ears broadly oval; nose pink; feet noticeably blackish. Eyeshine red. **Voice** Short sequence of sharp grunts or coughs, 'ke ke ke', not a rattling series as in Common Brushtail Possum. **Distribution, habitat and status** Common in cool-temperate wet forest and subtropical rainforest with a luxuriant understorey of non-sclerophyllous shrubs and ferns, through the ranges and coastal plains from inland of Gladstone, Qld to Mt Cole, w. Vic. **Behaviour** More terrestrial than other possums, often feeds on ground plants and dens in fallen logs as well as standing trees; eats wide range of plant material from the ground and shrub layers including leaves, fern fronds, buds, fruits, fungi and lichen. Dens in tree hollows or other sheltered cranny, sometimes at ground level. Long-lived, sedentary, forms long-term monogamous pairs, females do not breed until 3 years old; births occur mostly Mar–June, the single young remains in pouch for 5–6 months, weaned at about 8–11 months and may not disperse until 18 months old. **Similar species** See under Common Brushtail Possum.

Common Brushtail Possum *Trichosurus vulpecula*

Silver-grey Possum, Bushy-tail Possum

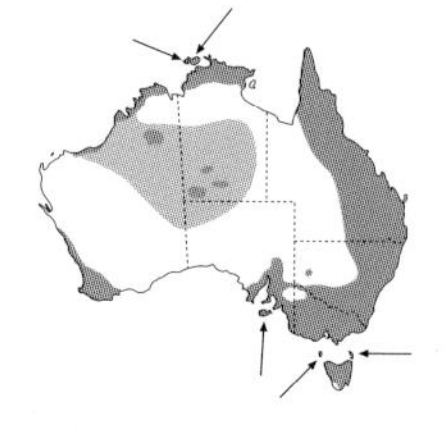

hb 350–500 mm; **t** 250–400 mm; **el** 50–60 mm; **wt** 1.5–4 kg
A cat-sized arboreal and terrestrial possum; ears obviously longer than broad, with narrowly rounded tip. Fur colour, length and density highly variable. In e. and s. Aust mostly uniform silver-grey above with cream underparts, including chin and partial collar; belly often stained yellow/orange; tail black, only slightly bushy; black around eyes, on sides of muzzle and narrow midline of face and head. Some individuals pale yellowish white, others have varying amounts of rufous. Tas animals, and some on Wilsons Promontory (Vic), are larger (to 4 kg), longer-furred and often blackish flecked rufous. Race *arnhemensis* (Top End and Kimberley) is smaller (1.5 kg), short-haired, rufous buff in colour with a sparsely furred blackish tail. In central e. Qld upland rainforest subspecies *johnstonii* is short-furred, uniformly bright copper-coloured with a black tail and may be a separate species. **Voice** A characteristic loud series of rattling nasal coughs and hisses. **Distribution, habitat and status** One of the best-known marsupials; found in most treed environments, including cities, towns and farmland; replaced in wet sclerophyll forest in se. by Mountain Brushtail. Common in much of se. and sw; declining and endangered in central Aust. **Behaviour** Eats mostly leaves of a wide variety of trees, shrubs and herbaceous plants, also flowers, fruits. Dens in tree hollow or other sheltered cranny, sometimes at ground level, commonly in building roofs. Breeds between autumn and spring, but most young born in autumn; the single young occupies the pouch for 16–20 weeks, then rides on mother's back for 4–8 weeks until weaned at 5–7 months old. N. Aust animals may breed in any month. **Similar species** Mountain Brushtail Possum is larger, more thick-set, darker furred, bushier tailed with blunter muzzle. Ears are not obviously longer than wide.

premolar outside line of molars

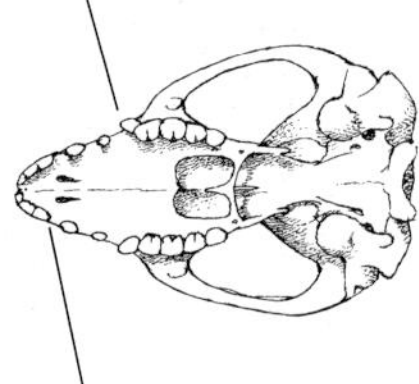

gap between C^1 and I^3

all black form
Mountain Brushtail Possum
subspecies
johnstonii
dark form
subspecies arnhemensis
Common Brushtail Possum

PLATE 22

Long-tailed Pygmy Possum *Cercartetus caudatus*

hb 100–110 mm; **t** 128–150 mm; **wt** 25–40g

Warm brown and grey above merging through buff flanks to whitish underparts. Extensive black patch around eye and along sides of muzzle. Tail clearly > hb, coiled, nearly naked apart from base, skin dark brown. **Distribution, habitat and status** Common in rainforest and fringing she-oak forest at altitudes above 300 m between Mt Spec and Daintree R., ne. Qld. Also in coastal rainforest and eucalypt–tea tree forest n. of Daintree to near Cooktown. **Behaviour** Nocturnal, mostly arboreal, agile climber readily jumping from branch to branch or to the ground; eats arthropods and nectar; shelters in spherical nest of leaves in tree hollow or other sheltered cranny. Births recorded Aug–Nov and Jan–Feb. Up to 4 young weaned at about 66–80 days old. **Similar species** Prehensile-tailed Rat is larger; lacks furred base to tail and black eye-patches; has Roman nose and smaller simple ears.

Shape of P_4

Western Pygmy Possum *Cercartetus concinnus*

Mundarda (WA)

hb 70–100 mm; **t** 70–95 mm; **wt** 8–18 g

Mouse-sized; bright cinnamon-grey above; clear demarcation of whitish underparts, including chin and lower jaw. Narrow dark-brown ring around large, black, bulging eye. Tail coiled, very sparsely furred, never fattened at base. Muzzle sparsely furred and pink; ears long, rounded, flesh-coloured. Female has 6 teats. **Distribution, habitat and status** Widespread in wheatbelt and s. coast of WA; Mallee areas of SA and Vic, s. to Edenhope area; Kangaroo Is. Generally uncommon in dry, heathy woodland, mallee shrubland and semi-arid heath. **Behaviour** Nocturnal, arboreal, terrestrial; eats arthropods and nectar. Shelters in spherical nest of bark and leaves in tree hollow or other cranny. Births recorded in most months; 2 or 3 litters of up to 6 young may be produced yearly. Becomes torpid in cold weather. **Similar species** Little Pygmy Possum has much less contrast in colour—uniformly grey above, grey rather than white below; tail can become fattened at base; has 4 molars and female has only 4 teats. Eastern Pygmy Possum is larger; more grey underparts; has 3 molars; tail often fattened. House Mouse lacks 2-tone colour, has musty odour, lacks opposable hindtoe, 2nd and 3rd hindtoes not fused.

Little Pygmy Possum *Cercartetus lepidus*

hb 63–70 mm; **t** 60–75 mm; **wt** 6–10 g

Smallest diprotodont marsupial. Upperparts uniformly grey with tinge of tan; underparts paler grey, whitish under chin. Tail very sparsely furred, flesh-coloured; can be strongly swollen at base. Ears large, rounded, often held forward. Presence of 4th molar distinguishes it from other pygmy possums. **Distribution, habitat and status** Found in confusing range of habitats; uncommon in wet and dry eucalypt forest in Tas; rare on Kangaroo I.; uncommon in mallee shrubland and semi-arid heath with spinifex in Murray Mallee of SA and Vic. **Behaviour** Nocturnal, arboreal, terrestrial; eats mainly arthropods, also small skinks and nectar. Shelters in spherical nest of bark and leaves in tree hollow or under spinifex hummock (in Mallee). On mainland births recorded in most months; in Tas breeds in summer. Litters of 4 young weaned at about 12 weeks. **Similar species** See under Western Pygmy Possum and Eastern Pygmy Possum.

P_4 with 2 small points

Eastern Pygmy Possum *Cercartetus nanus*

hb 70–105 mm; **t** 75–105 mm; **wt** 15–38 g

Upperparts grey tinged with fawn; underparts pale grey tipped white, including chin and lower face. Tail prehensile, very sparsely furred, brown; can be strongly swollen at base. Ears large, rounded, often held forward. **Distribution, habitat and status** Sparse to locally common in wide range of vegetation on the GDR, including western slopes, and coastal plains from se. Qld to se. SA; also Tas including Flinders and King Is. Found in wet and dry eucalypt forest, subalpine woodland, coastal banksia woodland and wet heath. **Behaviour** Nocturnal, arboreal, eats mainly nectar and pollen, also arthropods and fruit. Shelters in spherical nest of bark and leaves in tree hollow or other cranny. On mainland births recorded Aug–April; in Tas Aug–Oct. Two litters of 3–4 young can be raised per year. **Similar species** Little Pygmy Possum is under half the weight; has 4 molars, and P_4 has 2 points not 1. See also under Western Pygmy Possum.

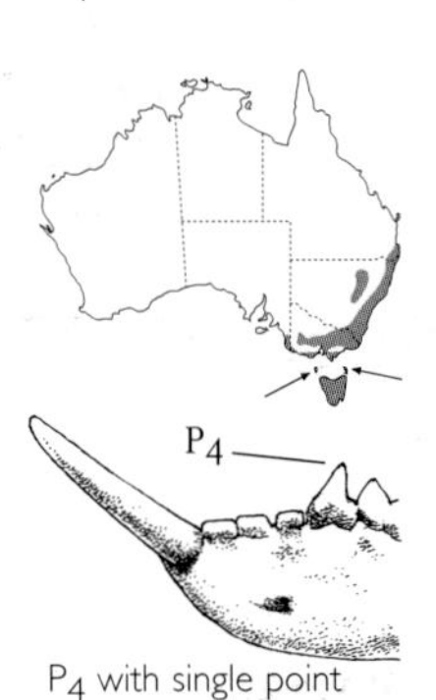

P_4 with single point

Long-tailed Pygmy Possum
Western Pygmy Possum
Little Pygmy Possum
Eastern Pygmy Possum

PLATE 23

Mountain Pygmy Possum *Burramys parvus*

Burramys

hb 100–115 mm; **t** 130–150 mm; **wt** 35–80 g

Upperparts uniform mid-grey with brown tinge; underparts and cheeks cream or fawn; flanks and underparts can be washed orange during breeding. Tail > hb; furred for first 2 cm then thin, naked, pinkish-grey, coiled at tip. P^3 and P_3 very large, with serrated cutting edge and grooved sides. **Distribution, habitat and status** The only Australian mammal restricted to alpine environments. Confined to tiny patches of alpine rock scree with heathy vegetation above 1400 m. Populations on Mt Buller, Mt Higginbotham – Mt Loch, Mt Bogong and scattered on Bogong High Plains (Vic) and Mt Kosciuszko – Mt Blue Cow (NSW). Endangered by commercial development. **Behaviour** Mostly terrestrial, active amid tumbled boulders, even beneath deep snow. Eats mostly arthropods and fruit during summer and caches seeds and fruit for winter. Short summer breeding season—4 young born Nov–Dec, weaned by March. After breeding, animals gain weight rapidly (may double) then hibernate between about April and Oct. **Similar species** Eastern Pygmy Possum is smaller; t not clearly > hb.

P^3 longer than each molar, grooved and serrated

Feathertail Glider *Acrobates pygmaeus*

hb 65–80 mm; **t** 70–80 mm; **wt** 10–14 g

World's smallest gliding mammal—size of small mouse. Tail unique among Aust mammals—almost hairless except for fringe of long stiff hairs on either side giving feather-like appearance. Gliding membrane between elbows and knees. Upperparts uniform mid-grey; often with white patch behind ears; blackish around eyes. Underparts whitish, including face below eyes. **Distribution, habitat and status** Widespread in cool-temperate and tropical eucalypt forests of e. Aust to se. SA. Needs high diversity of trees and shrubs to provide year-round nectar. More common in wet and old-growth forest than dry or regenerating ones. **Behaviour** Nocturnal; moves rapidly through canopy of trees and shrubs; makes controlled glides up to 25 m; when disturbed may spiral to ground then dash into cover. Eats nectar, manna, arthropods. Nests colonially in spherical leaf nest in hollow. In n. may breed in any month; in s. births occur late winter–summer. Two litters of 3–4 young may be raised per year.

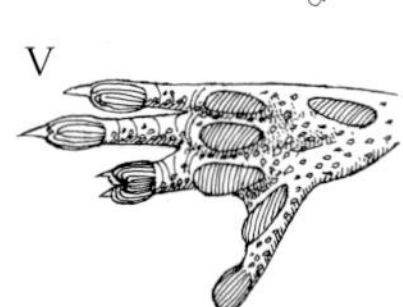

Honey Possum *Tarsipes rostratus*

Noolbenger

hb 65–85 mm; **t** 70–100 mm; **wt** 5–10 g

Unmistakable tiny animal with elongated muzzle and 3 longitudinal brown stripes. Fur grizzled grey-brown above grading to rufous on flanks and cream below. Distinct dark brown central stripe from behind ears to base of tail; outer 2 stripes less distinct and pale brown. Tail thin, slightly > hb. **Distribution, habitat and status** Locally common in floristically rich heath in coastal sw. WA from Kalbarri to e. of Esperance. Needs high diversity of shrubs to provide year-round nectar. **Behaviour** Nocturnal but partly diurnal; terrestrial and arboreal; feeds solely on nectar and pollen; agile and fast moving, darts between blossoms; confiding when feeding. Shelters in tree hollow, old bird's nest or other cranny; frequently becomes torpid in cold weather. Breeds throughout year; up to 4 litters of 2–4 young may be raised per year; young not left in nest at any stage.

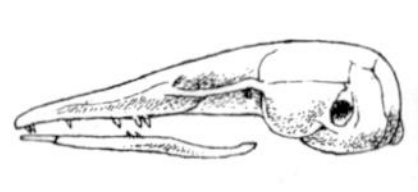

Common Striped Possum *Dactylopsila trivirgata*

hb 255–270 mm; **t** 320–350 mm; **wt** 245–520 g

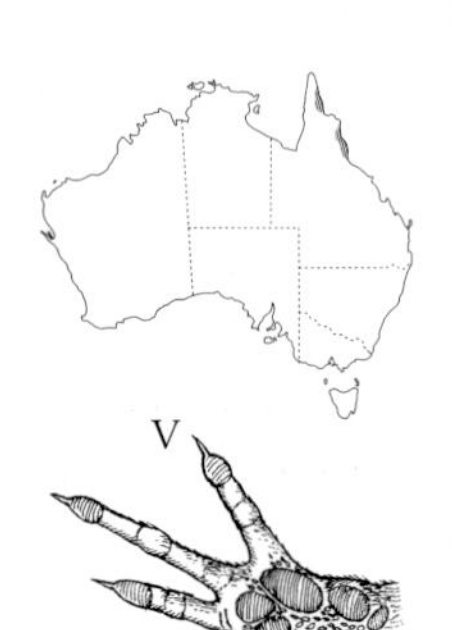

Mid-sized possum with distinctive black and white pelage. Fur coarse, woolly; yellowish-white with 3 variable black stripes—middle stripe from between eyes to tip of tail; lateral stripes from sides of muzzle through eyes and ears to rump with branches along outside of each limb. Front-on, face shows distinct white Y pattern. Tail > hb; bushy; greyish, darker on top, terminal third blackish, usually with white tip. **Voice** Loud, rolling, guttural shrieks—'gar-gair gar-gair'. **Distribution, habitat and status** Uncommon but widely distributed in rainforest and adjacent eucalypt and melaleuca woodlands of ne. Qld from Mt Spec to Iron Range. **Behaviour** Nocturnal; mostly arboreal; moves rapidly and noisily through canopy, leaping from branch to branch. Generalist insectivore, also eats fruit and honey. Obtains wood-boring grubs by gouging with lower incisors then extracting them with elongated 4th finger. Builds leaf nest in hollow or cranny amongst epiphytes. Breeds during dry season, usually 2 young raised.

elongate 4th finger

Mountain Pygmy Possum
Feathertail Glider
Honey Possum
Common Striped Possum

PLATE 24

Leadbeater's Possum *Gymnobelideus leadbeateri*

hb 150–170 mm; **t** 150–180 mm; **wt** 100–160 g
Small, fast-moving. Upperparts uniformly olive-grey; irregular blackish midline from nose to base of tail; cream cheeks and throat divided by narrow black stripe from ear to gape. Underparts cream. Tail pendulous, narrowest at base; similar colour to upperparts. **Voice** Distinctive alarm 'ch-ch-chir' like scolding bird, can be called up by imitation. **Distribution, habitat and status** Locally common within restricted distribution in montane wet sclerophyll forest ne. of Melbourne; outlying lowland population in swamp woodland at Yellingbo. Requires mixed-age eucalypt stands with dense acacia mid-storey. Threatened by decay of hollow stags and logging of old-growth forest. Endangered. **Behaviour** Nocturnal; constructs bulky nest of shredded bark in hollow; emerges shortly after dark and disperses rapidly through mid-storey vegetation in search of arboreal arthropods and acacia sap. Births occur in winter and spring; litter size usually 2. **Similar species** Sugar Glider is pale grey; lacks contrasting cream and black stripes below ears; has gliding membrane with distinctive white fringe; tail tapered.

Sugar Glider *Petaurus breviceps*

right forefoot, attachment of gliding membrane

hb 160–200 mm; **t** 165–210 mm; **wt** 90–150 g; smaller in n.
Upperparts pearl-grey, browner in n. (subspecies *ariel*); blackish midline from between eyes to lower back. Blackish patches around eyes; alternate black and cream patches at base of ears. Edge of gliding membrane blackish fringed in white. Underparts pale grey or creamy yellow (but often appears white in spotlight beam). Tail slightly tapered; grey then blackish for terminal quarter, frequently tipped white. Muzzle short, rounded, strong 'Roman nose'. **Voice** Distinctive shrill 'yip yip' sometimes repeated for long periods. **Distribution, habitat and status** Common, widespread in wet and dry sclerophyll forest and woodland from cool-temperate se. to wet/dry tropical n. Introduced to Tas. **Behaviour** Nocturnal, arboreal, can glide up to 90 m; constructs leaf nest in hollow; eats arboreal arthropods, nectar, pollen, manna. In s. births occur July–Aug, earlier in n.; litter size usually 2. **Similar species** Squirrel Glider is larger, especially in s.; has clear white underparts, tail never tipped white; tail fur long, flouncy; muzzle longer, slightly pointed. Mahogany Glider larger again; has buff to mahogany brown underparts.

Squirrel Glider *Petaurus norfolcensis*

hb 170–240 mm; **t** 220–300 mm; **wt** 190–300 g; smaller in n.
Upperparts pearl-grey; blackish midline of varying width from between eyes to mid back, darkest and broadest on head. Blackish patches around eyes; alternate black and cream patches at base of ears. Edge of gliding membrane blackish fringed with white. Underparts white, including sides of face to below eyes. Tail pale grey, blackish for terminal half to third; fur long, flouncy, especially at base, giving distinct taper. **Voice** Mostly silent. **Distribution, habitat and status** Along GDR from central Cape York to near Stawell, w. Vic; mostly in dry sclerophyll forest on inland slopes and nearby riverine corridors; in se. Qld and NSW n. of Sydney also in damp coastal eucalypt/banksia forest and woodland. Locally common but threatened in s. by habitat fragmentation. **Behaviour** Nocturnal, arboreal, can glide up to 90 m; constructs leaf nest in hollow; eats arboreal arthropods, nectar, pollen, manna, sap. Two young born May–Dec. **Similar species** See under Sugar Glider and Mahogany Glider.

PLATE 25

Mahogany Glider *Petaurus gracilis*

hb 218–265 mm; **t** 300–380 mm; **wt** 255–410 g
Upperparts vary from mahogany brown to smoky grey with patches of yellow-brown on shoulders, flanks and rump; blackish midline from between eyes to lower back; underparts vary from pale buff to honey or pale mahogany. Tail long (1.5 × hb), its fur is short, honey-grey, blackish for terminal third. **Voice** Not described, mostly silent. **Distribution, habitat and status** Restricted to narrow band (<12 km wide and 100 km long) of swampy coastal woodland below 200 m between Bambaru (s. of Ingham) and Tully ne. Qld. Endangered due to habitat clearance. **Behaviour** Nocturnal, arboreal; shelters in tree hollows. Omnivorous; eats nectar, pollen, sap (from eucalypts, acacias, grass trees), manna, arboreal arthropods. Silent, evasive of spotlight beams, dull red eye-shine. **Similar species** Squirrel Glider is smaller, grey and white without brown or honey tones; has shorter, fluffy tail. Yellow-bellied Glider is larger; predominantly olive-grey and pale yellow; highly vocal with diagnostic calls.

no large palatine vacuities

Yellow-bellied Glider *Petaurus australis*

Fluffy Glider

hb 270–320 mm; **t** 430–480 mm; **el** 52–63 mm; **wt** 450–700 g
Upperparts olive-grey; broad blackish areas on top of head and neck constricting to irregular black midline to base of tail; legs and terminal half of tail black. Underparts pale yellow or cream, sometimes tinged orange. Tail 1.5 × hb, fluffy. Ears long (50 mm), pink-grey, naked. Dull red eye-shine. **Voice** Highly vocal. Typical call 1 or 2 loud, high-pitched shrieks followed by long gurgle; audible over 500 m distance. Whirring, moaning calls given while gliding. **Distribution, habitat and status** Patchily distributed in wet sclerophyll forest from about Mackay, Qld s. to near Melbourne with isolated populations in Otway Range and far sw. Vic. Subspecies *reginae* occurs between Mt Windsor and Cardwell, ne. Qld. Uncommon. **Behaviour** Active, highly mobile, social, group cohesion maintained through frequent calls. Eats mostly plant and insect exudates—sap, nectar, honeydew and manna, with arthropods and pollen for protein. Obtains eucalypt sap by using teeth to create characteristic triangular incisions in trunk and main limbs of selected trees, and maintains these wounds by regular chewing. **Similar species** Mahogany Glider smaller, not vocal, lacks black on limbs. Dark-phase Greater Glider sooty above, whitish below, outside of ears heavily furred, silent, bright white eye-shine.

sap-feeding incisions

Greater Glider *Petauroides volans*

hb 350–450 mm; **t** 460–600 mm; **el** 41–52 mm; **wt** 900–1700 g
Largest gliding marsupial, size of cat. Gliding membrane attached to elbows and ankles; ears large, round, thickly furred on outside and fringed with long fur. Fur long, shaggy; colour variable—underparts whitish or pale grey, upperparts sooty brown, grey, cream, mottled grey and cream, or grey with white head and tail. Tail > hb, fluffy. Eye-shine brilliant white-yellow. **Voice** Silent. **Distribution, habitat and status** Locally common in wet sclerophyll forest on the ranges and coastal plains from near Mossman, ne. Qld to Daylesford, Vic. Requires large tree hollows for shelter. **Behaviour** Nocturnal, solitary; can glide over 100 m. Eats only eucalypt leaves and buds. Mating occurs Mar–Jun; the single young remains in pouch until about 4 months old then left in nest or rides on mother's back until independent at 9 months. **Similar species** See under Yellow-bellied Glider. Lemuroid Ringtail has small, naked ears, short fur.

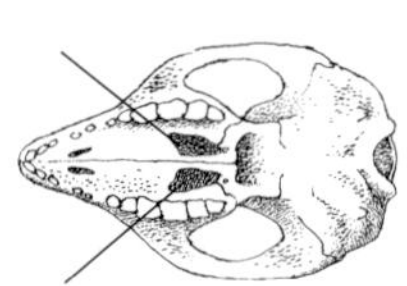

palatine vacuities present

Mahogany Glider
Yellow-bellied Glider
dark form
Greater Glider

PLATE 26

Rock Ringtail Possum *Petropseudes dahli*

hb 335–385 mm; **t** 200–270 mm; **wt** 1200–2000g
Stocky build, short legs, steep forehead and short, finely pointed muzzle. Tail diagnostic: fur thick and woolly at base; tapering steeply to thin, naked terminal half which is often carried pointed downwards. Upperparts mid-grey grizzled with brown and whitish hairs; rump and base of tail lightly washed rufous; narrow dark stripe from muzzle to mid-back; whitish eye-rings and patch behind ears. Underparts greyish white. Bright red eye-shine. Scats like slightly bent cigar, reddish brown, up to 25 mm long. **Distribution, habitat and status** Common in rocky escarpment country with eucalypt woodland or vine forest thickets in n. Kimberley, Arnhem Land and Gulf of Carpentaria e. to Lawn Hill NP, Qld; also Groote Eylandt. **Behaviour** Lives in monogamous pairs or family groups; agile climber on rocks and trees; shelters in rock crevices during daylight, emerges at night to feed on leaves, flowers, fruits; births of a single young occur in all months. **Similar species** Scaly-tailed Possum has only basal fifth of tail furred and lacks steep forehead, fine muzzle and white eye-rings.

Common Ringtail Possum *Pseudocheirus peregrinus*

Ringtail

hb 320–380 mm; **t** 300–380 mm; **wt** 660–900 g
Colour variable: usually grizzled grey-brown above with strong rufous tinge to limbs and flanks; white underparts and patches behind and below ears. Tail rufous-grey with white terminal quarter to half, short-haired, tapering, prehensile, often carried in coil. Populations in rainforest in se. Qld and ne. NSW (subspecies *pulcher*) have bright orange face, limbs and flanks, blackish flecked rufous back, rufous wash to underparts; those in Tas are darker grey-brown above. **Voice** Soft, high-pitched, insect-like chirruping, also harsher 'zip zip'. **Distribution, habitat and status** Common in open and closed forests, coastal scrub and gardens, especially where tall shrub layer is dense and diverse, from Cape York to se. SA and Tas. Also Flinders, King and Kangaroo Is. **Behaviour** Social, occurring as family groups; nocturnal; agile climber, leaps between clumps of foliage; eats leaves, flowers and fruits; in s. of range shelters in large spherical drey constructed of shredded bark, leaves and twigs in dense shrubbery, or in tree hollow lined with leaves; in n. mostly in tree hollows. Breeds Apr–Nov, litter of 2 young. **Similar species** Daintree River Ringtail lacks contrasting colour between upper body and limbs, confined to montane tropical rainforest. Cuscuses have terminal half to third of tail naked.

Western Ringtail Possum *Pseudocheirus occidentalis*

hb 320–400 mm; **t** 300–400 mm; **wt** 850–1000 g
Similar to Common Ringtail Possum but upperparts dark grey-brown without obvious rufous wash. **Distribution, habitat and status** Rare and localised in forest patches in sw. WA between Two Peoples Bay and the Collie R.; the most inland population is at Perup. Formerly widespread in sw. WA forests but declined due to clearing and fox predation. Common in gardens in Busselton and Albany. Weeping Peppermint is an important component of the habitat. Vulnerable. **Behaviour** Similar to Common Ringtail Possum but usually only 1 young is raised.

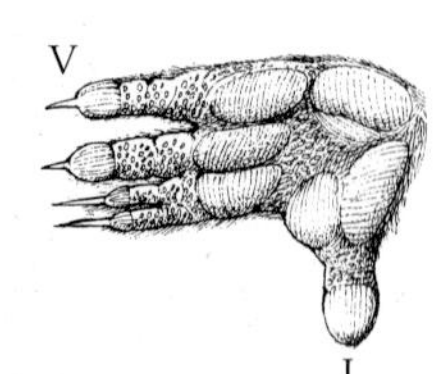

Rock Ringtail Possum
subspecies *pulcher*
Tas form
Common Ringtail Possum
Western Ringtail Possum

PLATE 27

Green Ringtail Possum *Pseudochirops archeri*

hb 300–380 mm; **t** 310–370 mm; **wt** 800–1300 g
Large, thickset; fur thick, woolly; tail strongly prehensile, completely furred, thick at base tapering to narrow tip which may be white. Upperparts olive grey grizzled with silver, yellow and black giving a peculiar greenish tinge; face grey; distinctive white patches above and below eyes and behind and below ears. Three brown stripes along back, of variable width and boldness. Underparts creamy white to buff. Eye-shine red. **Distribution, habitat and status** Sparsely distributed in montane tropical rainforest above about 300 m between Paluma and Mt Windsor Tableland (w. of Mossman), ne. Qld. **Behaviour** Nocturnal, arboreal, solitary, eats mostly leaves. Does not build nest or use hollows, but spends daylight hours sitting curled-up on branch. Births occur Aug–Nov and a single young is raised. **Similar species** Cuscuses have most of tail naked and do not overlap in range. Pelage colour and white facial patches diagnostic.

Daintree River Ringtail Possum *Pseudochirulus cinereus*

Cuscus (erroneous)

hb 330–370 mm; **t** 320–390 mm; **wt** 800–1250 g
Upperparts uniformly cinnamon or brown grading to creamy white underparts; dark brown midline from between eyes to lower back. Tail short-furred, similar colour to rest of upperparts, tapering to narrow, usually white tip. Bright red eye-shine. **Distribution, habitat and status** Confined to montane tropical rainforest above 400 m n. of Cairns—separate populations on Carbine Tableland, Mt Windsor Tableland and Thornton Peak massif. **Behaviour** Little known; probably similar to Herbert River Ringtail but breeding period more protracted—pouch young recorded July–Dec. **Similar species** Juvenile Herbert River Ringtail Possums similarly coloured, but no overlap in range.

Herbert River Ringtail Possum *Pseudochirulus herbertensis*

hb 300–400 mm; **t** 330–480 mm; **wt** 800–1500 g
Upperparts uniformly chocolate brown or black. Underparts usually pure white but chin and insides of lower limbs may be pale brown. Some males entirely dark brown. Tail narrow, naked underneath for terminal half, usually tipped white. Eye-shine orange. **Distribution, habitat and status** Locally common in montane tropical rainforest above about 350 m from Mt Lee (w. of Ingham) to Lamb Range (w. of Cairns), ne. Qld. **Behaviour** Nocturnal, arboreal, mostly solitary; moves slowly and carefully through the canopy; eats leaves from wide range of rainforest trees. Most births May–July, 2 young reared. **Similar species** Lemuroid Ringtail Possum has bushier, untapered tail, short muzzle, white eye-shine.

Lemuroid Ringtail Possum *Hemibelideus lemuroides*

Lemur-like Ringtail

hb 320–400 mm; **t** 300–370 mm; **wt** 750–1100 g
Upperparts, face, entire tail, limbs and feet rich chocolate; underparts greyish brown, sometimes with yellow tinge. Pale eye-rings. Tail bushy for entire length, not tapered. Muzzle short, forehead steep, ears short. Bright white eye-shine. Some individuals entirely creamy white with an orange tinge (most frequent on Carbine Tableland). **Distribution, habitat and status** Common in mature-age rainforest above 450 m between Ingham and Cairns, ne. Qld. Isolated population above 1100 m on Carbine Tableland. **Behaviour** Social, occurring as pairs or family groups; agile, leaps noisily from one clump of foliage to another. Shelters by day in tree hollow, emerges at night to feed on leaves, flowers, fruits. Birth of a single young occurs Aug–Nov. **Similar species** Other ringtails have short-haired, tapered tails, whitish underparts, longer muzzles. Greater Glider has long, heavily furred ears.

Green Ringtail Possum
Daintree River Ringtail Possum
juvenile
dark ♂
Herbert River Ringtail Possum
white form
Lemuroid Ringtail Possum

PLATE 28

Rufous Bettong *Aepyprymnus rufescens*

Rufous Rat-kangaroo

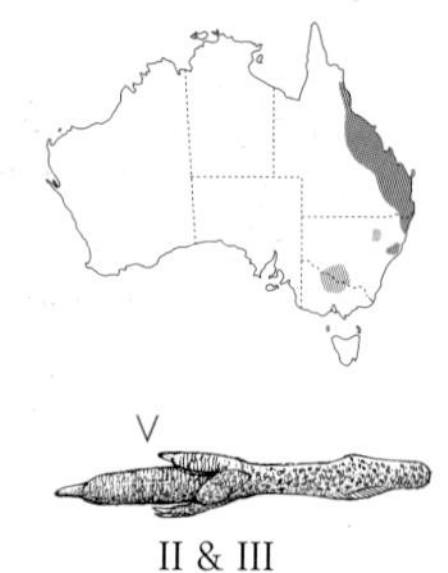

II & III

hb 375–390 mm; **t** 340–390 mm; **el** 48–57 mm; **wt** 2.5–3.5 kg
Largest of the rat-kangaroos. Fur shaggy and bristly; upperparts grey with distinct rufous tinge and emergent silvery hairs; underparts pale grey; faint pale hip stripe. Tail uniformly grey-brown sometimes with white tip. Muzzle short, lightly furred between nostrils; ears relatively long, triangular, pink inside, blackish outside with fringe of silvery hairs; bare pink skin surrounds eyes. **Voice** Low, hissing alarm call and regular grunts. **Distribution, habitat and status** Coastal and subcoastal NSW and Qld. from n. of Newcastle to Cooktown, Qld; inland to 300 km w. of Townsville. Formerly in Murray Valley of NSW and Vic. Locally common in grassy coastal forest, now rare in dry forest on inland slopes. **Behaviour** Strongly nocturnal; solitary, or male accompanying 1 or 2 females; shelters in woven nest built in scrape beneath dense cover. When feeding moves slowly using all 4 limbs plus tail for support; when relaxed, head is pulled into shoulders; stands tall with arched back to check surroundings; when alarmed hops rapidly with forelimbs tucked against body. Digs with strong claws for tubers and underground fungi, also eats some grass, sedge stem, seeds and insect larvae. Breeds year round; 2 or 3 young raised per year. **Similar species** Northern Bettong is smaller, has blackish tail, lacks shaggy appearance, rufous tones and pointed, triangular ears.

Northern Bettong *Bettongia tropica*

Tropical Bettong

hb 270–350 mm; **t** 320–365 mm; **el** 38–40 mm; **wt** 1.0–1.4 kg
Uniformly grey, paler below. Tail with low crest of longer black fur along top of outer half. **Distribution, habitat and status** Endangered; restricted to 4 areas of dry, grassy eucalypt open-forest and adjacent rainforest on the inland edge of the tropical rainforest zone in ne. Qld—Mt Windsor Tableland, w. of Mossman; Carbine Tableland; Lamb Range, e. of Atherton; Coane Range, near Paluma. Mostly above 450 m. Early specimens from Dawson R. valley, inland of Rockhampton. **Behaviour** Nocturnal, rests during day in nest of woven grass concealed under tussock. Breeding continuous. Eats mostly fungi, also tubers and grass. **Similar species** See under Rufous Bettong.

Southern Bettong *Bettongia gaimardi*

Tasmanian Bettong

hb 315–330 mm; **t** 290–345 mm; **el** 29–40 mm; **wt** 1.0–2.3 kg
Upperparts uniformly yellowish grey, tipped with silver and washed with buff, particularly on face, shoulders and flanks. Forefeet cream; underparts whitish. Tail grey-rufous, blackish for terminal third, often white-tipped. **Distribution, habitat and status** Now confined to eastern $^{2}/_{3}$ of Tas, including Maria and Bruny Is. Formerly se. mainland from se. Qld to se. SA. Inhabits dry grassy open-forest on infertile soils. Vulnerable; habitat threatened by timber harvesting and agriculture. **Behaviour** Nocturnal, solitary. Specialised feeder on underground fungi, also eats tubers, seeds, soil arthropods. Builds nest of woven grass and bark in a scrape beneath tussock or other cover. Nest material carried in slightly prehensile tail. Births occur in most months, 2 or 3 single young can be raised per year. When startled, zigzags away at great speed with long bipedal hops, head held low, body arched and tail extended. **Similar species** Long-nosed Potoroo is smaller, darker, tail noticeably < hb and tapered; muzzle longer, more pointed.

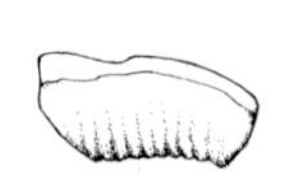

serrated and grooved premolar

Rufous Bettong

Northern Bettong

Southern Bettong

PLATE 29

Burrowing Bettong *Bettongia lesueur*
Boodie

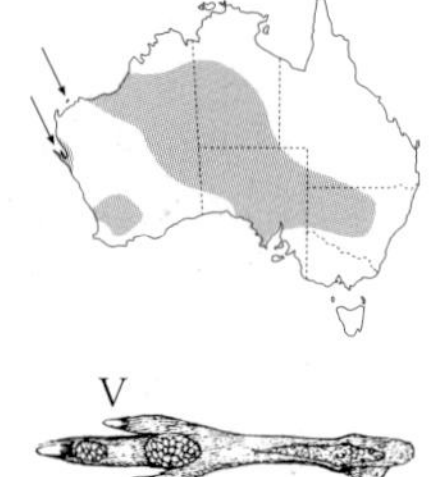

Bernier and Dorre Is: **hb** 360 mm; **t** 280 mm; **wt** 900–1600 g
Barrow Island: **hb** 280 mm: **t** 215 mm; **wt** to 980 g
Thickset with hunched posture. Fur dense, yellowish brown, paler below. Muzzle short and deep; ears short and rounded. Tail thick, < hb, not tapered, never tufted, yellow-brown but black dorsally for terminal third; tip sometimes white. **Voice** Highly vocal: a variety of grunts, hisses and squeaks. **Distribution, habitat and status** Once widespread across arid and semi-arid s., central and w. Aust. Now confined to 4 WA islands: Bernier and Dorre off Shark Bay, and Barrow and Boodie off Pilbara. Re-introduction programs to mainland under way. Inhabits hummock grassland and scrub. Warrens dug in firm, loamy soils rather than sand, or under layers of capping rock. Vulnerable. **Behaviour** Only macropod to regularly shelter in burrows. Nocturnal, gregarious; extensive warrens may support up to 100 animals. Omnivorous, eats tubers and underground fungi, leaves, flowers, seeds, fruits, soil arthropods and beach-washed carrion. Locomotion always bipedal. Breeds throughout year, up to 3 single young raised per year. **Similar species** Only bettong to live in burrows. Woylie has distinct crest of long, blackish hair on terminal half of tail, is grey above and whitish below with pale eye-rings, has longer, more pointed muzzle.

Woylie *Bettongia penicillata*
Brush-tailed Bettong

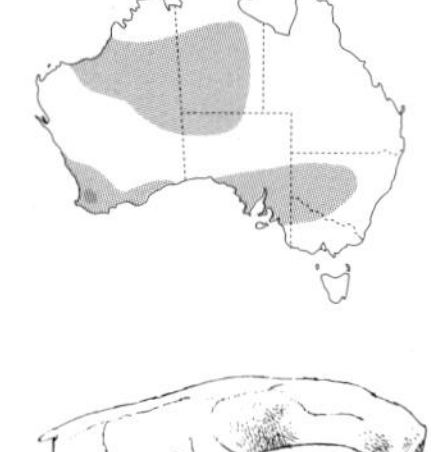

serrated and grooved premolar, I^2 and I^3 broad

hb 310–380 mm; **t** 290–350 mm; **el** 28 mm; **wt** 1.0–1.6 kg
Upperparts grey-brown grizzled with silver, washed with buff on lower face, flanks and thighs; Pale bare skin around eye; underparts cream. Tail rufous then blackish with crest of longer fur on top. **Distribution, habitat and status** Restricted to remnant habitat patches in sw. WA where population is expanding following widespread control of Red Fox, and re-introduction programs. Once widespread in dry habitats across s. and nw. Aust. Main populations inhabit dry sclerophyll forest with dense understorey at Perup, Dryandra and Tutanning. **Behaviour** Nocturnal, solitary. Specialised feeder on underground fungi, also eats tubers, seeds, soil arthropods. Builds nest of woven grass and bark in a scrape beneath tussock or other cover. Nest material carried in slightly prehensile tail. Births occur in most months, 2 or 3 single young can be raised per year. When startled zigzags away at great speed with long bipedal hops, head held low, body arched and tail extended. **Similar species** See under Burrowing Bettong. Quokka has short tail without crest, short muzzle, and is twice as heavy. Gilbert's Potoroo is restricted to dense heath; smaller; darker; t clearly < hb, tapered; muzzle longer and more pointed.

Desert Rat-kangaroo *Caloprymnus campestris*

hb 255–285 mm; **t** 300–380 mm; **hf** 110–120 mm; **el** 31–42 mm; **wt** 630–1080 g
Upperparts uniformly buff, tinged ginger, heavily flecked with long silver hairs, merging into creamy underparts, face and tail. Tail thin, > hb; hind feet long. Muzzle short and blunt, ears long for rat-kangaroo. **Distribution, habitat and status** Extinct. Known only from gibber desert and associated loamy flats in the Lake Eyre Basin of ne. SA and sw. Qld between Chambers Gorge and Birdsville, and possibly as far w. as Oodnadatta. Last recorded 1935. **Behaviour** Nocturnal, omnivorous, sheltered in nest of grass and twigs in scrape. **Similar species** Hare-wallabies (plate 33) larger, greyer with white moustache, shaggier fur, t ≤ hb.

Burrowing Bettong

Woylie

Desert Rat-kangaroo

PLATE 30

Long-nosed Potoroo *Potorous tridactylus*

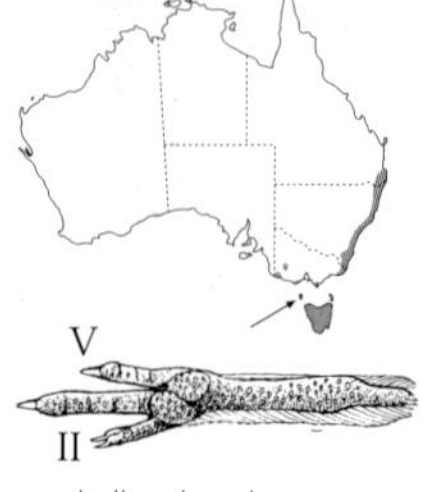

no hallucal pad

hb 340–400 mm; **t** 180–250 mm; **hf** 70–82 mm; **wt** 660–1600 g
Size of small rabbit. Fur brown-grey with rufous tinge on flanks; pale grey underparts. Tail sparsely furred, tapered, < hb, blackish, often white-tipped in s. but rarely so in n. Hindfoot length < head, has only 2 pads; ears short, rounded, dark grey on outer surface. Rhinarium and adjacent muzzle naked. **Distribution, habitat and status** Locally common in e. and n. Tas; unknown status on King and Flinders Is. Rare and patchy on mainland from coastal sw. Vic (including Grampians and French I.) to se. Qld. In Tas inhabits moist sclerophyll forest with dense shrub layer; in Vic mostly coastal heathy woodland; in n. of range rainforest, adjacent to wet sclerophyll forest and coastal wallum. Requires dense cover for shelter and adjacent, more open foraging sites. Vulnerable. **Behaviour** Not strictly nocturnal, mostly solitary; makes runways through dense ground vegetation. Builds rough squat of vegetation beneath dense cover. Omnivorous, eats underground fungi, tubers, soil arthropods, some seeds, fruits, green vegetation. Breeds throughout year, up to 2 single young may be raised per year. **Similar species** Southern and Northern Brown Bandicoots have brindled yellowish brown fur; tail is shorter and paler. Long-nosed Bandicoot is paler, tail shorter and pale brown, ears and muzzle much longer. Long-footed Potoroo difficult to distinguish in field: larger, hf > head, tail thicker at base and proportionally longer, has hallucal pad on hindfoot.

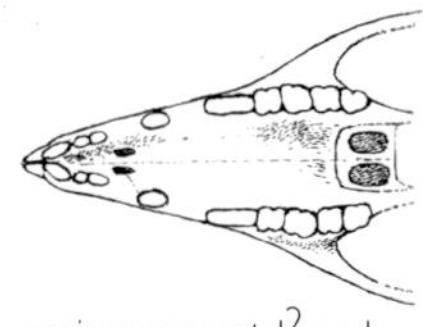

canines present, I^2 and I^3 small

Gilbert's Potoroo *Potorous gilbertii*

hb 270–290 mm; **t** 215–230 mm; **hf** 65-70 mm; **wt** 785–965 g
Similar in most respects to Long-nosed Potoroo but smaller, more rufous and diet more restricted to fungi. Critically endangered; known only from the Albany area, sw. WA where rediscovered in Dec 1994 after more than 100 yrs, in dense heath in Two Peoples Bay Nature Reserve.

Long-footed Potoroo *Potorous longipes*

hallucal pad

hb 380–410 mm; **t** 315–325 mm; **hf** 103–114 mm; **wt** 1.6–2.2 kg
Uniformly brown above, pale grey below and on inside of limbs; hindfeet large and flesh-coloured. Tail sparsely haired, blackish above, otherwise brown. Hallucal pad present on hindfoot. **Distribution, habitat and status** Restricted to 2 small areas of se. Aust: between Snowy and Cann Rivers in East Gippsland, Vic, and adjacent Bondi State Forest, NSW; and watersheds of the Buffalo and Wonnangatta Rivers, sw. of Mt Hotham, Vic. Inhabits temperate rainforest and wet sclerophyll forest with adjacent dry ridges. Extremely cryptic. Endangered. **Behaviour** Nocturnal, solitary, shelters in rough nest in dense ground cover such as sedges or wiregrass. Feeds by digging in surface soil; eats mostly underground fungi, also plant material, soil invertebrates. Births of a single young can occur in winter, spring or summer. **Similar species** See under Long-nosed Potoroo.

Broad-faced Potoroo *Potorous platyops*

hb 305 mm; **t** 178 mm; **hf** 54 mm; **wt** approx. 800 g
Upperparts grey-brown, tipped straw colour, giving streaked appearance, grading to greyer flanks and pale grey underparts, including feet. Head broad, muzzle short. **Distribution, habitat and status** Extinct. Known as a living animal only from sw. WA, inland of the tall forests. Last recorded before 1875. **Behaviour** Nothing known.

Long-nosed Potoroo
Gilbert's Potoroo
Long-footed Potoroo
Broad-faced Potoroo

PLATE 31

Musky Rat-kangaroo *Hypsiprymnodon moschatus*

hb 155–270 mm; **t** 125–160 mm; **wt** 360–680 g
The smallest macropod and only one with 5 toes, the first being opposable. Normal gait is quadrupedal bound, similar to bandicoot. Fur uniformly rich dark brown with reddish tinge, greyer on head and underparts. Whitish band from throat to belly of variable extent and intensity. Feet black; tail blackish, appears naked, scaly, < hb. Distinctive musky odour. **Distribution, habitat and status** Locally common in extensive patches of rainforest in ne. Qld between Mt Lee (w. of Ingham) and Mt Amos (s. of Cooktown). Inhabits montane rainforest on Atherton, Carbine and Windsor Tablelands as well as lowland rainforest, e.g. at Mission Beach and Cape Tribulation. **Behaviour** Diurnal, mostly solitary, terrestrial but can clamber through low branches. Forages in rainforest litter for fruit, fungi and insects. Sleeps at night in rough nest in sheltered site. Twins usual, mostly born Oct–Apr; pouch life about 21 weeks after which young left in nest until weaned. Sexually mature at 2 years old. **Similar species** Northern Brown Bandicoot is larger, paler, brindled yellowish brown, with short, pale tail.

Bennett's Tree Kangaroo *Dendrolagus bennettianus*

hb 700–750 mm; **t** 730–830 mm; **wt** 8–13.5 kg
General body colour grey-brown or buff with rufous crown, neck and shoulders, black forefeet and hindfeet, grey head with straw muzzle. Tail long, not tapered, upper surface blackish at base then pale grey/buff then black terminal third; underside entirely blackish. **Distribution, habitat and status** Restricted to a 75 km × 50 km area at n. end of Australia's wet tropics—from the Daintree R. n. to Mt Amos, s. of Cooktown, and inland to Mt Windsor Tableland. Sparsely distributed in lowland vine forest, montane rainforest and adjacent eucalypt forest where vines are plentiful. **Behaviour** Agile, competent climber; hindlegs can move independently, unlike those of other macropods; leaps from branch to branch or to ground, where it hops like typical kangaroo. Nocturnal, wary, extremely cryptic and territorial. Eats leaves of rainforest trees and vines, some fruit. Male mates with several females within its territory; most births occur just prior to wet season; females rear 1 young every second year. **Similar species** Lumholtz's Tree Kangaroo has underparts distinctly paler than upperparts, bicoloured head and uniformly coloured tail.

Lumholtz's Tree Kangaroo *Dendrolagus lumholtzi*

hb 520–650 mm; **t** 660–740 mm; **wt** 6.0–9.5 kg
The smallest tree kangaroo. Dark grey-black upperparts contrast strongly with pale yellowish buff underparts, blackish face contrasts strongly with pale grey-yellow band across forehead, crown, face behind eyes and neck. Black forefeet and hindfeet contrast with buff ankles and wrists. Tail long, cylindrical, blackish above, flecked with rufous in basal half, with yellow-buff underside. **Distribution, habitat and status** Now restricted to montane rainforest (> 800 m) in ne. Qld between Kirrama (nw. of Cardwell) and Mt Spurgeon (w. of Mossman). Extirpated from lowland rainforest between Innisfail and Cairns by clearing for agriculture and housing. Readily observed by spotlight at edges of rainforest on Atherton Tableland, including The Crater NP. **Behaviour** Nocturnal, cryptic, territorial. Movements as for Bennett's Tree Kangaroo. Sleeps during day crouched with head tucked between legs in dense foliage or on large branch. Eats leaves of rainforest trees. Births can occur in all months; a single young raised every 2 years. **Similar species** Bennett's Tree Kangaroo lacks contrasting colours between back and underside, and on head; not known to overlap in distribution.

Musky Rat-kangaroo
Bennett's Tree Kangaroo
Lumholtz's Tree Kangaroo

PLATE 32

Spectacled Hare Wallaby *Lagorchestes conspicillatus*

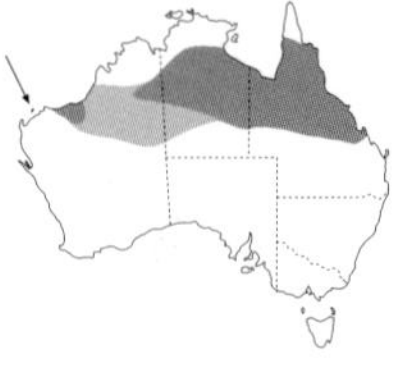

hb 480–520 mm; **t** 425–470 mm; **wt** 1.6–4.5 kg
Stocky, thickset, with short, broad muzzle. Fur shaggy, grey-brown broadly tipped with straw, giving highly grizzled appearance. Underparts pale buff or whitish. Distinctive orange eye-patch extends back to below ear. Ears pointed, edged with silvery hairs, orange patches at base. Distinct white moustache contrasts with black rhinarium. Indistinct pale hip stripe. Tail = hb; not strongly tapered, buff coloured. Barrow I. animals darker. **Distribution, habitat and status** Inhabits tropical tussock or hummock grassland with mid-dense or sparse tree and shrub cover. Has declined drastically in Great Sandy Desert, WA and in w. NT. Remains widespread and locally common through a broad swathe of n. NT and n. Qld and on Barrow I. **Behaviour** Solitary, nocturnal; spends day in tunnel in hummock grass or other dense vegetation. Moves in a slow walk until chased, when hops at speed. Feeds on green tips of grasses, herbs and shrubs. Breeds throughout year, births peak in Mar and Sept. **Similar species** Rufous Bettong lacks orange eye-patches and distinctive white moustache.

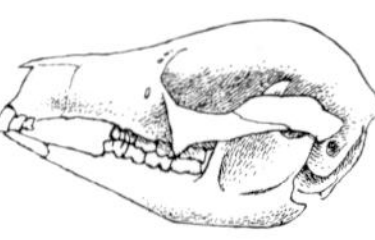

length of premolar < length of largest molar

Mala *Lagorchestes hirsutus*

Rufous Hare Wallaby

hb 310–390 mm; **t** 245–300 mm; **wt** 800–2000 g
Small, dainty; long, shaggy, rufous fur, greyer and flecked with silver on back and rump. Underparts and forearms pale yellow. Ears long, fringed with white; whitish moustache. Tail < hb, sand-coloured merging to grey tip. Animals on Bernier and Dorre Is larger and greyer dorsally. **Distribution, habitat and status** Endangered. Once widespread in deserts of central Aust and much of WA, now confined to Bernier and Dorre Is off Shark Bay, WA. Tiny colony in Tanami Desert, NT died out in 1991; re-introduction program under way. Inhabits hummock grass sand ridge country, and arid shrubland, particularly where patchy fires create mosaic of vegetation age classes. **Behaviour** Solitary, nocturnal; shelters in short burrow beneath hummock grass; when disturbed explodes from cover with rapid zigzag bounds. Feeds on green tips and seed heads of grasses, herbs, leaves of some shrubs. Timing of breeding determined mainly by rainfall patterns. **Similar species** Desert Rat-kangaroo smaller, less rufous; relatively longer hindfoot and tail. Burrowing Bettong darker brown, has short ears and thick muzzle. Banded Hare Wallaby grey with distinct darker dorsal bands.

Eastern Hare Wallaby *Lagorchestes leporides*

hb 450–500 mm; **t** 300–320 mm
Upperparts reddish-brown, heavily grizzled with silver and yellow, paler rufous on face, flanks, limbs and feet. Underparts pale grey; blackish patch on rear of upper forelimb. Tail < hb, grey-brown above, whitish below. **Distribution, habitat and status** Presumed extinct. Once common on the inland grassy plains of NSW, nw. Vic and se. SA, not recorded since 1891. **Behaviour** Nocturnal, solitary; sheltered in scrape under cover and when startled or pursued could flee with prodigious leaps and great speed.

Central Hare Wallaby *Lagorchestes asomatus*

No information on pelage or body dimensions is available.
Distribution, habitat and status Presumed extinct. A single animal was collected in 1932 between Mt Farewell and the n. end of Lake Mackay, w. of Yuendumu in NT. Unfortunately only the skull was kept. The habitat was desert sandhills.

Spectacled Hare Wallaby

Mala

Eastern Hare Wallaby

PLATE 33

Agile Wallaby *Macropus agilis*

hb males to 850 mm, females to 720 mm; **t** males to 840 mm, females to 700 mm; **st** to approx 850 mm; **wt** males to 27 kg, females to 15 kg

Yellow-brown or reddish above, paler below. Distinct white face stripe bordered above by black line through eye. Ears edged black; dark brown midline between eyes and ears; pale hip stripe. Tail long, pale sand colour. Hops with upright stance, head held high and forearms extended. Males twice weight of females. **Distribution, habitat and status** The most common kangaroo in tropical coastal Aust, occurring in a wide range of grassy forest and woodland communities on plains, often in riverine environments; rare in hilly country. **Behaviour** Grazes and browses on wide variety of plants, also eats fallen fruit. Groups shelter during day under dense vegetation and emerge in late afternoon to feed. Breeds year-round; single young remains in pouch for about 30 weeks, independent at 12 months. **Similar species** Antilopine Wallaroo larger, lacks facial stripe, has contrasting colours above and below. Northern Nailtail Wallaby (plate 41) has dark midline, terminal third of tail dark, longer ears, and characteristic head-low gait.

Antilopine Wallaroo *Macropus antilopinus*

hb males to 1.2 m, females to 840 mm; **t** males to 900 mm, females to 700 mm; **st** to approx. 1.1 m; **wt** males to 49 kg, females to 20 kg

Fur short; males reddish tan above, whitish below; females similar but greyer on head, neck and shoulders. Cheeks paler than top of muzzle, sometimes giving appearance of broad pale cheek stripe. Head of male mule-like, similar to Red Kangaroo. Ears noticeably pale inside and around edge. Tail pale, heavy, uniformly thick for most of length. Paws blackish. Naked black rhinarium. **Distribution, habitat and status** Locally common in tropical woodlands with perennial grasses, usually on plains and low hills, in central Cape York, Top End of NT and Kimberley of WA. **Behaviour** Gregarious; social groups controlled by dominant male. In dry season emerges from shelter in late afternoon to graze on grasses; in wet season can be active throughout day and night. Breeds year-round; births concentrated at end of dry season. **Similar species** See under Agile Wallaby. Euro can be similarly coloured but is stockier, has shorter, thick muzzle, is not gregarious, prefers rocky hillsides. Red Kangaroo is deeper red, has black and white flashes on muzzle.

Whiptail Wallaby *Macropus parryi*

Pretty-face Wallaby

hb males to 925 mm, females to 755 mm; **t** males to 1050 mm, females to 860 mm; **st** to approx. 1 m; **wt** males to 26 kg, females to 15 kg

Slender; long, narrow muzzle; distinctive pale colour. Upperparts pale grey-brown (greyer in winter); underparts, including limbs and all but tip of tail, whitish. Bold white face stripe contrasts with dark brown head. Ears long and distinctly tricoloured outside: blackish tip, then broad white stripe above brown base. Paws blackish. Distinct pale hip stripes merge into tail, forming U-shape when seen from behind. T > hb. **Distribution, habitat and status** Locally common in grassy open-forest in hilly areas of coastal eastern Aust from s. of Cooktown in Qld to near Grafton, NSW. **Behaviour** Gregarious, occurring in social groups controlled by dominant male; mobs of up to 50 not uncommon. Active at any time of day or night; grazes on grasses, herbs and ferns. Confiding, stands erect to look about; when not hurrying hops with upright posture and jumps high; alarm communicated by stamping of hind foot, causing group to flee in zigzag bounds with back and tail held horizontally. Single young vacates pouch at about 37 weeks, independent at about 15 months.

PLATE 34

Parma Wallaby *Macropus parma*

White-throated Wallaby

hb 450–525 mm; **t** 405–540 mm; **wt** 3.2–5.8 kg

The smallest *Macropus*. Dark reddish brown to grizzled grey-brown above, greyer on face; narrow blackish midline from forehead to mid back; white stripe along upper lip barely reaches to below eye; white fringe around base of ear. Clear white throat and upper chest, rest of underparts greyish. Tail sparsely furred, blackish; about half of animals have small white tip. Hops with head and body close to ground, forearms tucked into chest, tail held in shallow U shape. Scats characteristically square or rectangular and flattened. **Distribution, habitat and status** Rare and patchy in wet forest along GDR in NSW from Watagan Mountains n. to Gibraltar Range. Formerly as far s. as the Illawarra. **Behaviour** Nocturnal, solitary, cryptic; shelters in dense cover during day; grazes on edges of grassy clearings; makes runways through dense cover. Most births in Feb–June but can be in any month; single young weaned at about 10 months. **Similar species** Difficult to distinguish in field from pademelons, which have shorter, stiffer tails. Red-necked Pademelon has clear rufous-brown shoulders and lacks a facial stripe. Red-legged Pademelon has bright rufous patches at base of ears, on cheeks and hindlegs.

Tammar Wallaby *Macropus eugenii*

Dama Wallaby

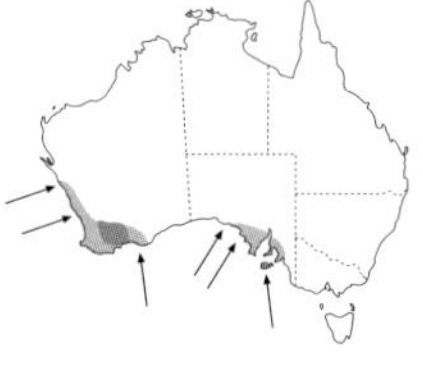

V

II & III

hb to 680 mm; **t** to 450 mm; **st** to approx. 650 mm; **wt** to 10 kg, mostly 4–7

Dumpy, thickly furred; grizzled grey-brown pelage variably washed with rufous on shoulders, flanks, thighs and forearms; grey underparts. White cheek stripe of variable intensity to beneath eye, accentuated above by blackish area between nostril and eye. Dark midline on forehead. Tail thickish, similar colour to body. Forearms held apart, especially when hopping. **Distribution, habitat and status** Formerly widespread in sw. WA and Eyre Pen., SA. Now reduced to tiny populations on mainland, e.g. Tutanning, Dryandra and Perup forests. Good populations remain on Kangaroo I., East and West Wallaby Is in Houtman Abrolhos, Garden I., and North Twin Peaks and Middle I. in the Recherche Archipelago, WA. Inhabits dense coastal heath and scrub, and some dry sclerophyll forests with dense patches of cover. **Behaviour** Nocturnal, mostly solitary; grazes in grassy clearings close to cover; makes runways through dense cover. On Kangaroo I. breeding is seasonal, most births occurring Jan–Mar. Pouch life lasts 8–9 months; most young independent at 12–13 months. **Similar species** Western Brush Wallaby is primarily grey, has black gloves and black tuft on final third of tail, white cheek stripe extends to base of ear, ears distinctly white, tipped black.

Western Brush Wallaby *Macropus irma*

Black-gloved Wallaby

hb to 900 mm; **t** 600–950 mm; **st** to approx. 800 mm; **wt** 7–9 kg

Generally gunmetal grey with brownish tinge to neck and back; dark brown midline on head and back; chest grey, belly buff. Bold white stripe from mouth to ear; ears blackish outside, whitish within, with clear black tip; distinctive black gloves, black toes and black crest to terminal half of tail. Some animals have a series of indistinct dark bars across back and rump. **Distribution, habitat and status** Locally common in dry sclerophyll forest and woodland in sw. WA, including some mallee areas, with a grassy understorey and thickets of shrubs. Has declined in recent decades, but populations recover after fox baiting. **Behaviour** More diurnal than many macropods; grazes on grasses and forbs; flees with head held low and tail horizontal. Most young born Apr–May, leave pouch in Oct–Nov. **Similar species** See under Tammar Wallaby.

Parma Wallaby

Tammar Wallaby

Western Brush Wallaby

PLATE 35

Eastern Grey Kangaroo *Macropus giganteus*

Great Grey Kangaroo, Forester

hb males to 1.3 m, females to 1 m; **t** males to 1.1 m, females to 0.84 m; **st** to approx. 1.4 m; **wt** males to 66 kg, females to 37 kg

Upperparts grey-brown, paler on hindlegs and feet; underparts pale grey or whitish, palest in females. Tail < hb; thick, heavy, similar colour to upperparts but blackish for terminal third. Ears long, dark brown outside, pale grey inside with whitish inner fringe. Males up to twice weight of females, more heavily developed in chest and forearms. Hops in upright posture with head held high and tail curved upwards. **Voice** Loud cough alarm call; clucking sounds between mother and young and courting adults. **Distribution, habitat and status** Widespread and common in e. Aust from southern Cape York Pen. to se. SA and n. and e. Tas including Maria I. where introduced. Mostly in higher rainfall (>300 mm/year) open-forest, woodland, farmland with patches of remnant vegetation, but extends into semi-arid w. NSW and parts of nw. Vic where overlaps extensively with Western Grey Kangaroo. Occurs at all altitudes up to subalpine woodland. **Behaviour** Nocturnal and crepuscular; gregarious, large mobs gather where food is abundant; grazes on grasses and herbs, some browsing of shrubs. Breeding continuous but most young born Mar–May; single young leaves pouch permanently at about 44 weeks; females capable of breeding at about 18 months. Undergoes delayed implantation, so another young may be born shortly after the pouch is vacated. **Similar species** Western Grey Kangaroo has finer build; darker, browner upperparts which contrast clearly with pale grey underparts; face and much of tail are blackish; ears are slightly shorter, sparsely furred, blackish inside giving sharp contrast with white fringing hairs. Males have a distinctive musty odour. Eastern Wallaroo has stockier build; longer, shaggy dark grey fur; mostly confined to rocky slopes.

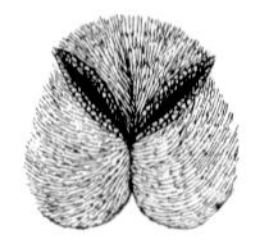

haired rhinarium

upper incisor shape and grooves

Western Grey Kangaroo *Macropus fuliginosus*

Black-faced Kangaroo, Mallee Kangaroo, Stinker

hb males to 1.2 m, females to 0.9 m; **t** males to 1.0 m, females to 0.8 m; **st** to approx. 1.3 m; **wt** males to 54 kg, females to 28 kg

Mainland animals brown-grey above, blackish on muzzle and face; underparts from upper chest to belly pale grey. Nominate subspecies on Kangaroo I. sooty brown throughout and heavily built. Hops in upright posture with head held high and tail curved upwards. **Voice** Loud cough alarm call; clucking sounds between mother and young and courting adults. **Distribution, habitat and status** Locally common across southern Aust, extending n. to the Bulloo and Paroo R. systems, sw. Qld, and to southern Shark Bay in WA. **Behaviour** Nocturnal and crepuscular, rests in shade during middle of day; gregarious, large mobs may form where food is locally abundant. Grazes on grasses and herbs, some browsing on shrubs; Births can occur in any month but most are in summer; the single young leaves pouch permanently at 42 weeks but continues to suckle until about 18 months old; females are reproductively mature at 18 months. Does not undergo embryonic diapause. **Similar species** See under Eastern Grey Kangaroo.

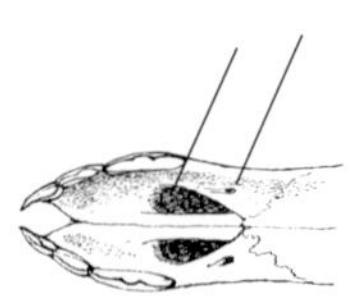

placement of palatine foramina relative to palatine vacuity

Eastern Grey Kangaroo
♀
♂
juvenile
Western Grey Kangaroo
♂
♀
subspecies *fuliginosus*
subspecies *melanops*

PLATE 36

Red Kangaroo *Macropus rufus*

Blue Flier (adult female)

hb males to 1.4 m, females to 1.1 m; **t** males to 1.0 m, females to 0.9 m; **st** to approx. 1.5 m; **wt** males to 85 kg, females to 35 kg

The largest living marsupial, with proportionally the longest hindlegs. Fur short. Males typically rust or brick-red above grading to pale buff below and on limbs, ears and tail; sometimes grey wash to head, shoulders and base of tail. Females about half the weight of males, blue-grey with a brown tinge above; whitish below. In central Aust both sexes are usually reddish. Diagnostic features are the mule-like head with long, pointed ears and heavy eyelids; white then black flashes on sides of muzzle; broad white cheek stripe from upper lip to below eye. Hops with body almost horizontal, tail held in shallow U shape. **Voice** Loud cough alarm call; clucking sounds. **Distribution, habitat and status** Widespread and common across semi-arid and arid Australia, e. to western slopes of ne. NSW. Inhabits semi-arid plains, grasslands, shrublands, woodlands and some dry open-forests on GDR. Avoids rocky country; sparse in desert country. Provision of bore water for stock has allowed populations in arid country to expand. **Behaviour** Nocturnal and crepuscular; rests during day in a shady scrape; gregarious in small groups but large mobs will gather where rainfall has allowed growth of green herbage; grazes on grasses and herbs. Births can occur in any month, mostly determined by availability of green feed, but most in spring–summer. Single young vacates pouch at about 9 months, independent at 12 months. Undergoes embryonic diapause. **Similar species** Euro is stockier, has shaggier fur, lacks black patch on side of muzzle and clear white cheek stripe. Antilopine Wallaroo lacks bold facial markings and occupies tropical, monsoonal woodlands.

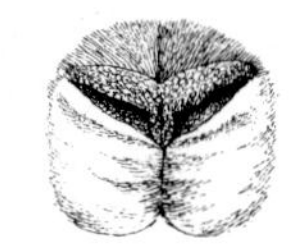

naked rhinarium

upper incisor shape and grooves

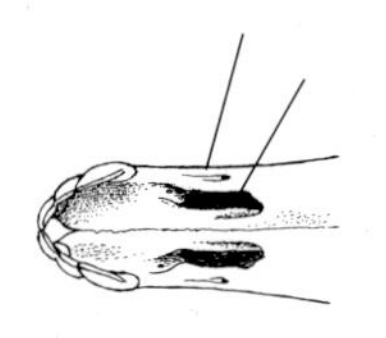

placement of palatine foramina relative to palatine vacuity

Red Kangaroo

PLATE 37

Euro *Macropus robustus*

Common Wallaroo, Eastern Wallaroo, Barrow Island Euro

hb males to 1.1 m, females to 0.8 m; **t** males to 0.9 m, females to 0.75 m; **st** to approx. 1.1 m; **wt** males to 55 kg, females to 25 kg

Shaggy, heavily-built kangaroos with wide geographic range and habitat tolerance but inhabits mainly hilly or mountainous terrain. Considerable variation in coat colour and texture between animals from wet versus arid areas, but all have blackish digits. Males 2 or 3 times heavier than females and always darker coloured. Subspecies *robustus* (Eastern Wallaroo) occurs in forests of the GDR from Cooktown in ne. Qld to Snowy R., Vic. Males have shaggy blackish coat, females are bluish-grey, paler below. Colour of subspecies *erubescens* (Euro) grades westwards from brownish animals on the inland slopes of NSW and Qld through roan on the inland ranges to orange-red on the w. coast; much shorter hair than the Eastern Wallaroo. Those in monsoonal woodlands of nw. WA and NT are paler grey-brown, and almost white below. On Barrow Island subspecies *isabellensis* is smaller and uniformly reddish brown. All forms have a characteristic upright stance with shoulders back, elbows raised and tucked into sides, and hop with body held semi-erect. **Voice** Loud hiss. **Distribution, habitat and status** Widespread and common over much of Aust in rocky ranges, isolated hills, scarps and plateaux. Vegetation varies from the most arid shrublands and grasslands to wet eucalypt forests and subalpine woodland. **Behaviour** Nocturnal and crepuscular, mostly solitary; rests during day in cave or beneath rock overhang. Moves after dark to surrounding flat country to graze on grasses and forbs. Births can occur in all months, in arid areas timing dependent on rainfall. Single young vacates pouch at about 9 months, independent at 16 months. Undergoes embryonic diapause. **Similar species** Eastern Wallaroo distinguished from Eastern Grey Kangaroo by shaggy coat, stocky build, lower legs and feet uniformly blackish not pale. Euro differs from Red Kangaroo by lacking obvious facial markings except for clear pale eyelids, has shorter muzzle, less pointed ears, proportionally shorter hindlimbs, and a larger area of bare black rhinarium. Antilopine Wallaroo is taller, more slender, has paler underparts, usually not found in rocky habitats.

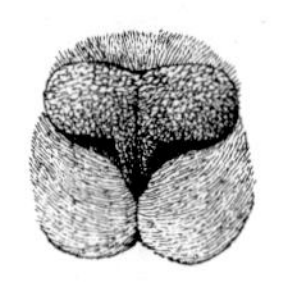

large naked rhinarium

upper incisor shape

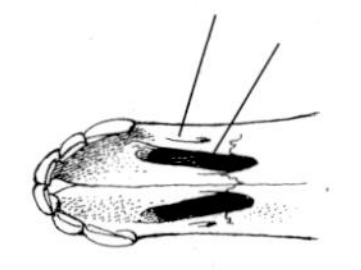

placement of palatine foramina relative to palatine vacuity

Black Wallaroo *Macropus bernardus*

hb to 730 mm; **t** to 640 mm; **st** to approx. 800 mm; **wt** males 19–22 kg, females about 13 kg

Smallest and most distinctive of the wallaroos; thickset and muscular; males uniformly dark brown to black with yellow-brown forearms, females mid-grey to grey-brown with blackish paws, feet and tail tip. Ears rather short and oval. Diagnostic groove in I^3. **Distribution, habitat and status** Confined to steep, rocky sandstone escarpment and plateaux of w. Arnhem Land between South Alligator R. and Nabarlek, including Mt Brockman and Nourlangie Rock. Common in monsoonal woodland, hummock grassland and sandstone rainforest. **Behaviour** Little studied. Nocturnal, solitary, shy. Shelters during day in cave or overhang, descends to foot of the escarpment after dark to graze on grasses and forbs. If disturbed moves rapidly to shelter of rocky escarpment.

distinct notches in I^3

subspecies *woodwardi*
♂
subspecies
erubescens
♂
subspecies *robustus*
♂
♀
Euro
♀
♂
subspecies *erubescens*
♂
♀
Black Wallaroo

PLATE 38

Toolache *Macropus greyi*

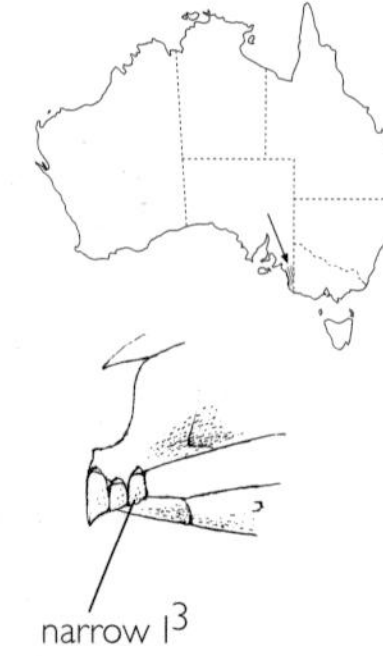

narrow I^3

hb 810–840 mm; **t** 710–730 mm; **hf** 212–216 mm; **el** 65–71 mm

Slender and lightly built. General colour of upperparts greyish fawn with slight ginger tinge, and grizzled silver; sometimes faint grey bars across rump; whitish underparts; tinge of rufous on back of ears, crown and shoulders; bold black stripe from rhinarium to eye bordered below by whitish cheek that has narrow brownish band beneath; ears whitish, tipped black; limbs obviously pale, digits black with sharp demarcation to pale grey paws and feet; tail very pale grey with slight terminal tuft. **Distribution, habitat and status** Extinct. Formerly common in hinterland of The Coorong in se. SA, in swampy sedgeland plains interspersed with stringybark woodland on rises. Possibly also inland to s. limit of mallee heath on 90 Mile Plain and Wimmera district, Vic, s. of Little Desert. Last wild specimens seen near Robe, 1927. **Behaviour** Nocturnal and crepuscular; gregarious, fleet-footed. **Similar species** See under Red-necked Wallaby.

Red-necked Wallaby *Macropus rufogriseus*

Bennett's Wallaby (Tas)

hb males to 920 mm, females to 780; **t** males to 875 mm, females to 790 mm; **st** to approx. 800 mm; **hf** 220–230 mm; **el** 76–78 mm; **wt** males to 27 kg, females to 16 kg

Upperparts grey-brown grizzled with cream; back of ears, neck and shoulders rusty brown; whitish stripe along upper lip not bold or clearly defined; digits black, merging to grey-brown limbs. Underparts whitish; Tail slightly < hb; pale grey, darker at tip. Tas and Bass Strait Island animals have long shaggy fur; mainland animals (subspecies *banksianus*) have short, coarse, less grizzled fur, more red on neck, shoulders and rump. **Distribution, habitat and status** Widespread and common in sclerophyll forest and coastal scrub in se. Aust from about Rockhampton, Qld to se. SA, including Tas and many Bass Strait Is. In Qld inland to Carnarvon NP. **Behaviour** Nocturnal and crepuscular; mostly solitary; grazes and browses on wide variety of green shoots. Tas animals have distinct birth period from Feb to April; on mainland births can occur in any month. Single young independent at 9–10 months. **Similar species** See under Black-striped Wallaby. Toolache is lightly built, particularly in forequarters, pale greyish fawn with very pale tail, bold black stripe from rhinarium to eye bordered below by white cheek, forearms are yellow-white not grey-brown. Black Wallaby is much darker, including wholly black tail.

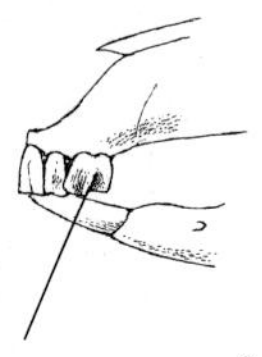

broad, notched I^3

Black-striped Wallaby *Macropus dorsalis*

Scrub Wallaby

hb males to 820 mm, females to 615 mm; **t** males to 830 mm, females to 615 mm; **st** to approx. 800 mm; **wt** males to 20 kg, females to 7.5 kg

Fur short and coarse; upperparts grey-brown with distinctly reddish forequarters including ears and arms; distinct, narrow black midline from forehead to mid back; obvious white cheek stripe from upper lip to below eye, with narrow dark borders above and below; cream horizontal hip stripe. Throat and chest whitish, rest of underparts pale grey. Tail about = hb; uniformly mid-grey, tinged rufous. Gait distinctive—short hop with head held low, back arched, rump tucked under body, arms extended forward and outward. Males 2–3 times heavier than females. **Distribution, habitat and status** Widespread but no longer common in sclerophyll forest and brigalow scrub with dense shrub cover from near Townsville and Longreach, Qld s. to about Narrabri, NSW. In NSW only w. of GDR. **Behaviour** Nocturnal, crepuscular, shy; gregarious, groups rest during day in thick scrub; at dusk move along well-defined paths in single file to open areas where graze on grasses and herbs. Births can occur in any month. **Similar species** Red-necked Wallaby lacks midline, hip stripe and clear cheek stripe; has less red in forequarters and grey not reddish arms.

Toolache
subspecies *rufogriseus*
subspecies *banksianus*
Red-necked Wallaby
Black-striped Wallaby

PLATE 39

Quokka *Setonix brachyurus*

hb 410–540 mm; **t** 250–310 mm; **wt** 2.5–4.2 kg

Small, stocky; short-tailed; short, broad head; rounded ears. Fur thick, coarse, grizzled brown above with rufous wash to cheeks, ears, neck and shoulders; underparts buff. Tail blackish, short, thick, sparsely furred. **Distribution, habitat and status** Rare and restricted in sw. WA from s. of Perth to Two Peoples Bay, common on Rottnest and Bald Is off Perth. On mainland requires dense, wet ground cover in forest or swampy flats. On fox-free Rottnest I. occupies wide range of semi-arid habitats, plus gardens and township land. Vulnerable. **Behaviour** Mostly nocturnal; gregarious, in large groups at good food sources, several may share a nest. Browses and grazes, eating grasses, sedges, succulents and foliage of shrubs. Can climb in low shrubs and trees. On Rottnest I. births occur in Feb–Mar; on mainland, throughout year.

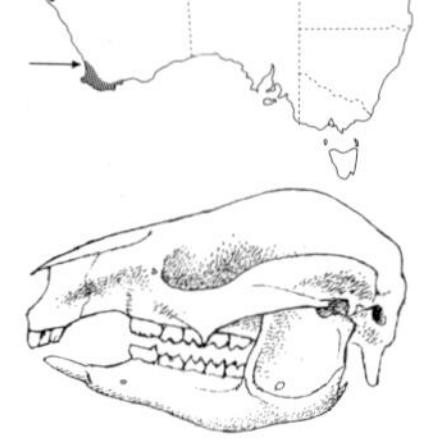

lacks canines

Black Wallaby *Wallabia bicolor*

Swamp Wallaby, Black-tailed Wallaby

hb males to 850 mm, females to 750 mm; **t** males to 860 mm, females to 730 mm; **st** to approx. 850 mm; **wt** males to 20 kg, females to 15 kg

Dark, stocky; fur coarse, dense, dark grey or blackish above flecked with pale grey, yellow and rufous; underparts pale yellow or rufous-orange; amount of rufous in pelage highly variable, but greater and richer in n. Top of muzzle mid-grey to blackish; forehead, crown and base of ears rich rufous; blackish area around and behind eyes; stripe from upper lip below eye towards ear varies from white, through pale grey or yellow to almost non-existent. Paws, feet and wrists black. Tail = hb, blackish, thick, not strongly tapered, sometimes with white tip. Characteristically hops with head and shoulders low, tail roughly horizontal. **Distribution, habitat and status** Common along entire e. coast from near tip of Cape York to Mt Gambier, SA; inland to Carnarvon Range, Qld. In wide range of forest, woodland, scrub and heath from tropical rainforest to dry brigalow, box–ironbark and some mallee associations. **Behaviour** More diurnal than most macropods; solitary, shy. A specialised browser, feeding on foliage of shrubs, ferns, sedges and some grasses. Births in any month, but in s. mostly winter; single young vacates pouch at about 8–9 months, independent at 16 months. **Similar species** Eastern Wallaroo is larger and lacks rufous and yellow tones. Red-necked Wallaby and Black-striped Wallaby are predominantly grey, have pale grey tails.

P^1 longer than longest molar

Banded Hare Wallaby *Lagostrophus fasciatus*

hb 400–450 mm; **t** 350–400 mm; **wt** 1.5–3.0, mostly <2 kg

Sole surviving species of sthenurine kangaroos. Fur thick, long and shaggy, grey above grizzled with pale yellow and silver; rufous tinge on flanks; pale grey below; head and face uniformly grey. Diagnostic series of transverse dark bars from mid-back to base of tail. Muzzle short. Tail < hb, grey, sparsely furred, black crest along terminal third. Rhinarium naked. **Distribution, habitat and status** Restricted to Bernier and Dorre Is in Shark Bay, WA. Formerly widespread through WA wheatbelt. Vulnerable. **Behaviour** Nocturnal; males aggressive; browses on foliage of low shrubs and spinifex. Births Dec–Sept; most females raise one young per year; independent at about 10 months. **Similar species** Burrowing Bettong and Rufous Hare Wallaby rufous not grey in general body colour, lack darker bands. Rufous Hare Wallaby has furred rhinarium.

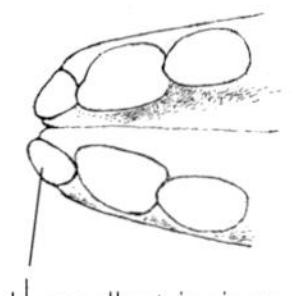

I^1 smallest incisor

Quokka

Black Wallaby

Banded Hare Wallaby

PLATE 40

Bridled Nailtail Wallaby *Onychogalea fraenata*

Merrin, Flash Jack

hb 450–700 mm; **t** 380–540 mm; **st** to approx. 650 mm; **wt** 4–8 kg
General body colour grey-yellow with rufous wash to neck, shoulders and upper back; prominent whitish stripes (bridle) from between ears along each side of dark midline across shoulders to underarm; lower section of bridle highlighted in front by chocolate brown crescent; prominent whitish cheek stripe from upper lip to beneath eye; edged above by blackish stripe from rhinarium to eye. Underparts cream. Tail < hb, tapering, pale grey, slightly crested with dark brown fur at tip. Ears long, pointed, grey outside, whitish within. When hopping fast, forearms extend down and out from body and move in circular motion. **Distribution, habitat and status** Endangered. Confined to acacia-dominated woodland and shrubland in Taunton Scientific Reserve near Dingo (w. of Rockhampton) and recently introduced to Idalia NP, central Qld. Formerly widespread in grassy woodland from Lake Hindmarsh in Vic to Charters Towers, Qld. **Behaviour** Partly diurnal; mostly solitary; shelters in scrape in thick shrubbery; grazes grasses, forbs and chenopods, some browsing on shrubs and possibly digging for tubers.

tail 'nail'

Crescent Nailtail Wallaby *Onychogalea lunata*

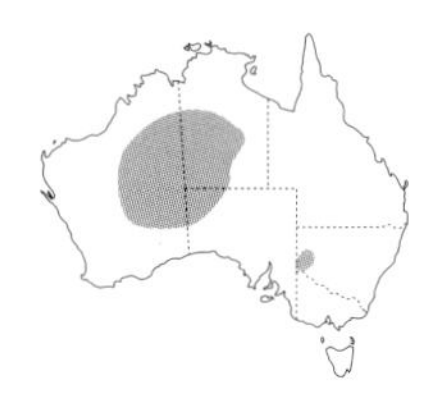

hb 370–510 mm; **t** 150–330 mm; **wt** approx 3.5 kg
Upperparts including tail ashy-grey; rufous wash across shoulders and onto flanks; distinct whitish crescents from shoulder blades down to chest; whitish stripe diagonally across flanks before thigh; another along thigh from knee to hip. Indistinct blackish stripe along top of muzzle to between ears; another along side of muzzle from rhinarium to eye. Underparts whitish. Tail grey with narrow, short, black crest along top. **Distribution, habitat and status** Extinct. Once widespread and quite common through a variety of arid and semi-arid shrublands and woodlands across much of central and southern Aust, including sw. NSW. Last definite records were from central Aust in 1940s. **Behaviour** Little known. When pursued sometimes sought refuge in hollow log or hollow base of tree.

Northern Nailtail Wallaby *Onychogalea unguifera*

Sandy Nailtail Wallaby

hb 500–700 mm; **t** 600–740 (650) mm; **st** to approx. 650 mm; **el** 80–92 mm; **wt** 5–9 kg
Largest and most widely distributed nailtail wallaby. Overall body colour sandy ginger; palest on head and neck with whitish patches above and below eyes; distinct rufous tinge to flanks; ears pale grey, very long, mobile; pale cream cheek stripe and hip stripe; brown midline from neck to base of tail; underparts including feet and lower limbs creamy white. Tail > hb; pale grey then blackish, sometimes banded blackish and grey for terminal third with black terminal tuft. Hops with head low, tail upturned, long arms held stiff with rotary action. **Distribution, habitat and status** Widespread and locally common across drier parts of n. Aust, from w. Cape York to Kimberley. Inhabits wide range of tussock grassland, savannah, grassy woodland, floodplain and riparian communities. Common around edges of blacksoil plains. **Behaviour** Nocturnal; solitary; rests during day in shallow scrape beneath dense cover; may hide by lying in long grass or crawling beneath low shrubs. Eats the most succulent forage available—herbs, green grass shoots, fruit, succulents. **Similar species** Agile Wallaby is larger, lacks dark midline and black on tail, has shorter ears and tail.

long, slender incisors

Bridled Nailtail Wallaby

Crescent Nailtail Wallaby

Northern Nailtail Wallaby

PLATE 41

Short-eared Rock Wallaby *Petrogale brachyotis*

hb 415–550 mm; **t** 320–550 mm; **el** 40–48 mm; **wt** 2.2–5.5 kg
Highly variable in colour and size. Fur short; upperparts uniformly grey-brown flecked with silver, imparting a glistening appearance; legs cinnamon; forequarters can be suffused reddish; whitish or buff side stripe from below shoulder to thigh with dark brown anterior border; buff cheek stripe with darker area between eye and nose; dark midline from forehead to mid back. Ears short—< half head length. Underparts pale grey. Tail short for a rock wallaby, cinnamon brown becoming blackish and tufted. Animals from Kimberley and Victoria R. area duller with indistinct stripes. **Distribution, habitat and status** Widespread but patchy across tropical n. Aust, n. of 600 mm rainfall isohyet, from Windjana Gorge (s. Kimberley) e. to near NT–Qld border. Also Groote Eylandt, Sir Edward Pellew, Wessel, and English Company Is groups. Inhabits low rocky hills, cliffs and gorges within savannah woodland. Locally common, e.g. in parts of Kakadu NP. **Behaviour** Mostly nocturnal but also diurnal, especially in cloudy weather; gregarious; sometimes shares habitat with Nabarlek. Feeds on grasses, sedges, seeds. **Similar species** Nabarlek is smaller, lacks pale side stripe, hops with body horizontal, tail curled high over back with tuft fluffed out.

Monjon *Petrogale burbidgei*

Warabi

hb 300–350 mm; **t** 265–290 mm; **hf** 80–92 mm; **el** 30–33 mm; **wt** 950–1400 g
Smallest rock wallaby; little-known, elfin. Upperparts marbled deep olive-buff; head forward of ears distinctly rufous and buff, palest on cheeks; eyes large and black; ears relatively short (<35 mm); rufous wash on upper limbs. Underparts, including chin whitish, tending to yellow on belly. Tail tawny, flecked black; terminal third with distinct dark brown tuft. **Distribution, habitat and status** Restricted to high rainfall (1200–1400 mm) areas of coastal Kimberley and some islands of Bonaparte Arch. (Bigge, Boongaree, Katers Is). Locally common in fractured King Leopold Sandstone country with open eucalypt woodland. **Behaviour** Timid; mostly nocturnal; remarkably fast and agile; shelters in crevices and caves; feeds on grasses and ferns. Births probably throughout year with peak in wet season. **Similar species** Nabarlek has distinct dark head stripe and pale cheek stripe, is generally more greyish, has ears >35 mm long; not known to occur with Monjon.

Nabarlek *Petrogale concinna*

Little Rock Wallaby

hb 310–365 mm; **t** 260–335 mm; **hf** 95–105 mm; **el** 41–45 mm; **wt** 1200–1600g
Tiny rock wallaby. Generally grey with rufous tinge to rump, limbs and tail which is tufted dark brown for terminal third. Clear whitish cheek stripe bordered above by blackish area between rhinarium and eye. Blackish midline from between eyes to neck; blackish armpit. When moving at speed carries body horizontally with tail arched high over back, terminal tuft fluffed up. **Distribution, habitat and status** Three discrete populations: Arnhem Land including Groote Eylandt; w. Top End between Mary and Victoria Rivers; coastal Kimberley including some islands of Bonaparte Arch. (Borda, Augustus, Long, Hidden Is). In sandstone cliffs and boulder screes, lateritic breakaways, granite boulder piles, usually where shrub cover is present. Locally common including parts of Kakadu and Litchfield NPs, e.g. Jim Jim Falls. **Behaviour** Mostly nocturnal; partly gregarious; timid. At night may travel some distance away from rock shelter to feed on grasses, sedges and ferns. Births can occur in any month but mostly in wet season. **Similar species** See under Monjon and Short-eared Rock Wallaby.

PLATE 42

Rothschild's Rock Wallaby *Petrogale rothschildi*

hb 470–600 mm; **t** 550–700 mm; **el** 56 mm; **wt** 3.7–6.6 kg
One of the largest rock-wallabies. Upperparts pale golden-brown with greyish wash to hindneck and shoulders, underparts buff. Top of head, muzzle and ears dark brown with sharp demarcation to pale buff-grey face below level of eyes and ears. Fur may have purplish sheen, specially around hindneck and shoulders. Tail long, brown, darker towards tip. **Distribution, habitat and status** Confined to Pilbara District, WA, including Hamersley and Chichester Ranges, Burrup Pen., and 5 Is of Dampier Arch. (Rosemary, Enderby, Dolphin, West Lewis, Burrup). Inhabits cliffs, rock piles, scree slopes covered in hummock grassland. Has declined in areas accessible to Red Fox. **Behaviour** Mostly nocturnal, shelters by day in cool crevices and caves. Feeds on green vegetation, especially grasses growing on sandplains between the rockpiles. **Similar species** Black-flanked Rock Wallaby has side stripes, dark midline, bicoloured ears.

Black-flanked Rock Wallaby *Petrogale lateralis*

Black-footed Rock Wallaby

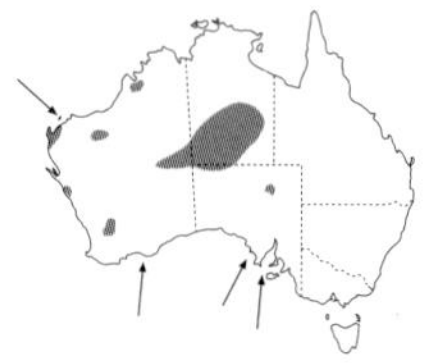

hb 465–570 mm; **t** 450–510 mm; **el** 43–62 mm; **wt** 5–7.7 kg
Highly variable. Back and rump rich brown flecked with silver; neck, shoulders and upper arms grey; forearms and base of ears brown; white side stripe bordered below by broad dark brown flanks; dark midline from forehead to mid-back. Head noticeably tricoloured: top dark brown, whitish or buff cheek stripe to ear, rest of head below level of ear greyish. Underparts whitish or buff. Tail brown, terminal third blackish. Distinctiveness of cheek and side stripes, and grey neck and shoulders, varies greatly. **Distribution, habitat and status** Widely scattered populations in ranges of central and w. Aust, plus Barrow I. and some islands in Recherche Arch. (WA) and Investigator Group (SA). Has declined drastically in WA and SA due to fox predation; some central Aust populations have declined, others seem stable. Vulnerable. **Behaviour** Timid, never venturing far from rock shelter. Mostly nocturnal but basks in sun in cooler weather. Shelters in crevices and caves; feeds on grasses and forbs. **Similar species** Rothschild's Rock Wallaby lacks side stripes and midline, has uniformly dark ears.

Yellow-footed Rock Wallaby *Petrogale xanthopus*

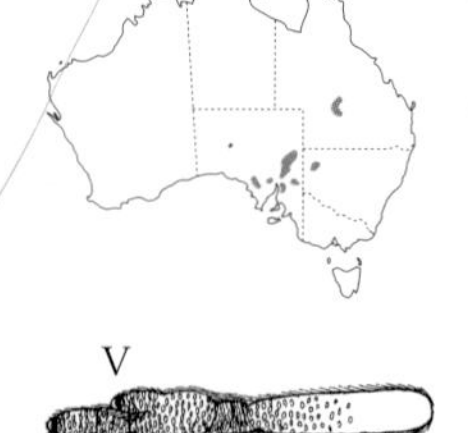

broad hindfoot; thick granulated sole

hb 480–650 mm; **t** 600–715 mm; **el** 70–90 mm; **wt** 6–11 kg
Upperparts generally grey; limbs yellow-orange; clear white side stripe bordered on both sides by chocolate patches; whitish hip stripe bordered above by brown patch; white cheek stripe bordered above by black patch from rhinarium to eye; dark brown midline from forehead to mid-back; ears long, yellow-orange outside, blackish fringed with white inside. Underparts white. Tail diagnostic: long, yellowish at base grading to grey with distinct dark rings becoming bolder towards tip, which is solidly dark brown. Digits black. Race *celeris* from sw. Qld is paler, hip stripe and tail rings indistinct. **Distribution, habitat and status** Locally common in Flinders and Gammon Ranges and Olary Hills, SA, and in the ranges of the Adavale Basin, sw. Qld (bounded by Adavale, Blackall and Stonehenge). Populations in Gap and Coturaundee Ranges, w. NSW, and Gawler Ranges, SA are endangered. Inhabits broken, rock scree and low cliffs in semi-arid ranges. Vulnerable. **Behaviour** Gregarious, in groups up to 100; mostly nocturnal but basks in sunshine in cool weather; feeds on grasses, forbs and browse. Births in any month.

Rothschild's Rock Wallaby

Black-flanked Rock Wallaby

Yellow-footed Rock Wallaby

PLATE 43

Eastern Queensland Rock Wallabies

In e. Qld a complex of 9 rock wallaby species forms a virtually unbroken chain from Pascoe R. on Cape York to the NSW border. The Proserpine Rock Wallaby is morphologically distinctive and the Brush-tailed Rock Wallaby has a wide distribution along e. coast; these 2 species are treated on Plate 45. The remaining 7 species are, on current knowledge, virtually indistinguishable in the field. They are separated most readily by analysis of chromosomes, or by geographic location in regions where genetic markers have been studied. For these species only one description is given, but knowledge of ranges, status and habitats is summarised separately for each.

There is much overlap in measurements of the 7 species. The largest, Herbert's Rock Wallaby, has **hb** males 500–615 mm, females 470–565; **t** males 510–660 mm, females 510–570; **wt** males 5.0–6.7 kg, females 3.7–5.0. The smallest, Sharman's Rock Wallaby, has **hb** males 490–530 mm, females 455–475 mm; **t** males 500–532 mm, females 435–480 mm; **wt** males 4.4 kg, females 4.1. kg

Typically brown above, but can be greyish; darker on muzzle and shoulders. Underparts buff or pale grey. Overall pelage colour may vary according to local rock type and to stage of moult—after autumn moult pelage is more grey, becoming progressively browner through the year. Blackish midline from forehead to hindneck; dark brown or black patch from armpits onto shoulders and flanks; pale cheek stripe and buff or whitish side stripe may be present; forearms, lower legs, hindfeet and base of tail distinctly pale, often buff; digits blackish. Tail long, cylindrical, brown grading to blackish and longer-haired towards tip. Northern species (Mareeba, Godman's, Cape York) can have dirty white tail tip of variable length. Amount of ornamentation to pelage tends to increase from s. to n. within a species range, and across the total range of the group. **Behaviour** Mostly nocturnal but some diurnal activity: gregarious, form long-lasting pair bonds and practise allogrooming; shelter by day in rock piles and caves, also climb sloping trees; feed on green vegetation, especially forbs and low shrubs. Breed probably throughout year. **Similar species** See Brush-tailed and Proserpine Rock Wallabies.

Distribution, habitat and status

1 **Cape York Rock Wallaby *Petrogale coenensis*** — confined to central e. Cape York from Musgrave to Pascoe R. Separated from Godman's Rock Wallaby by Hann R. catchment, a gap of 70 km.
2 **Godman's Rock Wallaby *Petrogale godmani*** — e. Cape York from Mt Carbine n. to Bathurst Head and inland to near Palmerville. Locally common but has declined at Black Mountain.
3 **Mareeba Rock Wallaby *Petrogale mareeba*** — in the ranges w. of Cairns from near Mt Garnet n. to the Mitchell R. and Mt Carbine, inland to Mungana.
4 **Sharman's Rock Wallaby *Petrogale sharmani*** — confined to the Seaview and Coane Ranges, e. of Ingham.
5 **Allied Rock Wallaby *Petrogale assimilis*** — widespread in hinterland of Townsville, from near Bowen inland to Hughenden, Mt Hope and Croydon, but excluding the Seaview Range to the n. of Townsville. Also Magnetic and Palm Is.
6 **Unadorned Rock Wallaby *Petrogale inornata*** — scattered colonies in coastal ranges from s. of Rockhampton to Home Hill, near Townsville. Also Whitsunday Island.
7 **Herbert's Rock Wallaby *Petrogale herberti*** — widespread and common in se. Qld from about 100 km nw. of Brisbane to the Fitzroy R., s. of Rockhampton; inland to near Clermont and Rubyvale.

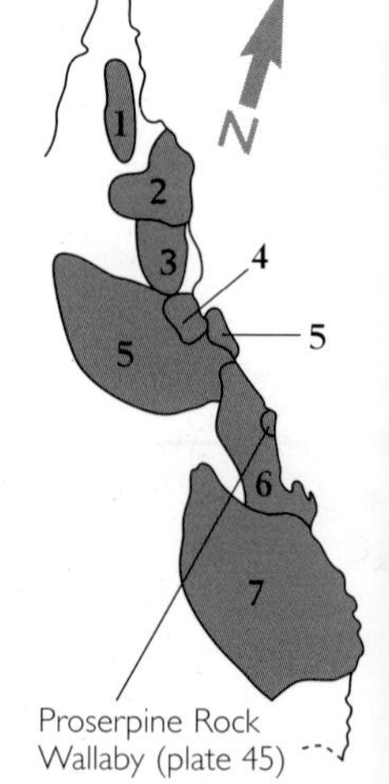

Cape York Rock Wallaby
Godman's Rock Wallaby
Mareeba Rock Wallaby
Sharman's Rock Wallaby
Allied Rock Wallaby
Unadorned Rock Wallaby
Herbert's Rock Wallaby

PLATE 44

Proserpine Rock Wallaby *Petrogale persephone*

hb 500–640 mm; **t** 520–670 mm; **wt** 4.2–8.8 kg
A large, short-furred rock wallaby with elongated, narrow muzzle. Upperparts dark grey-brown, greyer after summer moult; dark brown patch behind armpit. Underparts pale yellow or cream. May be faint pink or mauve tinge on shoulders and belly. Tail rich rufous at base, becoming blackish distally, often a small whitish tip. Ears uniformly brown-orange outside, blackish inside; pale cheek stripe from upper lip to below eye, pale grey face below this; narrow blackish midline from between eyes to mid-back. Feet and paws blackish. **Distribution, habitat and status** Endangered. Known only from about 26 sites, all in Whitsunday Shire, e. Qld, including the Clarke Range, Conway Range, Mt Dryander and Gloucester I. Restricted to boulder outcrops in pockets of semi-deciduous vine forest on foothills near grassy open-forest. **Behaviour** Timid, never venturing far from rock shelter. Partly diurnal; suns in cooler weather. Shelters in crevices and caves; feeds on grasses and forbs. **Similar species** Overlaps with Unadorned Rock Wallaby, which is smaller, browner, lacks bold facial pattern and rufous at base of tail and ears, never has white tail tip.

Brush-tailed Rock Wallaby *Petrogale penicillata*

hb 510–580 mm; **t** 500–700 mm; **wt** 5–11 kg
Fur thick, shaggy. Upperparts entirely rufous-brown; or grey-brown on shoulders and back, brown on rump and thighs. Paler below, sometimes with white blaze on chest. Forelegs, lower hindlegs, paws and feet blackish. Top of head dark brown with blackish midline from just behind eyes to hindneck; ears blackish outside, buff inside; cream or buff cheek stripe. Blackish patch behind forearms; pale side stripe may be present. Tail long, brown or black, shaggiest at tip. Animals from n. of range have shorter fur and are paler. **Distribution, habitat and status** Along GDR from 100 km nw. of Brisbane s. to upper Snowy R. in e. Vic and Grampians, w. Vic. Inland to Warrumbungle NP NSW. Inhabits rock piles and cliffs with numerous crevices and ledges in vegetation ranging from rainforest to dry sclerophyll forest. Has declined seriously in s. and w., still locally common in ne. NSW and se. Qld. Vulnerable. **Behaviour** Great agility on rock faces, also climbs sloping trees; gregarious in small groups; mostly nocturnal but basks in sunshine in cool weather; feeds on grasses, forbs and browse. **Similar species** Overlaps with Herbert's Rock Wallaby in Qld. Herbert's is greyer, less warm brown dorsally and buff or whitish below; forearms and hindfeet brown not black; less black on tail. Tail not shaggy. Black Wallaby is larger, short blackish fur on hindquarters and tail rather than shaggy brown.

Purple-necked Rock Wallaby *Petrogale purpureicollis*

hb 500–610 mm; **t** 450–600 mm; **wt** 4.7–7.0 kg
Upperparts pale grey-brown, greyer on shoulders, tending to rufous on rump. Characteristic purple wash over head, neck and shoulders varies in intensity and extent from faint pink to bright red or purple. Cheek stripe buff; brown midline from between eyes to hindneck; dark brown patch behind armpit may extend to thigh; no distinct side-stripe. Underparts sandy, chin whitish, feet pale grey-brown with dark brown digits. Tail brown, blackish towards tip. **Distribution, habitat and status** Locally common in rocky areas of central w. Qld from Kajabbi s. to Dajarra and from w. of Mt Isa to e. end of Selwyn Range. Also in Lawn Hill and Riversleigh limestone country where status unknown. Habitat includes hummock grassland and mulga. **Behaviour** Mostly nocturnal, shelters by day in cool crevices and caves. Feeds on green vegetation, especially grasses.

Proserpine Rock Wallaby
Brush-tailed Rock Wallaby
Purple-necked Rock Wallaby

PLATE 45

Rufous-bellied Pademelon *Thylogale billardierii*

Tasmanian Pademelon, Red-bellied Pademelon

hb 550–630 mm; **t** 320–415 mm; **wt** males to 9 kg, females to 5.8 kg
Small-medium size, squat; tail thick, short ($^{2}/_{3}$ hb); fur short, dense. Upperparts uniformly dark brown-grey flecked buff, bases of ears and narrow eye-rings tinged rufous. Underparts buff, including around mouth and chin, with rufous tinge especially on belly. When hopping, forelegs held close against chest. **Distribution, habitat and status** Common in Tas, including Bruny, King and Hunter Is and Furneaux Group. Extirpated from coastal Vic and se. SA. Inhabits dense vegetation adjacent to open patches, including paddocks and gardens, rainforest, wet sclerophyll forest, coastal heath and scrub, gullies in drier forests. **Behaviour** Mostly nocturnal; solitary, but aggregations may form at favoured feeding sites; never far from dense cover; feeds on grass, forbs, some browse including nectar-bearing flowers. Most young born Apr–June but births can occur in any month; pouch life about 200 days. **Similar species** Red-necked Wallaby is larger, has white cheek stripe, whitish underparts, black paws.

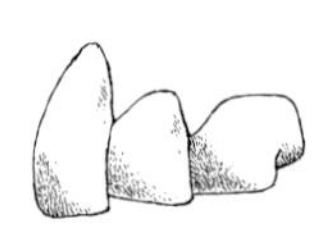

Posterior notch in I^3

Red-legged Pademelon *Thylogale stigmatica*

hb 390–535 mm; **t** 300–470 mm; **wt** males to 6.5 kg, females to 4.2 kg
A brightly coloured rainforest wallaby. Fur short, stiff; back and rump grizzled brown-grey; neck, shoulders and forelegs grey grading to rich rufous face, flanks and hindlegs; white cheek stripe; dark midline to forehead and crown; buff hip stripe. Underparts whitish. Tail grey above, buff below; short, thick and carried straight when hopping. **Distribution, habitat and status** Common in rainforest from near tip of Cape York to s. of Tamworth, NSW. Sometimes in adjacent wet sclerophyll forest. **Behaviour** Solitary, timid. Although active day and night, usually enters open areas only after dark. In s. of range feeds mostly in rainforest, mainly on fallen leaves and fruit; in n., forest edges are visited and grass forms a larger proportion of diet. Births year-round; single young in pouch for about 7 months, weaned at about 9 months. **Similar species** Red-necked Pademelon has rufous only on neck and shoulders; lacks facial colour and hip stripe. Parma Wallaby more petite, lacks obvious rufous, no hip stripe, carries tail in a shallow curve.

Red-necked Pademelon *Thylogale thetis*

hb 320–600 mm; **t** 300–510 mm; **wt** males to 7 kg, females to 4 kg
Upperparts brown flecked grey except for neck, shoulders and forehead, which are rich rufous. No obvious facial markings, but ring of reddish bare skin around eye. Chin, throat and chest whitish; rest of underparts cream. Tail short, thick, grey; carried close to horizontal when moving at speed. **Distribution, habitat and status** Common around fringes of rainforest and wet sclerophyll forest in far e. Aust from about Biloela in Qld to the Illawarra, NSW. **Behaviour** Mostly nocturnal but active within dense forest during daylight; shy, rarely far from cover; travels along well-defined runways from shelter to feeding areas; eats soft grasses, forbs and low shrubs. Births occur year-round. **Similar species** See under Red-legged Pademelon.

Rufous-bellied Pademelon

Red-legged Pademelon

Red-necked Pademelon

PLATE 46

Least Blossom Bat *Macroglossus minimus*
Northern Blossom Bat

fa 38–43 mm; **hb** 49–67 mm; **el** 13–16 mm; **wt** 10–19 (12.5) g
Like tiny flying-fox; long muzzle, long bristle-tipped tongue. Fur uniformly light russet-brown, paler below. Tail a tiny stub; no tail membrane, but narrow flap of skin along inside of legs. Ears naked, long, rounded. Mature males have reddish crescent gland on chest. Incisors tiny and separated. **Distribution, habitat and status** Locally common in rainforest, tropical riverine woodland and mangroves in Kimberley, Top End and e. Qld s. to Mackay. **Behaviour** Roosts alone under large leaves in dense cover, under loose bark or in entrance of large hollows. Feeds mostly on nectar and pollen, some soft fruit. Can fly slowly, manoeuvre skilfully and hover, but mostly lands beside flowers or fruit before feeding. Temporary feeding territories established around food sources with much vocalising, chasing and wing-clapping. Breeding probably continuous. **Similar species** Eastern Blossom Bat lacks a tail stub and skin flap along inside of legs (replaced by fringe of hairs); males have no chest gland; fur fawn rather than russet.

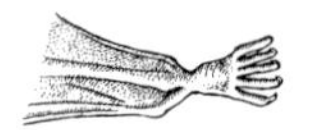
flap of skin along inside leg

Eastern Blossom Bat *Syconycteris australis*
Common Blossom Bat

fa 38–43 mm; **hb** 57–71 mm; **el** 13–17 mm; **wt** 13–17 (15.5) g
Similar appearance to Least Blossom Bat but lacks tail stub and flap of skin on inside leg—replaced by fringe of hairs; has shorter muzzle; I^2 clearly longer than I^1. **Distribution, habitat and status** Locally common in coastal e. Aust from tip of Cape York to about Kempsey, NSW; inland to rainforest patches in adjacent ranges. Roosts in rainforest and forages in variety of habitats including rainforest, tropical woodland, heathland. **Behaviour** Roosts alone or in small groups under broad leaves or amongst dense vine thickets. Feeds on nectar, pollen, soft fruits. Breeding probably continuous. **Similar species** See under Least Blossom Bat.

fringe of hair along inside leg

Eastern Tube-nosed Bat *Nyctimene robinsoni*

fa 65–70 mm; **hb** 82–93 mm; **el** 16–20 mm; **wt** 42–56 (48) g
Characterised by protruding tubular nostrils on short, broad, rounded muzzle, combined with small pale yellow or yellow-green spots scattered over wing membranes and ears. Body fur grey-brown to russet, greyest on head and face; distinct blackish midline from hindneck to rump. Lacks lower incisors; lower canines modified for eating fruit, bite against upper incisors. Bright red eye-shine. **Voice** High pitched 'seep' given in flight. **Distribution, habitat and status** East coast and adjacent ranges from islands of Torres Strait, s. to about Lismore, NSW. Inhabits rainforest, open-forest, woodlands and heath. Common in n. but uncommon s. of about Ingham, Qld. **Behaviour** Roosts in canopy of primary rainforest, mostly alone, sometimes in small groups. Eats rainforest fruits and some cultivated fruit; also nectar from banksia and other heathland plants. Highly manoeuvrable flight, hovers easily. Single young born Oct–Dec. **Similar species** Torresian Tube-nosed Bat has uniformly fawn rather than grey and russet body fur, and less distinct black midline.

Torresian Tube-nosed Bat *Nyctimene cephalotes*

fa 60–65 mm; **hb** 90–125 mm; **el** 14–18 mm; **wt** 40–50 g
Similar to Eastern Tube-nosed Bat but fur woolly fawn, with a narrow dark midline. **Distribution, habitat and status** In Aust known only from Moa Island in Torres Strait, where Eastern Tube-nosed Bat also occurs. **Behaviour** Not documented. **Similar species** See under Eastern Tube-nosed Bat.

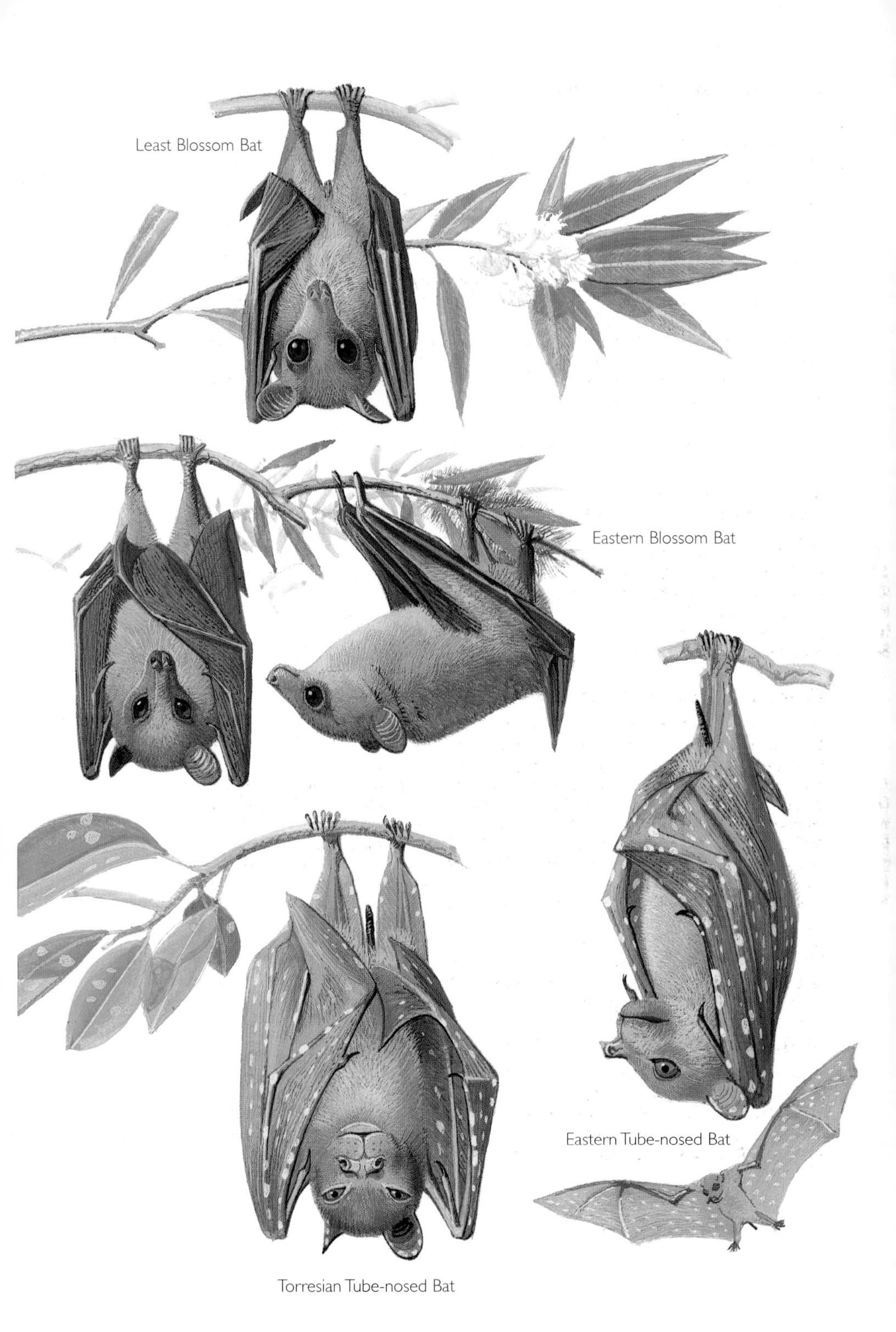
Least Blossom Bat
Eastern Blossom Bat
Eastern Tube-nosed Bat
Torresian Tube-nosed Bat

PLATE 47

Bare-backed Fruit Bat *Dobsonia moluccensis*

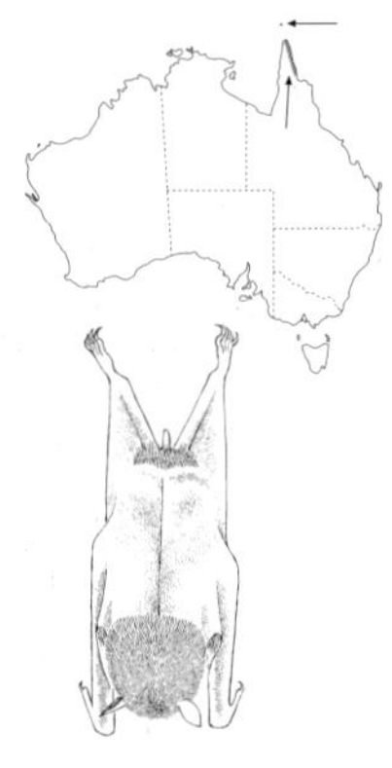

fa 140–155 mm; **hb** 190–230 mm; **el** 29–37 mm; **wt** 380–500 g
Immediately recognisable by 'bare back' due to wings joining along midline rather than attached to flanks (there is furred skin below the wings); claws on only the first digit of fa, and presence of tail stub. Fur dark brown, blackish on head; paler, sparsely furred underparts; all claws ivory colour. Only one pair of upper and lower incisors. Wingbeats make characteristic hollow 'pok-pok' sound. **Distribution, habitat and status** Uncommon and restricted to e. Cape York, s. to Silver Plains, in rainforest and tropical open-forest; also on Moa Is. Colonies mostly <100 individuals. **Behaviour** The only Aust fruit bat to regularly roost in caves or abandoned mines, always in twilight zone close to entrance; sometimes roosts in dense vegetation. Slow, highly manoeuvrable flight; hovers readily; forages below the canopy. Eats soft fruits and nectar from trees and shrubs. Single young born Sept–Nov.

Black Flying-fox *Pteropus alecto*

fa 155–190 mm; **hb** 230–280 mm; **el** 29–37 mm; **wt** 590–980 g
Large; all-black but sometimes with reddish-brown or yellow-brown fur on mantle; belly fur flecked with grey tips giving grizzled appearance. Lower leg unfurred. **Distribution, habitat and status** Common in tropical and subtropical forests and woodlands around n. coast from Exmouth Gulf in WA to Bowraville, n. NSW. Present on numerous offshore islands. Camps formed in mangroves, paperbark forest and rainforest. **Behaviour** Forms large permanent camps for daytime roosting; occasionally groups roost in cave entrances. Eats mostly nectar from eucalypts, melaleucas, turpentines, also fruits, including citrus and mangoes, some leaf material. May travel up to 50 km each night in search of flowering or fruiting trees. Wary. Single young born in camp; timing varies from Jan–Mar in n. Aust to Oct–Nov in se. Qld. **Similar species** Spectacled Flying-fox has straw-coloured patch on hindneck and may have conspicuous yellow-brown eye-rings. Little Red Flying-fox has uniformly reddish-brown body fur, transparent brown wing membranes and is smaller. Grey-headed Flying-fox has orange-brown collar around neck, greyish body fur.

Spectacled Flying-fox *Pteropus conspicillatus*

fa 160–180 mm; **hb** 220–240 mm; **el** 33–36 mm; **wt** 500–850 g
Similar to Black Flying-fox but large patch of straw-coloured fur on mantle. The broad eye-ring, which may extend forward along the muzzle, varies from dirty brown to straw. **Distribution, habitat and status** Confined to coastal Qld from about Tully to tip of Cape York and islands in Torres Strait. Camps in tall rainforest, gallery forest, mangroves or paperbark forests. Vulnerable due to habitat destruction and persecution by fruit-growers. **Behaviour** Usually roosts in single-species camps. Feeds mostly on rainforest fruits, some eucalypt nectar and pollen; territorial and aggressive at rich food resources. Mating activity common throughout first half of year but conception only in Mar–May; single young born Oct–Dec. **Similar species** See under Black Flying-fox.

Dusky Flying-fox *Pteropus brunneus*

Percy Island Flying-fox

fa 118 mm; **hb** 210 mm; **wt** approx. 200 g
Small, uniformly golden-brown flying-fox known from only one specimen collected on Percy I., n. of Shoalwater Bay in Qld, in 1874. Apparently died out late in 19th century before further details were recorded.

Bare-backed Fruit Bat
Black Flying-fox
Spectacled Flying-fox
Dusky Flying-fox

PLATE 48

Grey-headed Flying-fox *Pteropus poliocephalus*

fa 138–180 mm; **hb** 230–290 mm; **el** 30–37 mm; **wt** 600–1000 (700) g
The largest Australian bat; wingspan to 1 m. Body fur long, grizzled grey, paler on head; distinct, broad, complete collar of golden-orange fur. Legs furred to ankles. **Voice** Complex series of screeches and squeals. **Distribution, habitat and status** Eastern coastal Aust from Gladstone in Qld to S. Gippsland and Melbourne in Vic, rare influxes further w. and s. Rarely more than 200 km inland. Formerly bred n. to Mackay, now contracting s. Vulnerable. **Behaviour** In warmer months gathers in very large camps, usually in dense forest in gullies; population more dispersed in winter. Size of camps fluctuates in response to local food supplies; in s. numbers fluctuate in regular pattern, being highest in late summer–autumn and lowest in winter. Conception in Mar–Apr; single young born Oct–Nov. **Similar species** Little Red Flying-fox has uniformly reddish-brown body fur, transparent brown wing membranes, is smaller; lower leg not thickly furred. Spectacled Flying-fox is blackish rather than grey, pale colour on neck is confined to hind neck, and pale eye-rings are conspicuous.

Little Red Flying-fox *Pteropus scapulatus*

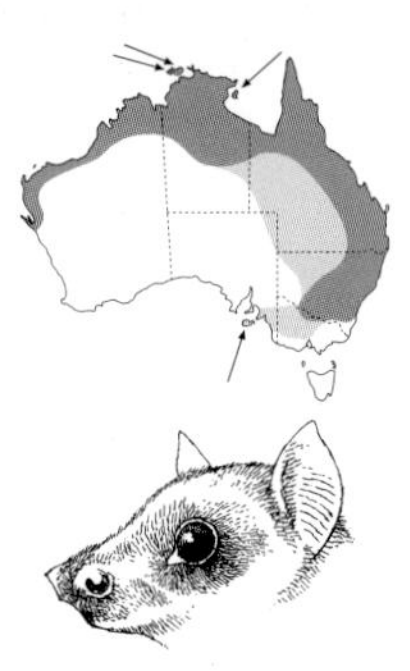

fa 120–150 mm; **hb** 125–200 mm; **el** 29–40 mm; **wt** 300–600 (450) g
Body fur entirely reddish brown, short; sometimes with paler yellowish mantle and creamy patches on shoulder; head dark grey to pale grey. Lower legs not furred. Wings distinctly pale brown and translucent in flight. **Voice** High-pitched twitterings, screeches, squeals; single sharp 'yap'. **Distribution, habitat and status** Found throughout coastal and subcoastal n. and e. Aust from Shark Bay in WA to n. Vic and, rarely, se. SA. Extends further inland than any other fruit-bat. Roosts by day in streamside trees and feeds in wide range of tropical, subtropical and temperate forests and woodlands. **Behaviour** During Oct–Nov forms large camps where mating occurs. Camps tend to disperse in Mar–Apr, before birthing. In camps individuals pack close together. Feeds mostly on eucalypt or melaleuca nectar, also native and cultivated fruit. Partially nomadic in response to unpredictable food supplies. In dry conditions skims surface of water, then licks wet fur. **Similar species** See under Grey-headed and Black Flying-foxes.

Large-eared Flying-fox *Pteropus macrotis*

fa 136–148 mm; **hb** 180–240 mm; **el** 30–37 mm; **wt** 315–425 g
Small dark brown flying-fox from New Guinea. Forms camps in mangroves on Boigu I. in Torres Strait. Identified by small size, uniformly dark brown fur tipped yellow on underparts, and relatively long, pointed ears. **Similar species** Could be difficult to distinguish from Little Red and Torresian Flying-foxes, but neither recorded from Boigu I.

Torresian Flying-fox *Pteropus banakrisi*

fa 128–141 mm; **hb** 160–200 mm; **el** 27–29 mm; **wt** 210–240 g
Similar to Black Flying Fox but much smaller, about 1/3 the weight and has proportionately longer ears. **Distribution, habitat and status** Known only from Moa I. in Torres Strait, where a camp of about 2000 individuals roosts in rainforest. **Behaviour** Nothing recorded.

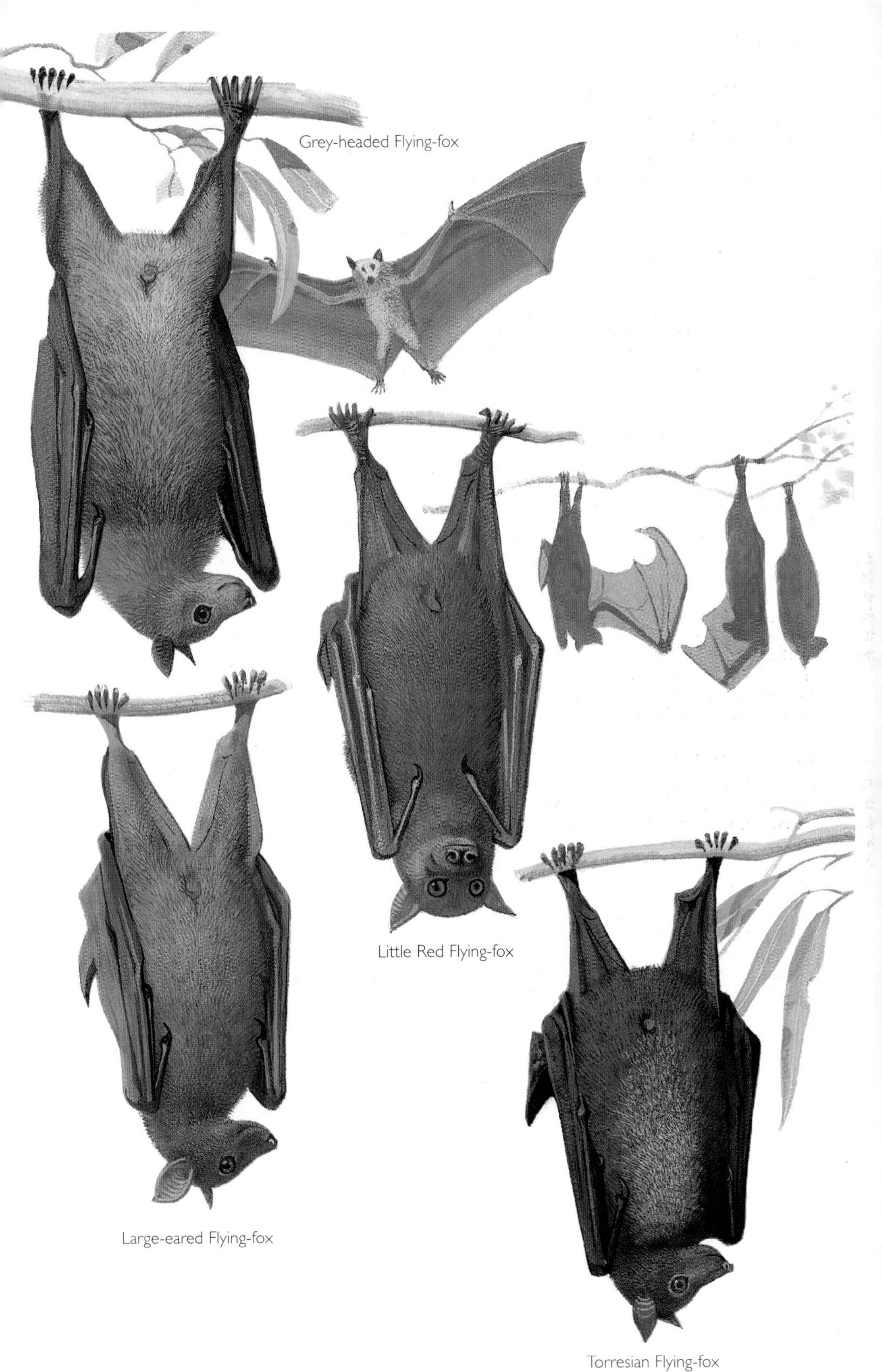
Grey-headed Flying-fox
Little Red Flying-fox
Large-eared Flying-fox
Torresian Flying-fox

PLATE 49

Ghost Bat *Macroderma gigas*

fa 96–112 mm; **hb** 98–120 mm; **el** 44–56 mm; **wt** 75–145 (105) g
The largest Aust microbat. Very long fluted ears joined along inner margin for about half their length; elongated but simple noseleaf. No tail, but a full tail membrane between long legs. Fur mid to pale grey above, whitish below and on head, dark eye-rings; bare skin of wings, ears and muzzle pale brown. **Voice** Audible calls include a 'dirrup dirrup' like a Fairy Martin, plus excited twitterings. **Distribution, habitat and status** Patchily distributed in small colonies in 3 areas of n. Aust: Pilbara and Kimberley in WA, Top End of NT, and ne. Qld. Extinct in central Aust. Requires undisturbed roost caves or mineshafts, usually complex systems with several openings. Vulnerable. **Behaviour** Eats large insects, geckoes, frogs, small birds, mammals including other bats. Kills made on ground or in flight, then taken to feeding perch, usually a rocky overhang or small cave. Single young born July–Sept.

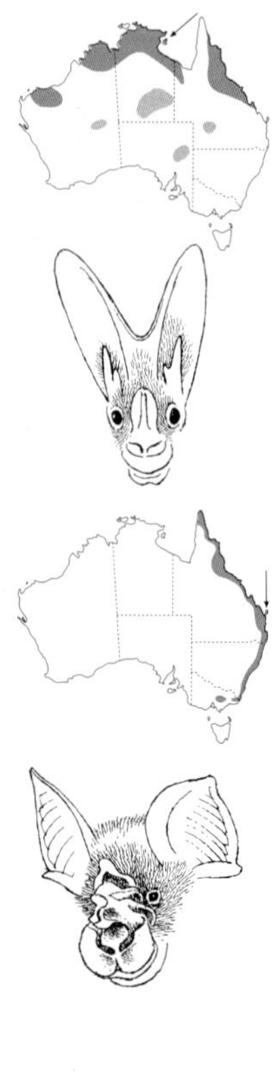

Eastern Horseshoe Bat *Rhinolophus megaphyllus*

fa 44–52 mm; **hb** 44–53 mm; **el** 17–21 mm; **wt** 7–13 g
Small with distinctive noseleaf; ears large, simple, pointed; eyes minute. Fur grey-brown, paler below; bare skin dark grey. In n. Qld some individuals have rufous-orange fur. Noseleaf pink, edged grey; comprises a lower horseshoe-shaped plate, a central fleshy projection, and triangular pointed upper leaf. Northern and southern forms may be separate taxa, but indistinguishable morphologically. **Distribution, habitat and status** Locally common in e. Aust seaward of GDR from tip of Cape York to central Vic. Roosts in small colonies (mostly 5–50 but up to 2000) in warm, humid sites within caves or mines. Availability of suitable roosting and maternity sites is main determinant of local distribution. **Behaviour** Captures insects in flight and by gleaning, at all heights within the canopy; flight slow, fluttery, highly manoeuvrable. Pregnant females move to special maternity caves in Sept–Oct to give birth and suckle single young; then return to non-maternity roost in Mar–Apr where mating occurs. In s. becomes torpid and inactive in coldest months. Hangs from cave ceiling, not walls, separately not in tight clusters, often with wings wrapped around body. **Similar species** Large-eared Horseshoe Bat has longer ears (>25 mm) and forearms (>50 mm), more complex noseleaf. Leaf-nosed bats have smaller, simpler noseleafs.

Large-eared Horseshoe Bat *Rhinolophus philippinensis*

fa 'large' form 53–59 mm, 'small' form 50–53 mm; **hb** 54–60 mm; **el** 'large' form 29–33 mm, 'small' form 25–27 mm; **wt** 8.5–15 g
Similar to Eastern Horseshoe Bat, but longer ears and forearms; larger, taller central projection on noseleaf. Some individuals have bright yellow noseleaf, anus, penis or pubic teats. Two forms differ in size and echolocation calls and may be separate taxa. **Distribution, habitat and status** Confined to e. Cape York, s. to Paluma, inland to Chillagoe. Inhabits rainforest, tropical eucalypt forest, melaleuca forest. Large form known only from McIlwraith and Iron Ranges and is endangered. **Behaviour** Little known. Captures insects in flight, often low in the understorey, also gleans insects from ground. **Similar species** See under Eastern Horseshoe Bat.

Orange Leaf-nosed Bat *Rhinonicteris aurantius*

fa 45–50 mm; **hb** 43–53 mm; **el** 12–13.6 mm; **wt** 7–11 g
Fur uniformly rich golden orange, some individuals brownish yellow. Ears short, finely pointed. Noseleaf without obvious forward projections; lower, horseshoe-shaped leaf deeply cleft at bottom; upper leaf scalloped. **Distribution, habitat and status** Found across n. Aust from about Derby in WA to Lawn Hill in Qld, with isolated population in e. Pilbara, WA (may be a separate taxon). During dry season roosts in deep, warm, humid caves or mines and forages nearby; in wet season is more widespread and may not require caves for roosting. Common in NT but scattered and localised in WA. **Behaviour** Forages low in open habitats, including grasslands and along roads. Captures insects in flight, feasts on flying termites. Births occur in wet season but maternity sites unknown. **Similar species** No other orange-coloured bat has short, pointed ears (<14 mm) and deeply cleft lower noseleaf without obvious forward projections.

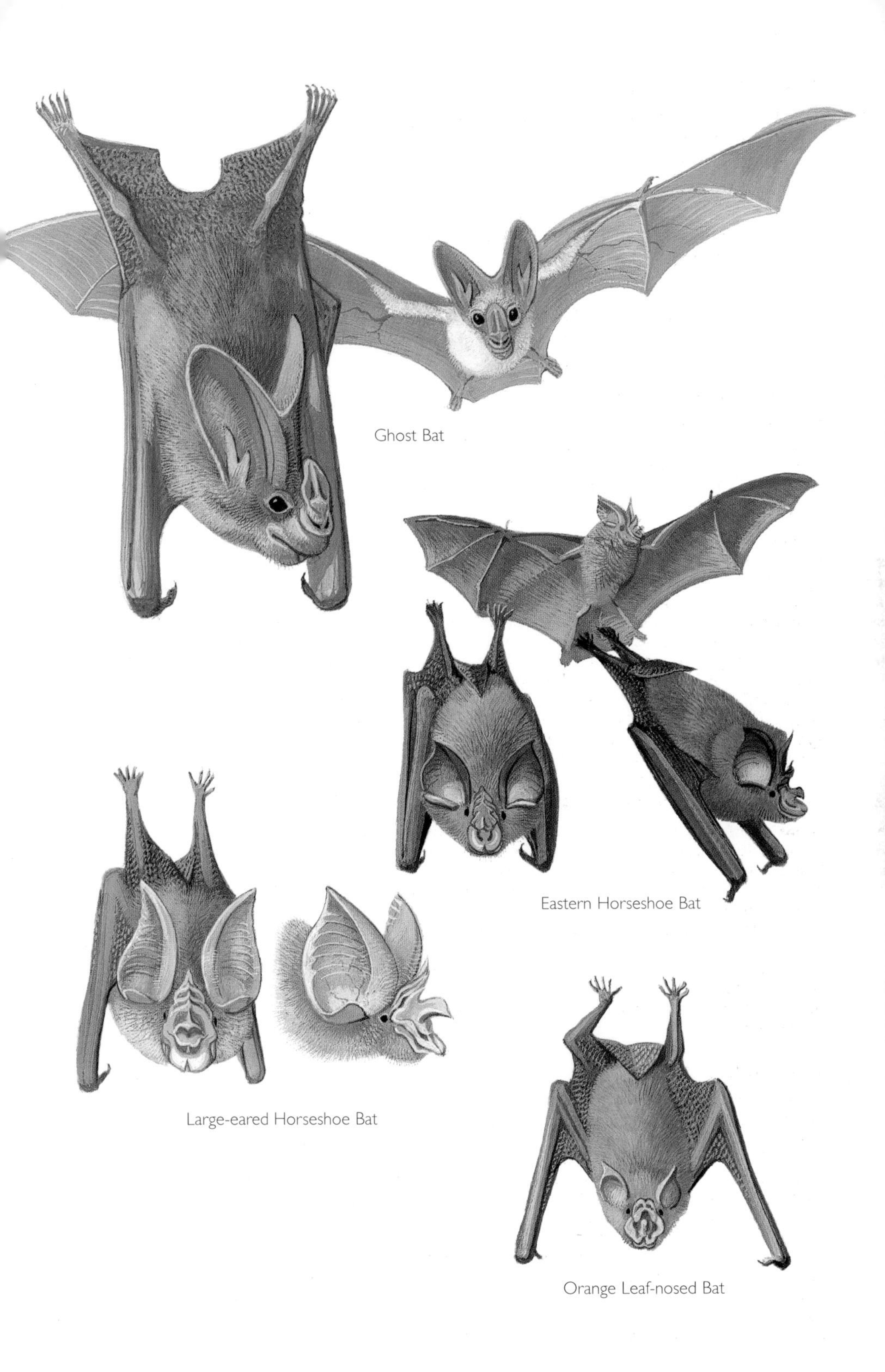
Ghost Bat
Eastern Horseshoe Bat
Large-eared Horseshoe Bat
Orange Leaf-nosed Bat

PLATE 50

Dusky Leaf-nosed Bat *Hipposideros ater*

fa 34.5–40.5 mm; **hb** 33–46 mm; **el** 17–21 mm; **wt** 3.4–5.6 (4.2) g
Tiny and delicate; ears extremely large, broad, rounded, length about 1/2 fa. Fur long, fluffy; upperparts mid grey-brown or mottled grey and ginger, some individuals uniformly bright orange. Noseleaf squarish, without adornment or secondary leaflets. **Distribution, habitat and status** Locally common where suitable roosting sites exist across n. Aust from about Derby in WA to Townsville, Qld. Southern limits of distribution unclear, with recent records from s. NT and central Qld. Also Melville and Bathurst Is and some islands of Buccaneer Arch. **Behaviour** Roosts by day in caves, mineshafts, sometimes tree hollows; prefers dark, warm, humid sites. Individuals hang separately from ceiling of cave. Forages for small flying insects, usually in dense vegetation such as rainforest and mangroves, but also eucalypt woodland. Flight slow, manoeuvrable, moth-like; hovers. Single young born Oct–Dec, weaned by mid Jan. **Similar species** Fawn Leaf-nosed Bat has 2 small secondary leaflets projecting beyond edge of lower noseleaf; shorter, more pointed ears. Other species much larger or have finely pointed ears.

noseleaf
(after Churchill)

Fawn Leaf-nosed Bat *Hipposideros cervinus*

fa 45–48 mm; **hb** 41–51 mm; **el** 13–15 mm; **wt** 5.6–8.5 (7.0) g
Noseleaf diagnostic: squarish, lower portion wider than upper, 2 subsidiary leaflets visible at sides of lower leaf below nostrils. Ears short, funnel-shaped with pointed tip. Fur varies from completely grey, through grey-brown with russet tinge, to bright orange. **Distribution, habitat and status** Known in Aust from only a few sites on e. Cape York and Torres Strait Is, s. to Coen. Roosts in caves or abandoned mines; forages in wide variety of vegetation, from rainforest to savannah woodland. **Behaviour** Forages for medium-sized flying insects low to ground, often over water or amongst low vegetation. Highly manoeuvrable. Single young born Nov–Dec. **Similar species** See under Dusky Leaf-nosed Bat.

noseleaf
(after Churchill)

Semon's Leaf-nosed Bat *Hipposideros semoni*

fa 44.5–50.5 mm; **hb** 39–49 mm; **el** 17–22 mm; **wt** 5.7–8.7 (7.2) g
Fur long, pale grey-brown above, paler below. Sparse pale brown or whitish fur on wing membrane near body. Ears long, narrow, acutely pointed. Noseleaf squarish with 2 wart-like projections, one 6–8 mm long projecting from top centre of lower leaf, level with eyes, the other from top of upper leaf; upper leaf divided into 4 depressions; 2 subsidiary leaflets at sides of lower leaf. **Distribution, habitat and status** Confined to eastern Qld, mostly n. of Cairns and at mid altitudes, with isolated southern records at Kroombit Tops and near Maryborough. Occupies range of tropical vegetation types including rainforest and savannah woodland. Endangered. **Behaviour** Roosts as solitary individuals in variety of sites—caves, mines, crevices, tree hollows, buildings. Slow, highly manoeuvrable flight; forages mostly <2 m above ground. Single young born in Nov. **Similar species** Northern Leaf-nosed Bat has smaller projections on noseleaf, not recorded e. of Mt Isa.

noseleaf
(after Churchill)

Northern Leaf-nosed Bat *Hipposideros stenotis*

fa 42–46 mm; **hb** 40–46 mm; **el** 17–21 mm; **wt** 4.6–6.4 (5.5) g
Very similar to Semon's Leaf-nosed Bat but the 2 wart-like projections on noseleaf are less prominent. **Distribution, habitat and status** Rare in w. Kimberley, including some islands of Buccaneer Arch., Top End of NT, and Gulf Country e. to Mt Isa. Roosts only in sandstone caves, boulder piles and disused mines. Forages in variety of vegetation types including rainforest, eucalypt woodland and spinifex-covered hills. **Behaviour** Wary, rarely captured. Feeds on flying insects taken low down among shrubby vegetation. Slow, fluttery flight. Single young born Oct–Jan.

noseleaf
(after Churchill)

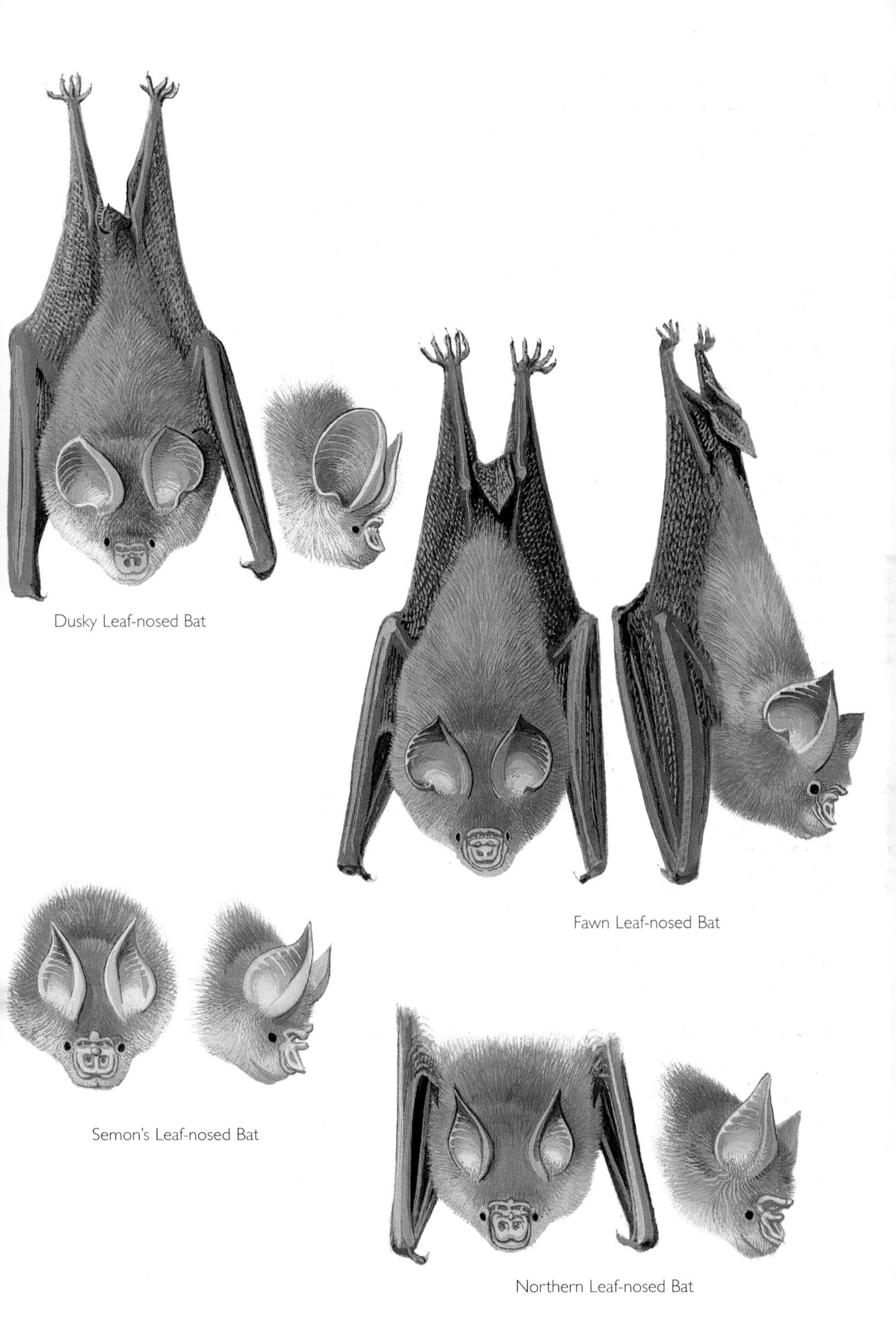
Dusky Leaf-nosed Bat

Fawn Leaf-nosed Bat

Semon's Leaf-nosed Bat

Northern Leaf-nosed Bat

PLATE 51

Diadem Leaf-nosed Bat *Hipposideros diadema*

Qld: **fa** 77–85 (82) mm; **hb** 74–96 (84) mm; **el** 25–30 (26) mm; **wt** 34–53 (44) g
NT: **fa** 68–73 (71) mm; **hb** 75–79 (77) mm; **el** 23–27 (25) mm; **wt** 22–35 (26.5) g
Larger than all other Aust leaf-nosed bats. Upperparts usually pale brown; underparts whitish or pale brown. Qld animals can have distinct white patches on shoulders and back. Some individuals reddish brown or orange. Noseleaf distinctive: upper portion wide, protruding, and divided into 4 depressions; lower portion with 3 or 4 subsidiary leaflets. Ears broad and acutely pointed. Eyes tiny, partly hidden by fur. **Distribution, habitat and status** Two distinct forms (may be separate species): subspecies *inornatus* from sandstone escarpments of Arnhem Land, NT; subspecies *reginae* from coastal n. Qld, s. to Townsville. Uncommon. Roosts in caves, mines, buildings and culverts. Forages in variety of vegetation types from rainforest to open eucalypt woodland. **Behaviour** Feeds on flying insects captured on the wing by perch and sally or direct pursuit. Flight more direct and stronger than other leaf-nosed bats. Single young born Nov–Dec.

noseleaf
(after Churchill)

Yellow-bellied Sheathtail Bat *Saccolaimus flaviventris*

fa 66–82 mm; **hb** 72–92 mm; **el** 17–23 mm; **wt** 30–60 g
Readily distinguished by size, glossy black upperparts and white or creamy yellow underparts. Head flat; muzzle long, unadorned; skin of face and muzzle naked, black. Ears long, ribbed, leathery; tail bristly. Wings long and narrow for fast, direct flight. Wing pouches absent; males have well-developed throat pouch, females a bare area outlined by ridge of skin. **Voice** Echolocation calls are audible to some people—a high-pitched, rapid-fire 'ting ting ting' with about 10–12 pulses per second. **Distribution, habitat and status** Common in n. Aust but rare late summer–autumn visitor to s. Occurs in most environments from wet forest to deserts. **Behaviour** Flies fast and direct above canopy so rarely captured in traps. In n. Aust readily observed in flight by spotlight. Roosts singly or in small groups in tree hollows; in treeless areas known to roost in burrows of terrestrial mammals. The single young is born Dec–Mar. **Similar species** Papuan Sheathtail Bat is smaller, has grizzled grey-brown not black upperparts, has a wing pouch lined with whitish hairs and females have a throat pouch rather than a simple bare area.

Papuan Sheathtail Bat *Saccolaimus mixtus*

Cape York Sheathtail Bat

fa 62–68 mm; **hb** 72–77 mm; **el** 18 mm; **wt** 24 g
Upperparts grizzled grey, darkest on head and shoulders; underparts grey-buff. Face, muzzle and ears naked, dark brown. Wing pouch present, lined with whitish hairs; throat pouch well developed in males, slightly less so in females. **Distribution, habitat and status** Known in Aust from only a few sites on n. Cape York Pen.; apparently prefers open-forest habitats. **Behaviour** Little known; forages for flying insects above the canopy and lower where clearings such as streams or roads provide flightlines. Probably roosts in tree hollows. **Similar species** See under Yellow-bellied Sheathtail Bat.

wing pouch

Bare-rumped Sheathtail Bat *Saccolaimus saccolaimus*

Qld: **fa** 72–77 mm; **hb** 81–96 mm; **el** 16–18 mm; **wt** 49–55 (53) g
NT: **fa** 77–80 mm; **hb** 88–96 mm; **el** 22 mm; **wt** 61 g
Rich reddish brown above with scattered irregular white patches; pale grey-brown underparts; skin of face and ears pale brown, skin of wings olive brown. Fur of lower back becomes increasingly sparse and is absent beyond hips. Wing pouches absent; throat pouch well developed in males, represented by naked circle of skin in females. NT form larger, darker brown, few whitish patches. **Voice** Echolocation clicks are audible. **Distribution, habitat and status** Uncommon and scattered in 2 separate populations: coastal Qld between about Cooktown and Bowen; w. Arnhem Land, NT. Found mostly in sclerophyll open-forest. Qld form critically endangered. **Behaviour** Roosts in tree hollows; forages above the canopy or lower in forest clearings, flying fast and direct. Agile on hard substrates, scurrying rapidly on all 4 limbs.

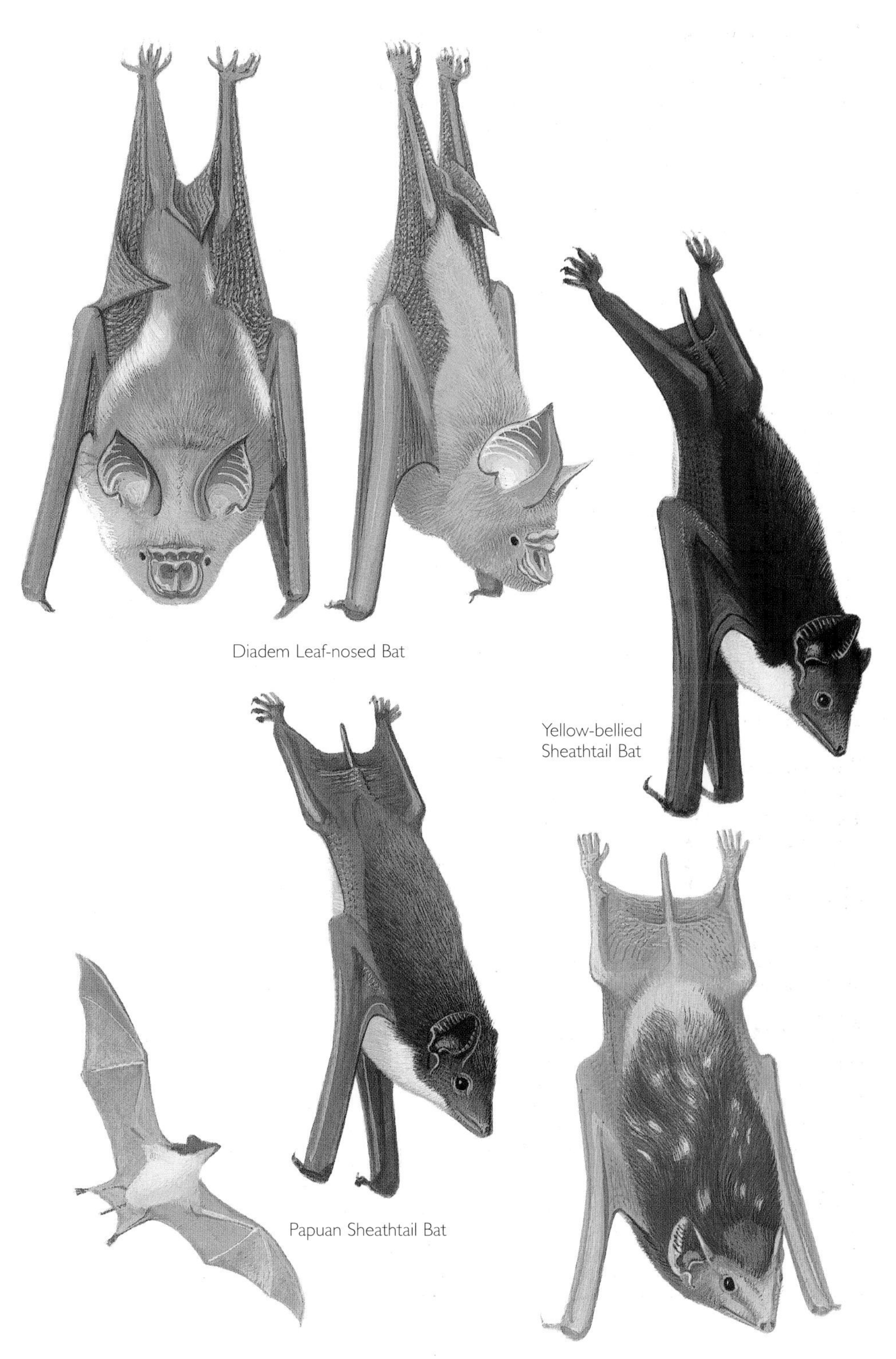
Diadem Leaf-nosed Bat
Yellow-bellied
Sheathtail Bat
Papuan Sheathtail Bat
Bare-rumped Sheathtail Bat

PLATE 52

Common Sheathtail Bat *Taphozous georgianus*

wing pouch

fa 66–75 mm; **hb** 75–89 mm; **el** 17–24 mm; **wt** 19.5–51 (32) g
As in all sheathtail bats, face is pointed with prominent nostrils on tip of muzzle; ears very broad and prominently ribbed; eyes large. When at roost hangs from vertical walls with characteristic posture: forearms propping head and shoulders out from substrate. Fur of upperparts uniformly dull brown, underparts slightly more grey. Skin of face mid-brown, wing membranes sepia, translucent. Throat pouch absent in both sexes; wing pouch present. **Distribution, habitat and status** Widespread and common in n. Aust from near Carnarvon, WA to coastal Qld excluding Cape York. **Behaviour** Roosts singly or in small groups in caves, crevices, mineshafts. Often roosts in twilight region of a cave but will readily fly or scurry to darker recesses if disturbed. Catches insects, mostly beetles, on the wing. Forages high over canopy or water. The single young is born Nov–Dec. **Similar species** Coastal Sheathtail Bat is paler, Hill's Sheathtail Bat is richer brown; in both, males have a throat pouch, females have a rudimentary ridge of skin. The canine teeth of Hill's Sheathtail Bat are longer (2.4–3.2 mm cf 2.2–2.6 mm). Troughton's Sheathtail Bat is larger (fa 73–76 mm, skull >24 mm, digit 3 metacarpal >68.0 mm cf <66.0 mm) and has darker fur.

Coastal Sheathtail Bat *Taphozous australis*

fa 63.5–67.5 mm; **hb** 61–75 mm; **el** 19–24 mm; **wt** 19–23 (21) g
Fur of upperparts uniformly fawn or grey-brown, underparts slightly more grey; some individuals greyer than others. Bare skin of face and limbs reddish brown. Males have throat pouch, females a vestigial ridge of bare skin. Wing pouch present. **Distribution, habitat and status** Rare and restricted to coastal e. Qld and islands in Torres Strait, s. to Shoalwater Bay. Always within a few km of the sea; often roosts on rocky islands; forages over most coastal vegetation types: open-forest, mangroves, scrub, heath, swamps. **Behaviour** Roosts in caves and rock crevices along the coast. Single young born Oct–Nov. Otherwise as for Common Sheathtail Bat. **Similar species** See under Common Sheathtail Bat.

Hill's Sheathtail Bat *Taphozous hilli*

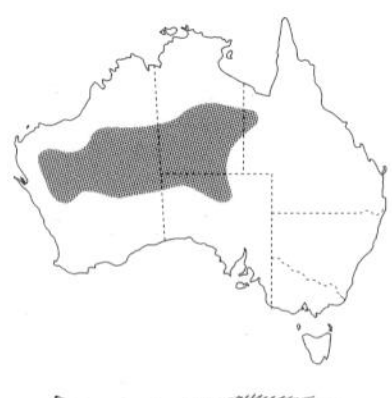

fa 61–72 mm; **hb** 65–81 mm; **el** 18–24 mm; **wt** 20–29 (25.5) g
Difficult to distinguish from Common Sheathtail Bat. Fur of upperparts rich brown; underparts slightly paler; wings grey-brown. Males have small, shallow throat pouch, females a bare patch. Wing pouch present. **Distribution, habitat and status** Widespread and common over much of n. arid WA, NT, and nw. SA (Everard and Musgrave Ranges). Occurs in rocky ranges and breakaway country; forages over arid eucalypt woodland, acacia shrubland, and hummock grassland. **Behaviour** Roosts in caves and disused mines; in WA often shares these with Common Sheathtail Bat. Single young born Sept–Mar in WA, earlier in central Aust. **Similar species** See under Common Sheathtail Bat.

Troughton's Sheathtail Bat *Taphozous troughtoni*

fa 72.5–75.5 mm; **hb** 79–86 mm; **el** 22–27 mm; **wt** N/A
A large sheathtail bat that lacks a throat pouch. Fur uniformly olive-brown. Bare skin of wing membranes and face pale brown. **Distribution, habitat and status** Known only from a few sites near Mt Isa and Cloncurry, nw. Qld, where roosts in caves and abandoned mines. Critically endangered. **Behaviour** Nothing recorded. **Similar species** Combination of large size (fa >72 mm, el >22 mm, digit 3 metacarpal >68.0 mm) and lack of throat pouch eliminates all but Common Sheathtail Bat, which has greyer fur and digit 3 metacarpal <66.0 mm.

Arnhem Sheathtail Bat *Taphozous kapalgensis*

fa 59–63 mm; **hb** 69–74 mm; **el** 16–18 mm; **wt** 26 g
Fur uniformly warm mid-brown above, buff below with orange tinge to chin; a band of long whitish fur extends along each flank beneath the wing, may be visible in flight. Throat pouch present in males, represented by a ridge of skin in females. **Voice** Loud, shrill calls when feeding. **Distribution, habitat and status** Rare. Known only from n. Arnhem Land, NT. Forages over open eucalypt woodland, melaleuca forest, swamps, streams. **Behaviour** Roosts in tree hollows and perhaps under the base of pandanus leaves. Forages for flying insects above the canopy and low over water. **Similar species** Pale flank stripes diagnostic.

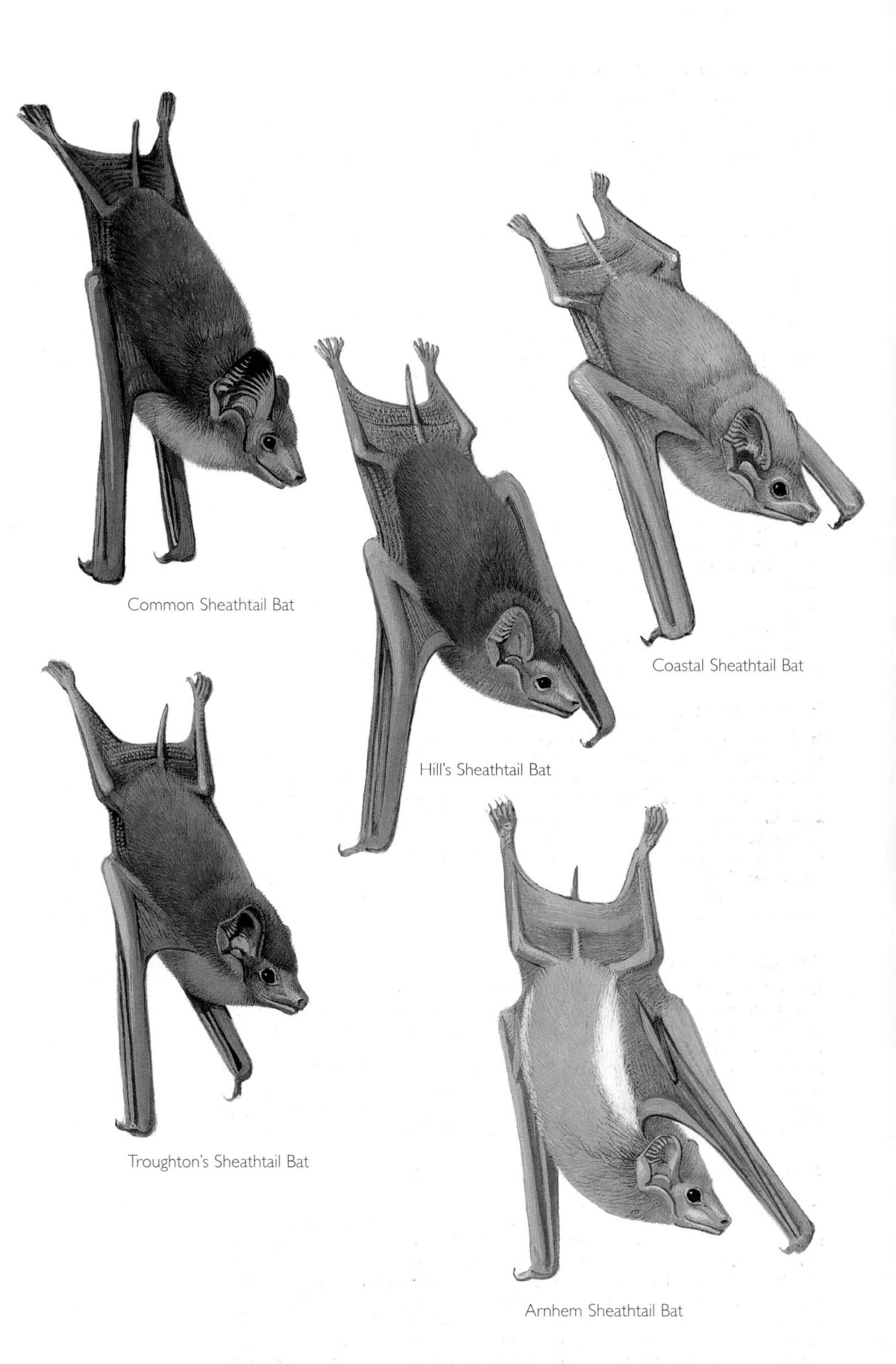
Common Sheathtail Bat
Coastal Sheathtail Bat
Hill's Sheathtail Bat
Troughton's Sheathtail Bat
Arnhem Sheathtail Bat

PLATE 53

White-striped Freetail Bat *Tadarida australis*

White-striped Bat, White-striped Mastiff Bat

fa 59–63 mm; **hb** 85–100 mm; **t** 40–55 mm: **el** 20–25 mm; **wt** 33–41 (37) g

Large, robust; fur chocolate brown with bold white stripe along flanks under wings. Sometimes also white patches on chest. Ears very broad and ribbed; touch but do not join above head. Lips distinctly wrinkled. Both sexes have throat pouch. Bare skin blackish, tinged pink. **Voice** Echolocation call audible as regular metallic 'ting-ting-ting' at a frequency of 1–2 per second. **Distribution, habitat and status** Widespread and common across southern $^2/_3$ of Aust, except Tas. Possibly migrates n. from s. areas in coldest months (June–Aug). Occupies very wide range of habitats including urban areas. **Behaviour** Roosts singly or in small groups in tree hollows. Flies fast above the canopy or through clearings. Agile on the ground and other hard substrates. Mating occurs in Aug and the single young is born Dec–Jan. **Similar species** Northern Freetail Bat lacks white stripes, has ears joined across top of head; lacks throat pouch; occurs primarily n. of Tropic of Capricorn. Yellow-bellied Sheathtail Bat also has audible echolocation call but gives 10–12 pulses per second.

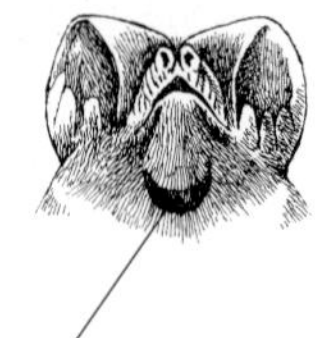

throat pouch

Northern Freetail Bat *Chaerephon jobensis*

Northern Mastiff Bat

fa 45–52 mm; **hb** 58–65 mm; **t** 32–37 mm; **el** 16–22 mm; **wt** 20–30 (26) g

Pig-like muzzle with greatly wrinkled upper lip. Upper jaw strongly overhangs lower. Inner margins of ears joined across forehead. Throat pouch absent. Fur uniformly chocolate brown above, paler and greyer below; bare skin dark brown. **Voice** Lower frequencies of echolocation call are audible. **Distribution, habitat and status** Found over most of Aust n. of Tropic of Capricorn in wide variety of habitats, including semi-arid. Occurs further s. in WA, to Shark Bay. **Behaviour** Roosts in tree hollows, caves, buildings. Forages above the canopy with fast direct flight; lower in open habitats or over water. Single young born Nov–Dec. **Similar species** See under White-striped Freetail Bat. Larger than all *Mormopterus* species.

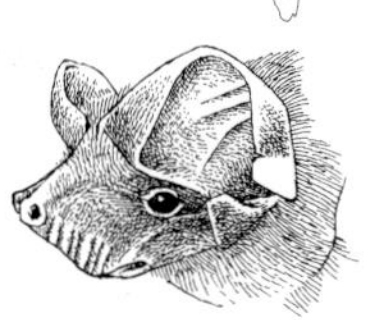

Beccari's Freetail Bat *Mormopterus beccarii*

fa 37–40 mm; **hb** 54–63 mm; **t** 21–32 mm; **el** 12–17 mm; **wt** 12–18 (15) g

The heaviest *Mormopterus*. Ears triangular, pointed; flattened, muscular body. Fur of upperparts mid dull brown with whitish bases to hairs; underparts distinctly paler. Bare skin dark grey-brown. Short, narrow, pointed wings. P^1 minute. **Distribution, habitat and status** Widespread and common over much of n. Aust. Occupies wide range of habitats from rainforest to savannah woodland, arid shrublands and grasslands. Common along inland watercourses. **Behaviour** Mostly roosts in tree hollows, also roof cavities, under peeling bark, etc. Flies fast with very rapid wingbeats. Forages above canopy or lower in open areas, often over water. Agile on ground and may obtain some prey there. Single young born Oct–Jan. **Similar species** Inland Freetail Bat overlaps in s. Qld, has paler skin and is smaller (fa 32–36 mm cf 37–40 mm).

East-coast Freetail Bat *Mormopterus norfolkensis*

fa 36–40 mm; **hb** 50–55 mm; **t** 35–45 mm; **el** 15 mm; **wt** 7–10 (8.5)

Fur of upperparts longish, dark brown or reddish brown; underparts slightly paler. P^1 well developed. **Distribution, habitat and status** Little-known. Found in coastal e. Aust from the Illawarra in NSW n. to near Brisbane. Inhabits dry and wet sclerophyll forests, coastal woodland. **Behaviour** Roosts in tree hollows and buildings; little else known. **Similar species** Beccari's Freetail Bat is more thickset, heavier (>12 g). Eastern Freetail Bat has t <25 mm and fa <34 mm.

White-striped Freetail Bat

Northern Freetail Bat

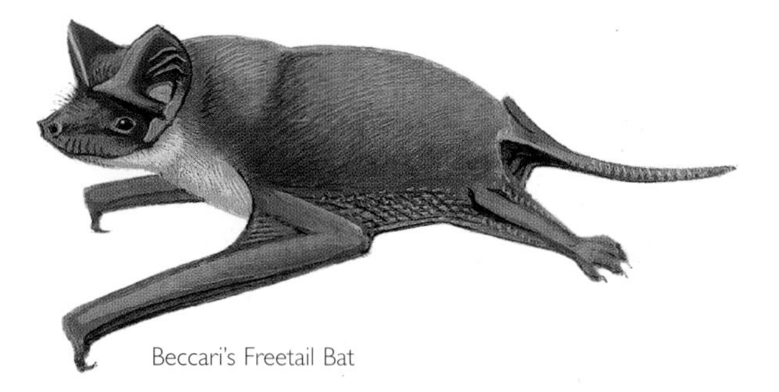

Beccari's Freetail Bat

East-coast Freetail Bat

PLATE 54

Undescribed Freetail Bats *Mormopterus species*

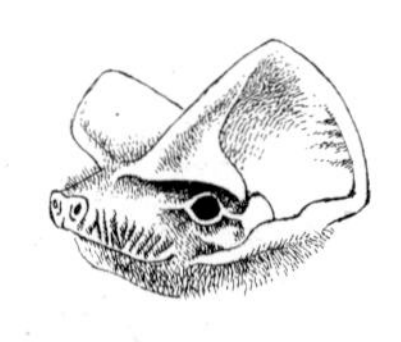

The species composition of the genus *Mormopterus* in Aust is unclear. Current taxonomic studies suggest that Beccari's Freetail Bat and the East-coast Freetail Bat are clear species (plate 54), but at least 7 species await scientific description and have no valid scientific name at present. Identification of these undescribed species by external morphology is difficult. However, most appear to have discrete geographical ranges, so locality can be a useful clue to identification.

Bats of the genus *Mormopterus* have a naked face; wrinkled lips; triangular ears compared to the broad circular ears of *Tadarida* and *Chaerephon*; short, narrow, pointed wings; very rapid wingbeats and fast, direct flight. They mostly roost in tree hollows, but also in buildings or other man-made cavities, and give birth to a single young during summer, or the wet season in n. Aust.

Little Northern Freetail Bat *Mormopterus loriae ridei, M. l. coboungensis (in part)*

fa 31–34 (33) mm; **hb** 42–55 (48) mm; **t** 26–31 (28) mm; **wt** 6.0–8.2 (7.4) g

Fur mid-brown, paler underparts. Bare skin dark brown. **Distribution, habitat and status** Cape York Pen., coast of Gulf of Carpentaria and Top End, NT. **Similar species** Beccari's Freetail Bat is larger. Eastern Freetail Bat has shorter tail.

Mangrove Freetail Bat *Mormopterus loriae cobourgensis (in part)*

fa 32.5–35 (34) mm; **hb** 47–55 (50.5) mm; **t** 30–36 (33) mm; **wt** 6.2–9.0 (7.3) g

Upperparts mid-brown to grey-brown; underparts greyish-buff, throat and chin grey-lemon. **Distribution, habitat and status** Restricted to mangroves and adjacent vegetation in narrow coastal strip between Derby and Exmouth Gulf, WA. **Similar species** Beccari's Freetail Bat is larger (fa > 36 mm, wt > 12 g).

Eastern Freetail Bat

fa 30.6–34.2 (33) mm; **hb** 45–48 (47) mm; **t** 18–25 (22) mm; **wt** 6.8–11.5 (8.8) g

Fur uniformly rich brown. **Distribution, habitat and status** Eastern Aust from Cooktown to sw. Vic; inland to w. slopes of GDR. Common in wet and dry sclerophyll forest. **Similar species** Beccari's and East-coast Freetail Bats are larger (fa >34.5 mm). Little Northern and Southern have longer tail (>25 mm).

Inland Freetail Bat *Mormopterus planiceps (short penis form)*

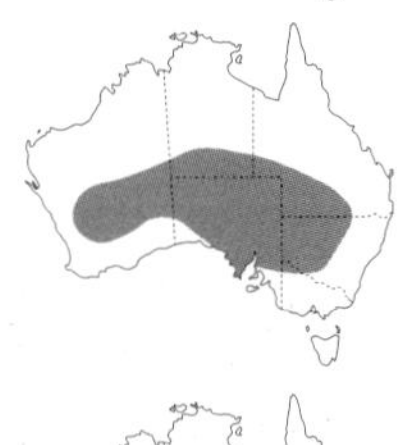

fa 32–40 mm; **hb** 47 mm; **t** 25–26 mm; **wt** 8–14.8 g (10.3 central Aust; 9.5 nw. Vic)

Fur short, pale brown or grey-brown, slightly paler underparts. Bare skin pink or grey. Penis short (<5 mm). Considerable geographic variation in size; smallest towards s. **Distribution, habitat and status** Common in arid and semi-arid s. Aust, reaching s. NT and nw. Vic. **Similar species** Beccari's is larger and darker. Hairy-nosed is smaller (<6 g). Male Western and Southern Freetail Bats have long penis (> 9 mm), fur darker, shaggier, bare skin is dark grey.

Southern Freetail Bat *Mormopterus planiceps (long penis form, in part)*

fa 30.6–35.7 (33.6) mm; **hb** 54–55 mm; **t** 25–33 (29.2) mm; **wt** 6.8–13.0 (8.9) g

Fur long, dark grey-brown with cream base, paler underparts. Bare skin dark grey-brown. Males have long penis (>9 mm). **Distribution, habitat and status** Inland of GDR in NSW and Vic, w. to about Spencer Gulf, SA. Common in dry sclerophyll forest and woodland, and mallee scrub. **Similar species** Combination of dark, long fur, long tail (>25 mm), long penis diagnostic in e. Aust.

Western Freetail Bat *Mormopterus planiceps (long penis form, in part)*

fa 31–37 (34) mm; **hb** 50–65 (57) mm; **t** 30–40 (33) mm; **wt** 10–14 (11) g

Fur dark grey-brown, slightly paler underparts. **Distribution, habitat and status** SW WA including most of wheatbelt. Found in dry sclerophyll forest, mallee shrubland and dry heath. **Similar species** Inland Freetail Bat has paler, shorter fur and males have short penis (<5 mm).

Hairy-nosed Freetail Bat

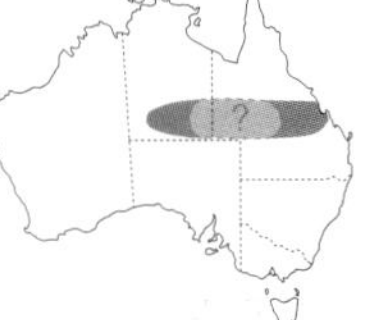

fa 32–35.3 (34.3) mm; **hb** 41–50 (45) mm; **t** 28–32 (30) mm; **wt** 5–6 g

Very small and lightly built (<6 g) with characteristic bristles on the muzzle, which is elongated and narrow. Fur pale grey or sand coloured. **Distribution, habitat and status** The few specimens assigned to this species were collected in s. NT and central Qld.

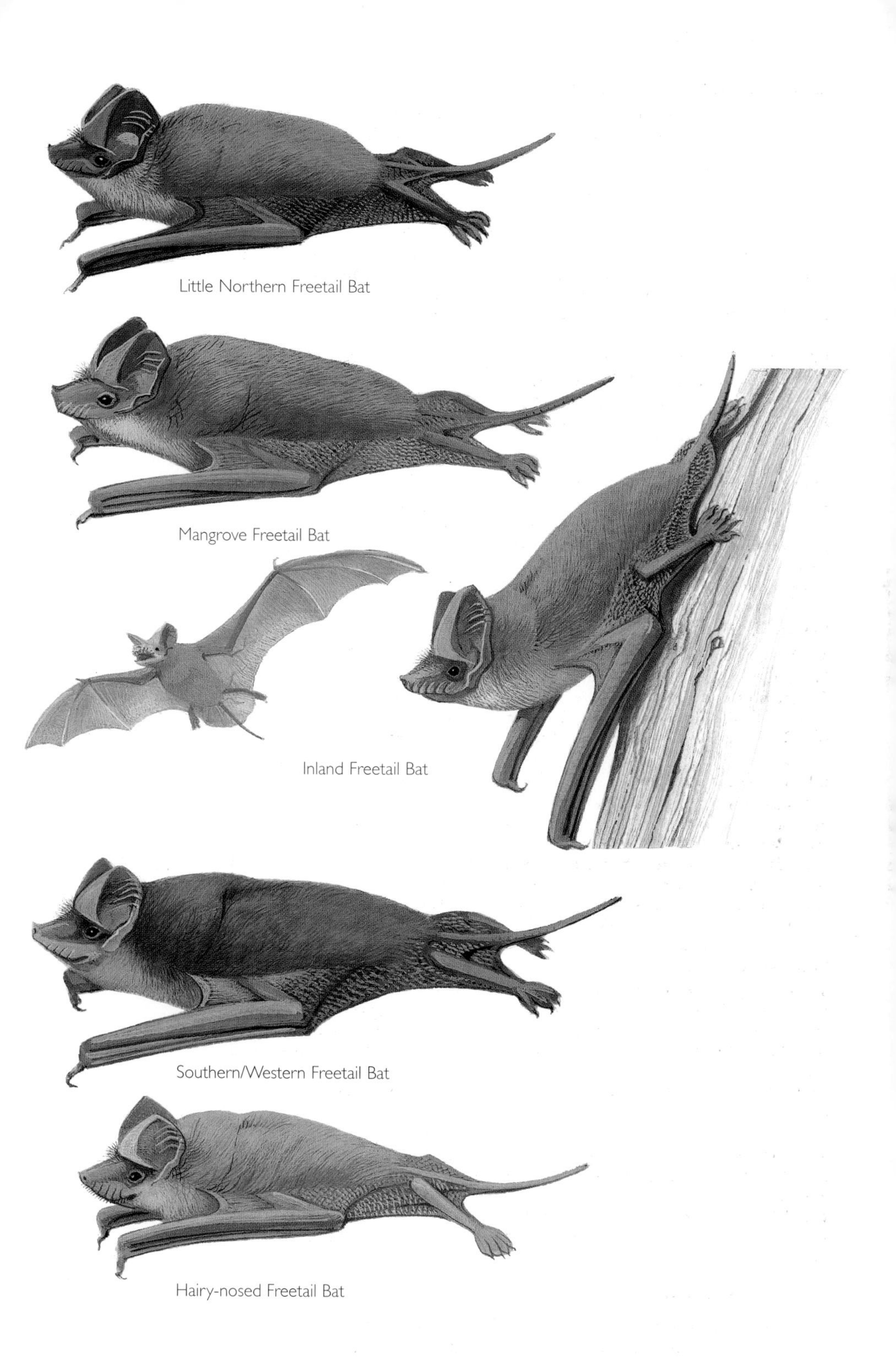
Little Northern Freetail Bat
Mangrove Freetail Bat
Inland Freetail Bat
Southern/Western Freetail Bat
Hairy-nosed Freetail Bat

PLATE 55

Large-eared Pied Bat *Chalinolobus dwyeri*

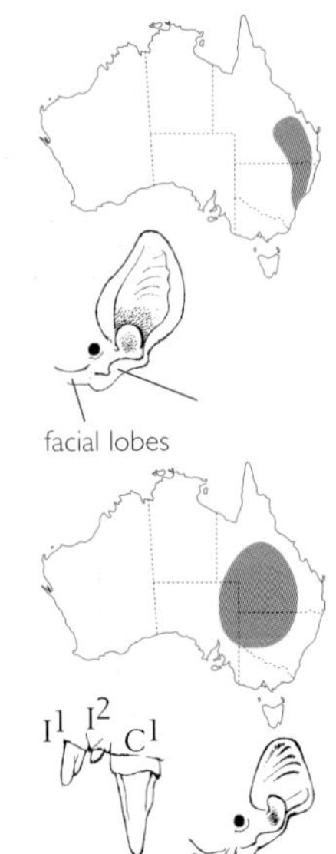

(after Parnaby)

fa 38–42 mm; **hb** 47–53 mm; **t** 42–47 mm; **el** 17 mm; **wt** 6.8–9.7 (8) g
Ears meet easily when pressed flat across crown; well-developed facial lobes, one between lowest point of ear and corner of mouth, another along lower lip. Fur uniformly glossy black, browner underneath with a variable strip of white along each flank beneath wings. These strips converge in pubic region to form V-shape. **Distribution, habitat and status** Uncommon in dry and wet eucalypt forests from Blackdown Tableland (central e. Qld) s. to near Wollongong, NSW and inland to Carnarvon NP. Vulnerable. **Behaviour** Roosts in small groups in caves, mines, usually in twilight zone near entrance. Births in Nov–Dec, usually twins. **Similar species** Little Pied Bat is smaller; ears do not meet when pressed across crown; Hoary Wattled Bat smaller, fur pale-tipped giving a frosted appearance.

Little Pied Bat *Chalinolobus picatus*

fa 32–36 mm; **hb** 45–49 mm; **t** 29–42 mm; **el** 8.6–11 mm; **wt** 4.3–7.1 (5.4) g
Small version of Large-eared Pied Bat, but with very short ears that do not project above fur of the domed forehead. Longitudinal notch on inner surface of I^1. **Distribution, habitat and status** Sparsely distributed in dry sclerophyll forest, woodland and scrub in inland NSW, ne. SA, and s. half of Qld. Often forages along watercourses. Not recorded s. of Murray R. **Behaviour** Little known. Roosts in caves, mineshafts, tree hollows, sometimes abandoned buildings. Two young born in spring. **Similar species** See under Large-eared Pied Bat.

Hoary Wattled Bat *Chalinolobus nigrogriseus*

fa 32–37 mm; **hb** 40–48 mm; **t** 29–39 mm; **el** 7–11 mm; **wt** 4–10 (6) g
Fur shaggy, dark grey with pale tips giving frosted appearance. Underparts grey-brown. Lobes poorly developed. Some specimens from ne. NSW have fringe of white fur on flanks, but not as dramatic as in the pied bats. WA animals smaller, more obviously 'frosted'. No notch in I^1. **Distribution, habitat and status** Common across n. Aust from about Derby in WA to Cape York, then increasingly uncommon southwards to ne. NSW. Occupies range of vegetation including vine forest, tropical savannah, dry sclerophyll forest, coastal scrub. **Behaviour** Roosts in tree hollows, also rock crevices. Flight relatively slow, manoeuvrable. Two young born Sept–Oct. **Similar species** Little Pied Bat always has clear white bands along flanks and notch in I^1.

Gould's Wattled Bat *Chalinolobus gouldii*

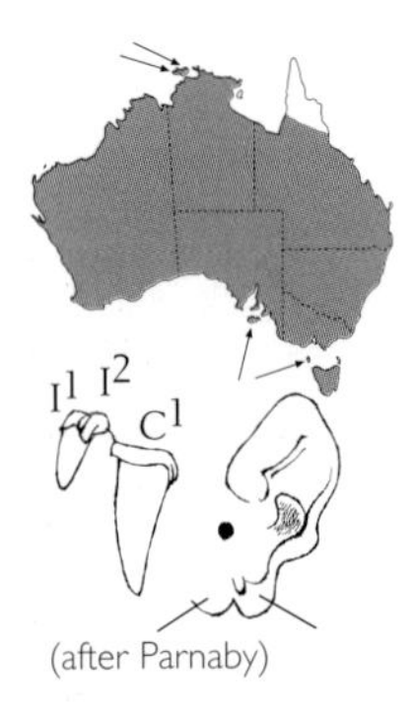

(after Parnaby)

fa 36–47 mm; **t** 31–46 mm; **el** 8–13 mm; **wt** s. Aust 10–20 (14), n. Aust 7–16 (10) g
Largest lobe-lipped bat in Aust. Fur of upperparts distinctly bicoloured: blackish head and shoulders sharply demarcated from brown back and rump; underparts mid-brown. Northern subspecies *venatoris* more uniformly black. Distinct fleshy lobe at base of ear near corner of mouth, distinct lobe on lower lip. **Voice** When handled emits a continuous buzz. **Distribution, habitat and status** Common throughout mainland except for Cape York Pen. Also Tas, King and Kangaroo Is. Found in most habitats except treeless deserts. Common in many towns and cities. **Behaviour** Roosts in tree hollows and buildings. Mates in Mar–June; births in Nov–Jan; twins common. Enters torpor in cold weather. **Similar species** Common Bentwing Bat has elongated terminal phalanx on 3rd finger, wing membrane attached to ankle, not base of outer toe.

Chocolate Wattled Bat *Chalinolobus morio*

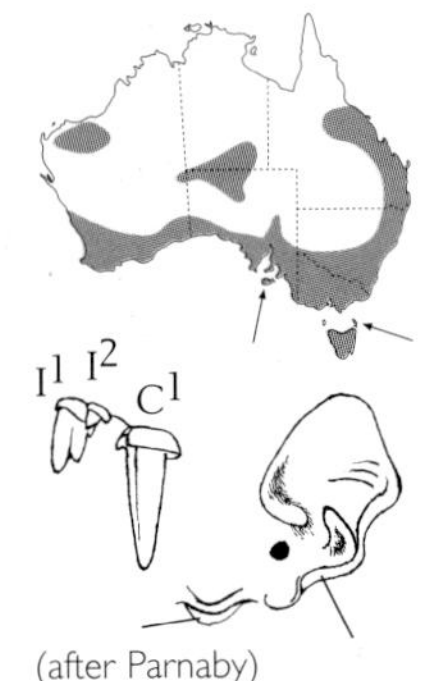

(after Parnaby)

fa 35–39 mm; **t** 39–49 mm; **el** 9–10 mm; **wt** 5.5–10.3 (8.0) g
Fur uniformly rich brown, but inland populations may be paler with contrasting pale grey underparts. Ears short, rounded; tragus short, broad; muzzle short with distinct furred ridge between eyes. Lobe on lower lip small, rounded, not obvious. **Distribution, habitat and status** Occurs across s. Aust including Tas, Flinders and Kangaroo Is, and along e. coast n. to about Townsville, Qld. Isolated inland populations in Pilbara of WA and central Aust ranges. Common in wide variety of habitats from subalpine woodland to arid plains. **Behaviour** Roosts in tree hollows, variety of artificial cavities, rarely in caves. Forages for flying insects below canopy. One or sometimes 2 young born Oct–Nov. **Similar species** Large Forest Bat similar in size and colour but lacks lobe on lower lip, and ridge between eyes; has long narrow tragus.

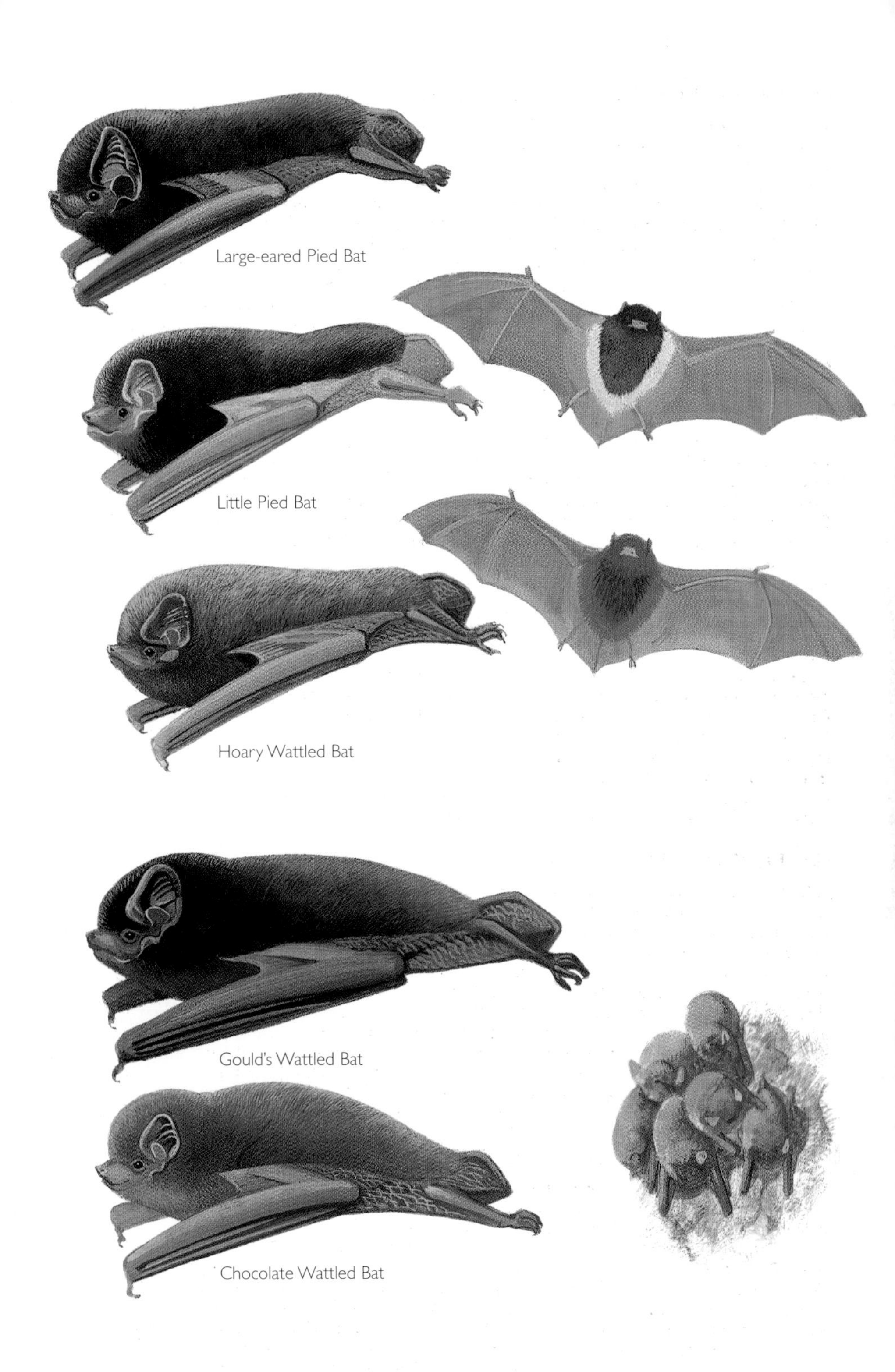
Large-eared Pied Bat
Little Pied Bat
Hoary Wattled Bat
Gould's Wattled Bat
Chocolate Wattled Bat

PLATE 56

Mangrove Pipistrelle *Pipistrellus westralis*

Northern Pipistrelle

fa 27.4–30 mm; **hb** 34.5–42 mm; **t** 29–37 mm; **el** 8–11 mm; **wt** 2.7–3.3 g
Tiny and delicate. Fur of upperparts uniformly reddish brown with blackish underfur; underparts buff with dark grey underfur. Bare skin blackish. Ears moderately long, broadly triangular with narrow, rounded apex. Tragus about half ear length, broadest at base. Outer canine width <3.8 mm. **Distribution, habitat and status** Confined to a narrow coastal strip across n. Aust from s. of Broome in WA to Karumba on s. coast of Gulf of Carpentaria, Qld. Feeds along tidal creeks and other openings in mangrove forests, melaleuca swamps and pandanus thickets. **Behaviour** Rapid, shallow wingbeats and highly manoeuvrable flight. Single young born in June–July. **Similar species** Cape York Pipistrelle is larger (fa >30 mm); outer canine width >3.9 mm; ear broader, more rounded, tragus broadest in middle, not at base; glans penis has a single distal fleshy tongue rather than small fleshy spines. Forest bats (*Vespadelus*) have only one pair of upper premolars.

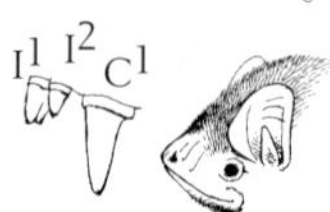

(after Parnaby)

Cape York Pipistrelle *Pipistrellus adamsi*

fa 30–32.5 mm; **hb** 35–44 mm; **t** 28–35 mm; **el** 8.5–11.5 mm; **wt** 3–5 (4.2) g
Small and nondescript. Fur of upperparts dark grey-brown to rust, slightly paler below. Colour of bare skin varies from mid-brown to blackish. Ear broad and rounded; tragus has large lobe at base and strongly convex rear edge, appears broadest in middle rather than at base. Glans penis with single distal fleshy tongue. **Distribution, habitat and status** Locally common in monsoon forest, melaleuca swamps, and along streams through savannah woodland on Cape York. Also a few records from coastal NT. **Behaviour** Probably roosts in trees; single young born in Oct–Nov. **Similar species** See under Mangrove Pipistrelle.

Flute-nosed Bat *Murina florium*

Tube-nosed Insectivorous Bat

fa 33–36 mm; **hb** 47–57 mm; **t** 31–37 mm; **el** 14–15 mm; **wt** 6–9 (8.4) g
Diagnostic tubular nostrils pointing sideways from tip of muzzle. Ears broad and rounded with distinct notch low on rear edge; tragus short and narrow. Fur long, dense; upperparts rufous-brown to grey-brown; underparts mid-grey. Fur covers much of tail membrane. **Voice** A loud, drawn-out, high-pitched whistling contact call is given in flight. **Distribution, habitat and status** Confined to tropical rainforest and adjacent wet scherophyll forest 200–1000 m asl in ne. Qld. Recorded from Shiptons Flat (s. of Cooktown) s. to Paluma. A small specimen from Iron Range may represent a second species in Aust. **Behaviour** Forages in mid to upper canopy where it gleans arthropods from rainforest trees. Slow fluttering flight, can hover. Roosts beneath hanging clusters of leaves or in suspended nests of scrub-wrens or fern-wrens. Said to wrap wings and tail around body during rain to form 'umbrella'.

Golden-tipped Bat *Kerivoula papuensis*

fa 36–40 mm; **hb** 50–60 mm; **t** 37–43 mm; **el** 14–16.5 mm; **wt** 5.7–6.5 (6.2) g
Fur long, woolly, slightly curly. Upperparts dark brown to blackish tipped with gold; underparts brown; bright golden fur on forearms, backs of legs and onto tail membrane. Ears funnel-shaped, pointed; tragus very slender, about 10 mm long. High domed crown. **Distribution, habitat and status** Localised and uncommon in rainforest, sometimes wet and dry sclerophyll forest, in e. Aust from McIlwraith Range (ne. Qld) to near Bega, se. NSW. Usually captured in dense vegetation, often close to creeklines, from sea level to 1000 m. **Behaviour** Feeds mostly by gleaning spiders from webs, and other arthropods from foliage. Highly manoeuvrable, slow fluttering flight interspersed with long glides. Roost sites mainly in abandoned domed and suspended nests of scrub-wrens and warblers (*Gerygone* spp.), also tree hollows and dense vegetation.

Mangrove Pipistrelle
Cape York Pipistrelle
Flute-nosed Bat
Golden-tipped Bat

PLATE 57

Northern Cave Bat *Vespadelus caurinus*

fa 26.6–31.7 mm; **hb** 32–40 mm; **t** 24–35 mm; **el** 8–12 mm; **wt** 2.3–4.2 (3.1) g
Tiny cave-roosting species. Fur grey-brown, warmer on rump, blackish at base. Bare skin dark brown or blackish. Head of glans penis laterally compressed. **Distribution, habitat and status** Throughout the rocky escarpment country of the Kimberley in WA (including many islands of the Bonaparte Arch.) and Top End of NT, in vegetation ranging from monsoon forest and vine thickets to open woodland and savannah. Locally common. **Behaviour** Roosts in cracks and crevices in caves, boulder piles, disused mines, culverts. Birth of 1 or 2 young Oct–Feb. **Similar species** Kimberley Cave Bat is larger, has greyish fur tinged yellow on shoulders and head. Inland Cave Bat difficult to distinguish except on length of fa (>32 mm) but not known to overlap in distribution. In pipistrelles the outer upper incisor is readily visible, not minute, and a small P^2 is present inside the line of the tooth row.

Kimberley Cave Bat *Vespadelus douglasorum*

Yellow-lipped Bat

fa 34.3–37.8 mm; **hb** 35–44 mm; **t** 35–38 mm; **el** 11–12.5 mm; **wt** 4.5–6 (5.3) g
Fur pale grey to olive buff with yellowish wash on head, shoulders, feet and forearms. Skin of lips pale cinnamon or orange-buff. Bare skin pale brown. **Distribution, habitat and status** Confined to w. Kimberley region, mostly where rainfall exceeds 800 mm/year. Forages in tropical woodland, often along streams. **Behaviour** Roosts in small colonies in caves. Single young born in first half of wet season (Nov–Dec). **Similar species** Northern Cave Bat smaller with darker fur and skin, lacks yellow tinge.

Inland Cave Bat *Vespadelus finlaysoni*

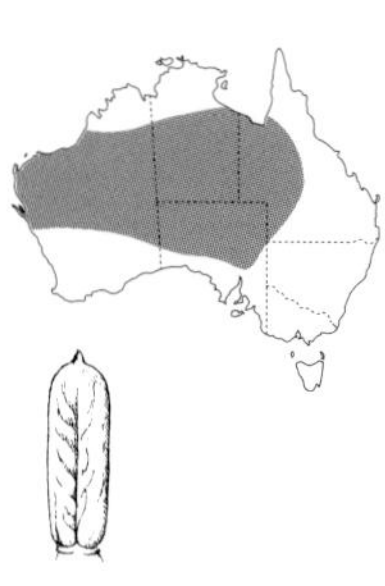

glans penis
(after Churchill)

fa 29.8–36.7 mm; **hb** 34–46 mm; **t** 31–42 mm; **el** 9–13 mm; **wt** 2.8–6.3 (4.3) g
Fur of upperparts dark brown to blackish with reddish tinge; underparts paler brown. Bare skin blackish. Glans penis rod-shaped with pointed tip. **Distribution, habitat and status** Common over much of arid Aust from coastal WA through the central deserts to tropical savannah around Gulf of Carpentaria. Always near rocky terrain. **Behaviour** Roosts colonially in caves, crevices and disused mines. Often forages around water. In n. of range births can occur year-round; further s. births mostly in Nov–Dec. One or 2 young are born. **Similar species** In all but n. of range, readily distinguished by dark reddish-brown fur and glans penis shape. In n., Northern Cave Bat is smaller (fa <32 mm), the glans penis is laterally compressed, with a deep longitudinal groove. In pipistrelles the outer upper incisor is readily visible, not minute, and a small P^2 is inside the line of the tooth row.

Eastern Cave Bat *Vespadelus troughtoni*

fa 33.0–36.4 mm; **hb** 38–44 mm; **t** 31–38 mm; **el** 10–13 mm; **wt** 4.6–6.7 (5.7) g
Fur of upperparts brown with gingery tips on face and head; underparts paler; bare skin dark grey. Identification best achieved by penis morphology: penis pendulous with swollen head, lacks distinct angular downturn; glans penis straight, blunt without pointed tip. **Distribution, habitat and status** Along GDR, inland slopes and coastal plains from top of Cape York to about Kempsey, NSW. Mostly in drier open-forest and woodland. Uncommon and poorly known. **Behaviour** Roosts in well-lit parts of caves and mineshafts, usually in small groups but colonies of up to 500 are known. **Similar species** Large Forest Bat is similar in size but has smaller, sharply angled penis without swollen head. Inland Forest Bat is smaller, has paler fur and skin, funnel-shaped tip to glans penis. Southern and Little Forest Bats are smaller (fa <33 mm), have complex vase-shaped or funnel-shaped glans penis. Eastern Forest Bat is smaller (fa <33 mm), has distinct angular bend in penis.

glans penis
(after Churchill)

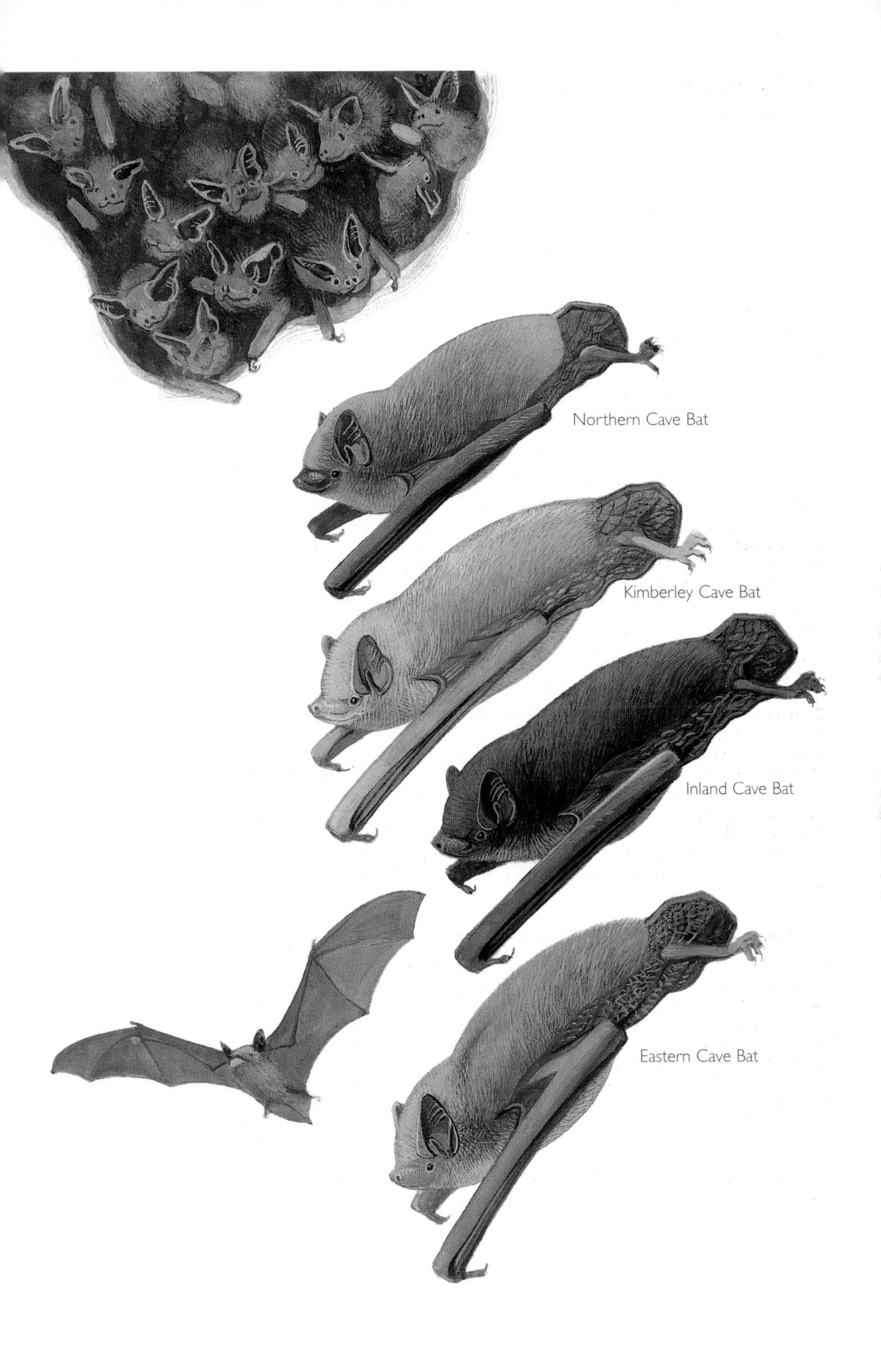
Northern Cave Bat
Kimberley Cave Bat
Inland Cave Bat
Eastern Cave Bat

PLATE 58

Inland Forest Bat *Vespadelus baverstocki*

fa 26.5–31.4 mm; **hb** 35–43 mm; **t** 26.5–34 mm; **el** 9–11 mm; **wt** 3.6–7.0 (4.8) g

Small, pale; upperparts usually sandy-brown, underparts cream or pale brown but some individuals darker, uniformly grey-brown. Tragus paler than rest of ear. Penis pendulous, without angular bend; glans penis funnel-shaped, not flattened. **Distribution, habitat and status** Widespread across inland southern Aust. Utilises most arid and semi-arid woodlands and shrublands. **Behaviour** Small groups or colonies of up to 60 roost in tree hollows, crevices in buildings. **Similar species** Combination of fa <32 mm, pale colour, non-angular penis and funnel-shaped glans penis identifies males. In s., females difficult to separate from Southern Forest Bat but usually smaller and paler. Female Little Forest Bat is even paler, and slightly smaller.

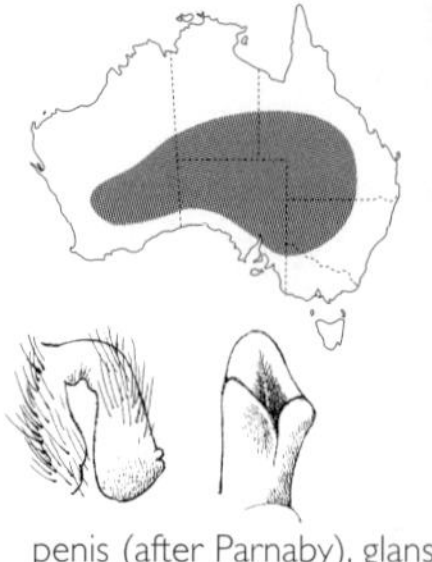

penis (after Parnaby), glans penis (after Churchill)

Large Forest Bat *Vespadelus darlingtoni*

fa 32.5–37.2 mm; **hb** 38–49 mm; **t** 29–38 mm; **el** 10–13 mm; **wt** 6.0–8.3 (7.2) g

Fur thick, long, rich dark brown; little contrast above and below. In Tas can be entirely blackish. Blackish triangle at point of lower lip. Short penis with distinctly angular bend, without swollen tip. On 3rd finger, ratio of lengths of 3rd to 2nd phalanges >0.84. **Distribution, habitat and status** Common in forest and woodland along GDR from se. Qld to Adelaide Hills and Tas; also Kangaroo, Flinders, Lord Howe Is. In n. restricted to higher, cooler areas; elsewhere found at all altitudes. **Behaviour** Roosts in tree hollows, forages in and below canopy. Single young born Nov–Dec. **Similar species** Among *Vespadelus* the combination of uniformly dark fur, black patch on lower lip, angular penis without swollen tip, and fa >33 mm is diagnostic. Separable from Eastern Forest Bat only on size (fa usually <33 mm). Chocolate Wattled Bat has a lobe on lower lip and ridge of fur between eyes.

penis shape (after Parnaby)

Eastern Forest Bat *Vespadelus pumilus*

fa 28.1–32.9 mm; **hb** 35–44 mm; **t** 28–34 mm; **el** 9–12 mm; **wt** 3–6 (4) g

Fur long, dense, dark brown, slightly paler below. Bare skin dark brown to blackish. **Distribution, habitat and status** Recorded at scattered sites along GDR from Atherton Tableland in n. Qld, to about Newcastle, NSW; also Lord Howe I. Found in wet forest at all altitudes. **Behaviour** Not documented. Roosts in tree hollows. **Similar species** Small size, dark skin, penis with angular bend and no swollen tip are strong clues. See under Large Forest Bat and Little Forest Bat.

Southern Forest Bat *Vespadelus regulus*

fa 28.0–34.4 mm; **hb** 36–46 mm; **t** 28–39 mm; **el** 9–13 mm; **wt** 3.6–7.0 (5.2) g

Upperparts brown, underparts greyish. Hairs bicoloured with dark brown base. Skull very flat, head triangular when viewed from above. Penis pendulous, not sharply bent; glans penis not compressed but funnel shaped with distinct lateral fold. On third finger, ratio of lengths of 3rd to 2nd phalanges <0.84. **Distribution, habitat and status** Common in coastal and subcoastal southern Aust from se. Qld to Eyre Pen. in SA, including Tas and Kangaroo I., and sw. Aust from near SA–WA border to n. of Perth. Habitats range from wet sclerophyll forest to semi-arid woodland, mallee. **Behaviour** Roosts in tree hollows, colonies often of a single sex. **Similar species** Little Forest Bat is smaller (fa <31 mm), has raised forehead, tragus often whitish. Inland Forest Bat paler, smaller (except in far s. of range), lacks lateral fold on glans penis. Eastern and Large Forest Bats have dark brown skin, angular bend in penis.

ear, glans penis (after Churchill)

Little Forest Bat *Vespadelus vulturnus*

fa 26.2–32.8 mm; **hb** 35–48 mm; **t** 28–34 mm; **el** 9–12 mm; **wt** 3–6.8 (4.3) g

Small, difficult to distinguish. Fur mid-brown to grey-brown; paler grey below. Hairs darker at base. Tragus often whitish. Head profile shows distinct raised forehead. Only *Vespadelus* with bulbous, rounded end to glans penis. **Distribution, habitat and status** Common in most of e. Aust s. of about Mitchell in Qld and e. of Adelaide, including Tas and Flinders I. Found in variety of forest and woodland types from wet sclerophyll to semi-arid woodland and mallee. **Similar species** Inland Forest Bat is usually paler furred, males have funnel-shaped glans penis. Southern Forest Bat is browner, tragus never whitish, has flatter head profile and more triangular head plan, glans penis funnel-shaped with distinct lateral fold. Eastern Forest Bat has distinct angular bend in penis, darker fur, usually longer forearm.

glans penis (after Churchill)

Inland Forest Bat
Large Forest Bat
Eastern Forest Bat
Southern Forest Bat
Little Forest Bat

PLATE 59

Common Bentwing Bat *Miniopterus schreibersii*

Large Bentwing Bat

fa 43–51 mm; **hb** 47–63 mm; **t** 43–58 mm; **el** 7–13 mm; **wt** N Aust 8.6–16 (11.3); Vic 10.5–19.5 (14.6) g

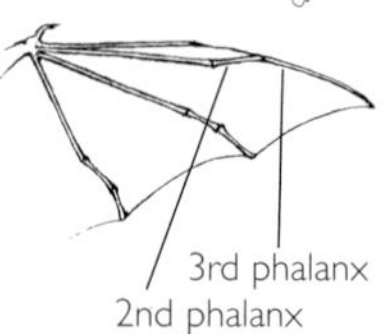

Characterised by 3rd (terminal) phalanx of 3rd finger being 3–4 times as long (>25 mm) as 2nd phalanx. Head profile distinctive: very short muzzle, high domed forehead, broad rounded ears with short rounded tragus. Fur dark brown, sometimes with ginger tinge or patches of orange fur; pale grey-brown below. Bare skin pale brown. **Distribution, habitat and status** Occurs in nw. Aust in the Kimberley and Top End (subspecies *orianae*), and coastal e. Aust from Cape York to central Vic (subspecies *oceanensis*), and w. Vic to se. SA (subspecies *bassani*). Absent from Tas. Populations centred on areas containing caves with temperature, humidity and conformation suitable for breeding. Locally common but vulnerable to disturbance at colonies. **Behaviour** Roosts in colonies in caves, old mines, road culverts. During Oct–Nov congregates at maternity caves, in colonies sometimes >150 000 animals. In late summer young form densely packed masses on cave ceiling. In March, when young are independent, depart maternity caves and form scattered smaller colonies, mostly within 300 km of maternity cave. Southern populations hibernate during coldest months but in n. remain active year-round. Forages above canopy and eats mostly moths. **Similar species** Little Bentwing Bat is very similar but smaller (fa <42 mm). Gould's Wattled Bat similar in head profile and size but lacks elongated 3rd phalanx on 3rd finger, has conspicuous lobes at base of ears and lower lips, wing membrane attaches at base of toe rather than ankle.

Little Bentwing Bat *Miniopterus australis*

fa 37–41 mm; **hb** 40–45 mm; **t** 39–47 mm; **el** 7–11 mm; **wt** 5.2–8.3 (6.7) g

Very similar to Common Bentwing Bat but smaller (fa <42 mm). **Distribution, habitat and status** East coast from Cape York to s. of Newcastle, NSW. Locally common in variety of habitats including lowland rainforest, wet and dry sclerophyll forest, paperbark swamps. Vulnerable to disturbance. **Behaviour** Roosts in caves and mine tunnels. Gathers in large maternity colonies in summer, e.g. Mt Etna near Rockhampton, disperses to smaller colonies after young become independent in March. Mating occurs in July–Aug; births in Dec. In s. shares maternity caves with Common Bentwing Bat. Forages beneath canopy. **Similar species** See under Common Bentwing Bat.

Large-footed Myotis *Myotis adversus*

fa 36–41 mm; **hb** 35–50 mm; **t** 33–42 mm; **el** 9–15 mm; **wt** 5.0–10.4 (8.3) g

Characterised by proportionately large feet (8.6–12.3 mm long, > than 1/2 tibia length) with long curved claws, and calcar extending 3/4 of distance to tail tip. Ears large, funnel-shaped, wide-set and not connected to corner of mouth. Tragus long (about 1/2 el), straight. Fur of upperparts grey-brown to russet; underparts cinnamon tipped with pale grey, particularly on chest. Bare skin of flight membranes and ears pale brown. **Distribution, habitat and status** Coastal and subcoastal areas from w. Kimberley across n. and e. Aust to lower Murray R. and se. SA. Always close to water, from small creeks to large lakes and mangrove-lined estuaries. Populations s. of se. Qld may be a different species. **Behaviour** Utilises a range of roost sites: caves, mineshafts, culverts, dense foliage, and (in s.) tree hollows. Forages mostly low over water taking flying insects; also captures aquatic insects and tiny fish by raking claws across the surface. Often 2 or more forage in unison. In n. up to 3 single births in a year; in s. a single young born in summer.

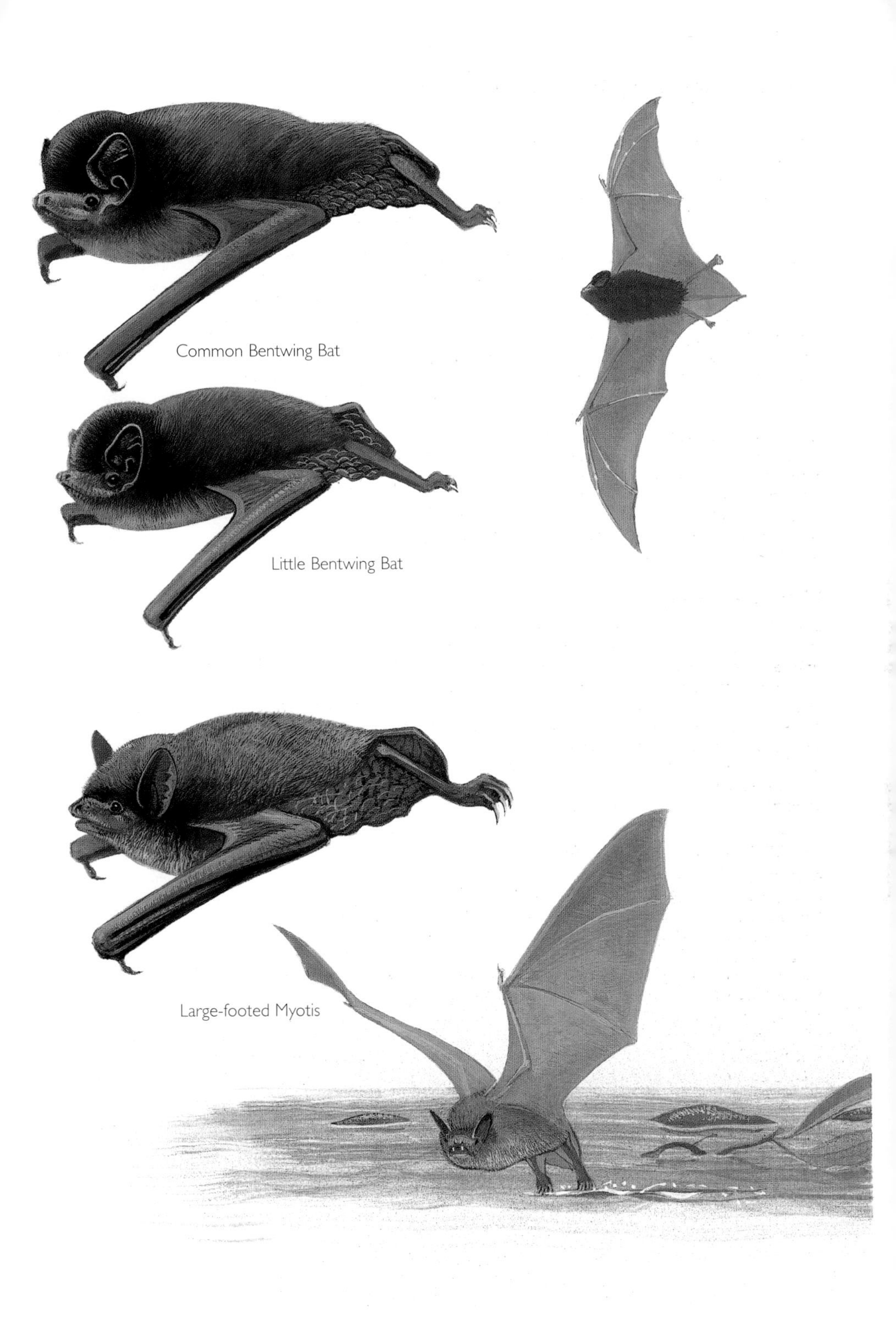
Common Bentwing Bat
Little Bentwing Bat
Large-footed Myotis

PLATE 60

Eastern False Pipistrelle *Falsistrellus tasmaniensis*

Great Pipistrelle, Tasmanian Pipistrelle

fa 48–54 mm; **hb** 55–70 mm; **t** 40–51 mm; **el** 14–19 mm; **wt** 14–26 (22) g

Large and robust. Fur of upperparts uniformly reddish brown, underparts paler. Face forward of ears naked and pale brown, as are ears; skin of flight membranes, lips, forearms and feet blackish. Ears long, narrow with rounded tips, distinct notch on upper rear margin; ears overlap when pressed over crown. Tragus long, narrow and strongly curved forward. I^2 present but minute; distinct gap between upper incisors and canine. **Distribution, habitat and status** Uncommon in tall forest on the GDR and adjacent coastal plains from se. Qld to se. SA, also widespread in Tas. **Behaviour** Roosts communally, in single-sex groups, in tree hollows or rarely in caves (Jenolan Caves, NSW). Forages above or within the canopy with swift direct flight. Single young born in Dec. **Similar species** Greater Broad-nosed Bat has somewhat shorter, broader ears that barely touch when pressed across crown; lacks minute I^2; no gap between upper incisors and canines; penis long, without bulbous tip.

penis shape
(after Parnaby)

Western False Pipistrelle *Falsistrellus mackenziei*

fa 48–54 mm; **hb** 55–66 mm; **t** 40–53 mm; **el** 14–18 mm; **wt** 17–26 (21) g

Similar to Eastern False Pipistrelle but more rusty brown above and cinnamon below. **Distribution, habitat and status** Confined to sw. WA, s. of Perth and e. to the wheatbelt. Most records from mature Karri forest but also known from wetter stands of Jarrah and Tuart, and woodland on Swan Coastal Plain. **Behaviour** Little known. Roosts in tree hollows, forages at canopy level. Births in spring, early summer. **Similar species** The largest vespertilionid in WA; should not be confused with any other species.

Greater Broad-nosed Bat *Scoteanax rueppellii*

fa 51–56 mm; **hb** 63–73 mm; **t** 44–58 mm; **el** 16–18 mm; **wt** 21–35 (30) g

Large and robust with broad head and short, squarish muzzle. Ears widely spaced, short, rounded apex with concave rear edge immediately below apex; barely touch when pressed together over crown. Upperparts vary from mid-brown to dark cinnamon-brown, paler on crown; underparts tawny-olive. Bare skin of face, ears, flight membranes pinkish-brown. Only one pair of upper incisors, abutting canines. **Distribution, habitat and status** Coastal e. Aust from Carbine Tableland in ne. Qld to se. NSW. Usually in tall wet forest, extending into drier forest along gullies. In s. rarely above 500 m, in far n. restricted to uplands. Nowhere common. **Behaviour** Roosts in tree hollows. Flies relatively slowly with limited manoeuvrability along forest edges or streams to capture large insects. May also eat small bats. Single young born in Jan. **Similar species** Eastern False Pipistrelle is similar in size but has longer ears which overlap by at least 5 mm when pressed together over crown; small I^2 present, with a gap between it and the C^1; penis short, with bulbous tip.

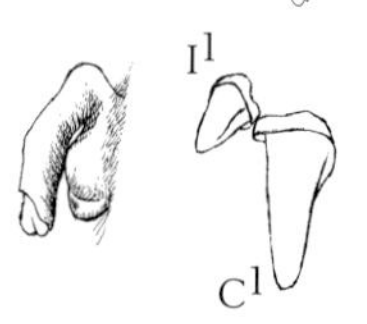

penis shape
(after Parnaby)

Eastern False Pipistrelle
Western False Pipistrelle
Greater Broad-nosed Bat

PLATE 61

Inland Broad-nosed Bat *Scotorepens balstoni*

Western Broad-nosed Bat

fa 32–40 (35.5) mm; **hb** 42–60 mm; **t** 20–42 mm; **el** 11–14 (12.5) mm; **wt** 6.3–12.5 (8.5) g
Slender body; short, squarish, hairless muzzle with distinct swollen glandular area on each side. Body colour variable: upperparts tawny olive to sandy, hairs unicoloured; underparts paler, hairs bicoloured, grey-brown basally with whitish tips. Size varies geographically; larger in e. and n. **Distribution, habitat and status** Common across much of arid and semi-arid Aust in wide range of habitats, including open woodland, shrubland, mallee, tree-lined watercourses. **Behaviour** Roosts in tree hollows, old buildings. Often emerges before dusk, forages among and below canopy. Flies with fast, flickering wingbeats. May forage on ground or other surfaces. 1 or 2 young born in Nov. **Similar species** Inland, Little and Northern Broad-nosed Bats are difficult to identify; may require biochemical methods such as protein electrophoresis. On average, Inland is larger than Little and Northern, but all measurements overlap. Eastern Broad-nosed Bat differs in ratio of forearm length to tibia length (>2.5 cf <2.4 in others); glans penis has 8 spines in circle rather than numerous spines in 2 rows, fur of underparts not obviously bicoloured.

Little Broad-nosed Bat *Scotorepens greyii*

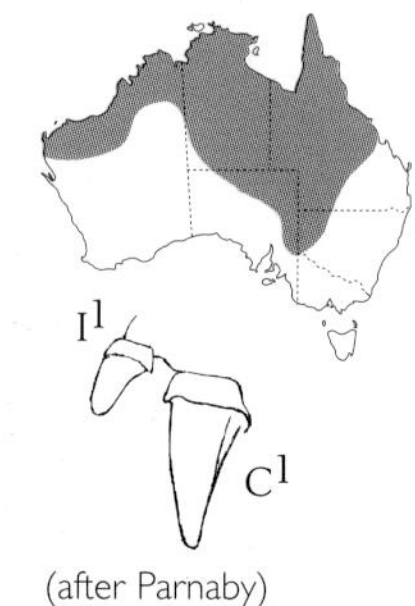

(after Parnaby)

fa 27–35 (31.3) mm; **hb** 37–53 mm; **t** 25–48.5 mm; **el** 10–13 (11.4) mm; **wt** 4.0–11.0 (6.6) g
Similar to Inland Broad-nosed Bat but smaller. Fur of upperparts yellow-brown to reddish-brown; underparts pale grey. Conspicuous glandular tubercule between corner of mouth and base of ear. **Distribution, habitat and status** Widespread across n. and e. Aust, w. of GDR, s. to sw. NSW. Locally common in variety of arid and tropical habitats, particularly along streams through open woodland and savannah, but also monsoon forest and paperbark forest. **Behaviour** Similar to Inland Broad-nosed Bat. During hot weather requires regular access to water. **Similar species** See under Inland Broad-nosed Bat. Forest bats (*Vespadelus*) have heavier build, more pointed muzzle, 2 pairs of upper incisors, not one, lack spines on glans penis.

Northern Broad-nosed Bat *Scotorepens sanborni*

nw. Aust: **fa** 28–34 (31) mm; **hb** 37–48 mm; **t** 27–36 mm; **el** 9–12 (10) mm; **wt** 5.7–7.3 (6.5) g
Qld: **fa** 31–36 (33) mm; **hb** 40–52 mm; **t** 29–39 mm; **el** 10–13 (11) mm; **wt** 5.7–9.1 (7.3) g
On average slightly larger in all dimensions than Little Broad-nosed Bat, but indistinguishable externally. **Distribution, habitat and status** Two separate populations: Kimberley and w. Top End, and coastal Qld from Cape York s. to about Rockhampton. The w. population is common in forests, especially those fringing waterways, including mangroves. In Qld inhabits monsoon forest, open woodland, heath. **Behaviour** Roosts in tree hollows, buildings; often forages over water; eats mainly small insects such as mosquitoes, midges. Young born Sept–Nov. **Similar species** See under Inland Broad-nosed Bat.

Eastern Broad-nosed Bat *Scotorepens orion*

fa 32–37 (34.8) mm; **hb** 44–53 mm; **t** 26–38 mm; **el** 10.6–13 (12) mm; **wt** 7.0–15.0 (11.0) g
Stockier and darker furred than other broad-nosed bats. Fur warm brown, duller below. Hairs not obviously bicoloured. Forearm length >2.5 tibia length. **Distribution, habitat and status** Confined to GDR and adjacent coastal plains from se. Qld to Melbourne, Vic. Inhabits mainly wet and dry sclerophyll forest; nowhere common. **Behaviour** Poorly known; roosts in tree hollows, occasionally buildings. Mates in autumn but ovulation and fertilisation delayed until spring; births in Nov–Dec. **Similar species** See under Inland Broad-nosed Bat.

Broad-nosed Bat (undescribed)

Similar to Eastern Broad-nosed Bat but smaller (fa < 34 mm, wt 6–8 g). Ratio of forearm length to tibia length <2.5, cf >2.5 in Eastern Broad-nosed Bat. Known from coastal ne. NSW and se. Qld, in dry sclerophyll forest, coastal woodland and heathland. Natural history not documented.

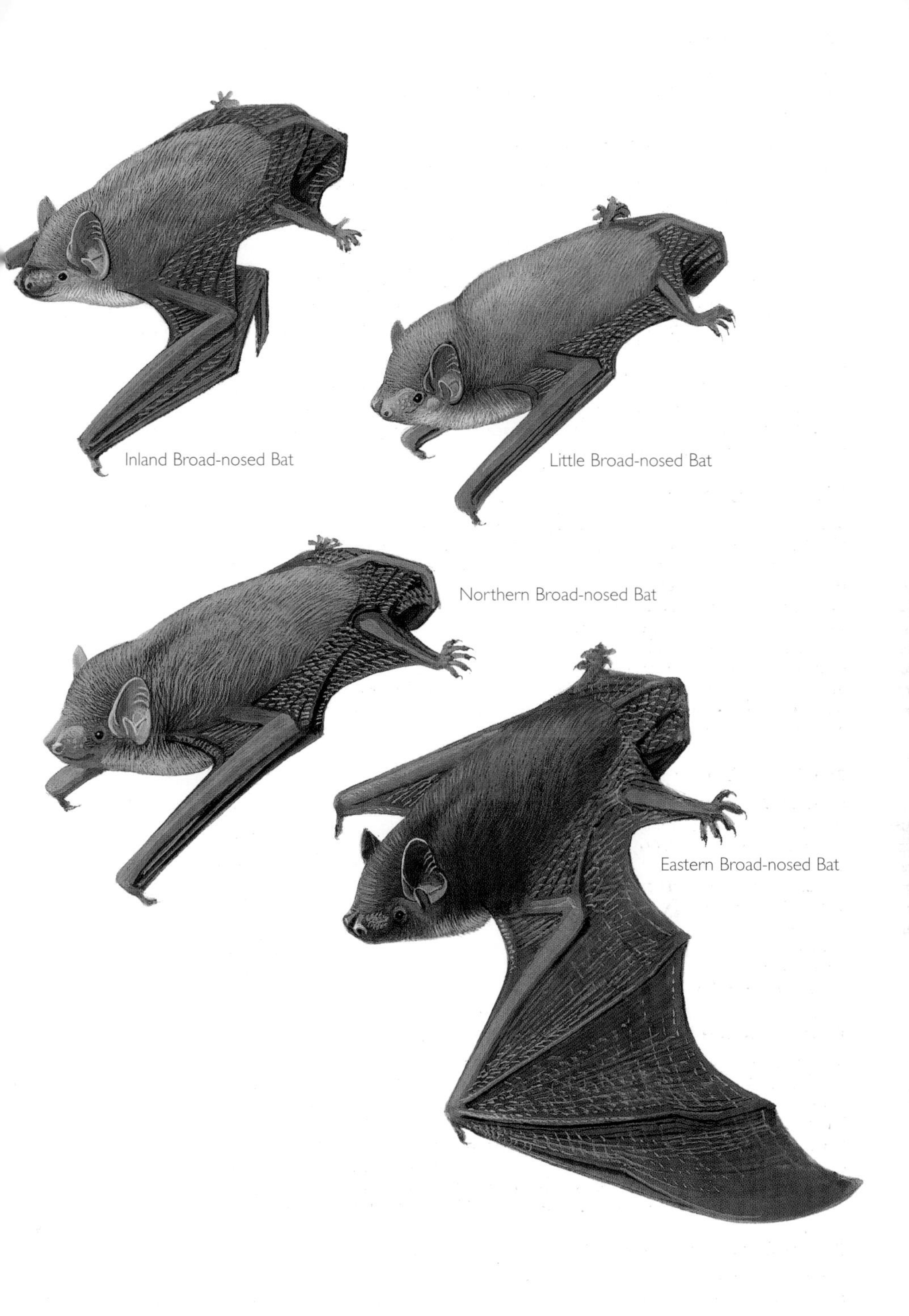
Inland Broad-nosed Bat
Little Broad-nosed Bat
Northern Broad-nosed Bat
Eastern Broad-nosed Bat

PLATE 62

Arnhem Long-eared Bat *Nyctophilus arnhemensis*

fa 36–40 mm; **hb** 40–58 mm; **t** 36–41 mm; **el** 16–20.5 mm; **wt** 5.2–7.7 (6.6) g
Post-nasal ridge across muzzle low and short with slight central depression and not wider than nasal exfoliations, which are well developed for a long-eared bat. Fur colour variable: upperparts from pale russet to dark brown, underparts slightly paler without obvious demarcation. Flight membranes mid-brown not blackish. **Distribution, habitat and status** Coastal and subcoastal n. Aust from Exmouth Gulf in WA to the w. shores of Gulf of Carpentaria, NT. Common in wide range of tropical vegetation including mangroves, pandanus along waterways, monsoon forest and melaleuca forest. **Behaviour** Roosts in dense vegetation or under peeling bark, occasionally in buildings. Forages below canopy with slow, fluttery flight. Births Oct–Feb, usually twins. **Similar species** Northern Long-eared Bat tends to be larger (wt >8 g, el >20 mm); has a wider post-nasal ridge (wider than nasal exfoliations, which are less developed) with a shallow central vertical groove. Pygmy Long-eared Bat is smaller (fa < 36 mm, el < 15 mm). Lesser Long-eared Bat has very high post-nasal ridge with distinct Y-shaped central groove; grey-brown upperparts clearly demarcated from whitish underparts.

post-nasal ridge

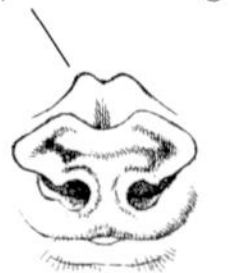

Northern Long-eared Bat *Nyctophilus bifax*

fa 38–45 mm; **hb** 35–56 mm; **t** 34–46 mm; **el** 20–25 mm; **wt** 8.0–13.2 (9.3) g
Post-nasal ridge low and broad, wider than nostrils, with shallow central dip. Fur of upperparts variable, rich reddish brown to pale brown, underparts pale grey-brown. Bare skin of face, ears and flight membranes mid grey-brown. **Distribution, habitat and status** Occurs in 3 separate regions: Pilbara; Kimberley and Top End, e. to near Burketown; and along e. coast s. to n. NSW. Locally common. Usually found in wet forest including rainforest, monsoon forest and riparian forest, sometimes in dry sclerophyll forest and woodland. In NSW known only from rainforest. WA, NT populations (subspecies *daedalus*) probably a separate species to the e. population (subspecies *bifax*). **Behaviour** Roosts in a range of situations including under peeling bark, in dense foliage and tree hollows. Has a more direct and rapid flight than other long-eared bats and tends to forage along forest edges rather than within the forest, but can hover and glean insects from foliage or the ground. Mates in May, births in Oct–Nov, twins normal. **Similar species** See under Arnhem Long-eared Bat and Greater Long-eared Bat. Gould's Long-eared Bat has grey rather than red-brown fur; a higher post-nasal ridge with distinct vertical groove; and on average has longer ears (>24 mm).

post-nasal ridge

Pygmy Long-eared Bat *Nyctophilus walkeri*

fa 30–36 mm; **hb** 38–44 mm; **t** 26–36 mm; **el** 11–15 mm; **wt** 3.3–7.0 (4.4) g
Smallest of Aust's long-eared bats, with shortest ears. Fur of upperparts pale orange-brown or fawn, underparts cream or buff, flight membranes distinctly blackish. Post-nasal ridge well defined, partially divided by central groove. **Distribution, habitat and status** Confined to the n. Kimberley and Top End, e. to near Burketown. Usually around permanent water and riparian vegetation, especially melaleuca and pandanus, also monsoon forest and savannah. Locally common, e.g. Mitchell Plateau and Drysdale R. NP. **Behaviour** Roosts in dense vegetation such as dead fronds of palm trees, also in tree hollows. Forages for flying insects low to the ground (normally below 3 m), often amongst dense vegetation or over water, with a slow, fluttery flight. Births of twins take place in Oct–Nov. **Similar species** See under Arnhem Long-eared Bat. Lesser Long-eared Bat has very high post-nasal ridge with a distinct Y-shaped groove; whitish underparts, el >17 mm.

post-nasal ridge

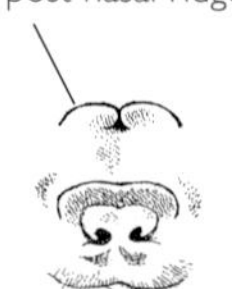

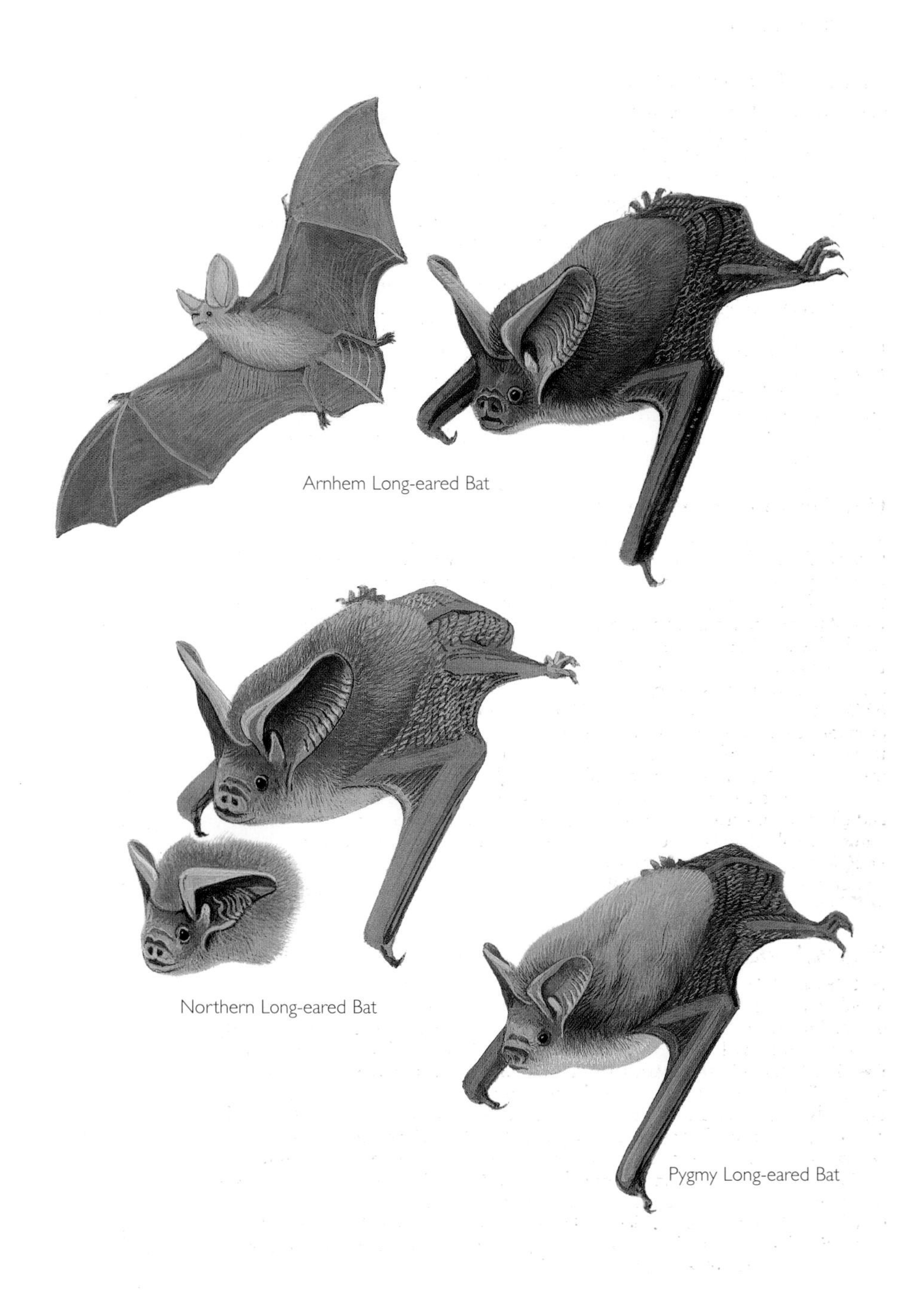
Arnhem Long-eared Bat
Northern Long-eared Bat
Pygmy Long-eared Bat

PLATE 63

Lesser Long-eared Bat *Nyctophilus geoffroyi*

fa n. Aust 31–39 (35); s. Aust 32–42 (37) mm; **hb** 38–50 mm; **t** 31–40 mm; **el** 18–25 mm; **wt** n. Aust 4.0–8.5 (5.8); s. Aust 6.0–10.2 (8.2) g

Post-nasal ridge well developed, about 1.5 mm high, split to give Y-shaped central groove. Nasal exfoliations obvious. Ears long (>17 mm). Fur of upperparts pale grey-brown, contrasting with uniformly whitish underparts. Hairs of dorsal fur clearly bicoloured, very dark at base. **Distribution, habitat and status** One of Australia's most widespread and abundant mammals. Common over most of mainland, except for Cape York and e. coast of Qld. Inhabits almost all habitats from deserts to wet forests, including urban areas. **Behaviour** Will use almost any hole or crevice as roost site, including tree hollows, rolled-up blinds, gaps in timber, etc. Forages low amongst shrubs with fluttery, highly manoeuvrable flight; catches insects in flight or gleans them from foliage or ground. Twins born in Oct–Nov in s. Aust, earlier in n. **Similar species** Only Gould's Long-eared Bat has post-nasal ridge approaching level of development in this species, but it is lower and lacks clear Y-shaped groove; Gould's is larger (fa >36 mm in n., >40 in s., el >24 in s.) and underparts are grey not whitish. Separation of these 2 species is most difficult towards inland edge of distribution of Gould's, where it tends to be smaller and paler.

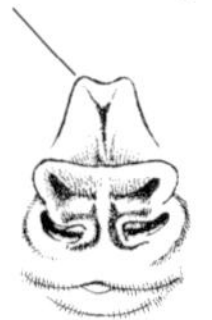

Gould's Long-eared Bat *Nyctophilus gouldi*

fa n. Aust 36–42 (38.5); s. Aust 40–48 (44) mm; **hb** 44–52 mm; **t** 39–41 mm; **el** 24–28 mm; **wt** n. Aust 5.2–7.5 (6.0); s. Aust 9.0–16.5 (12.3) g

Post-nasal ridge well-developed, with central groove that is usually T-shaped. Fur of upperparts grey-brown to grey; underparts pale grey or buff. Outer canine width <5.6 mm. Individuals on inland side of GDR tend to be smaller and paler than those from the ranges or coast. **Distribution, habitat and status** E. and s. Aust from about Cairns in Qld to se. SA; also in sw. WA. Inland limits correspond to limits of eucalypt open-forest. **Behaviour** Roosts in tree hollows or under peeling bark. Forages below canopy among vegetation. Mates in April; sperm stored by female until ovulation in Sept; births in late Oct–Nov. In s. hibernates during coldest months (May–Sept). **Similar species** See under Lesser Long-eared Bat. Northern Long-eared Bat is larger (wt >8 g), has shorter ears (<25 mm) and lower post-nasal ridge without central groove. Greater Long-eared Bat is more robust (wt >11 g) with broader muzzle (outer canine width >5.6 mm) and lower, undivided post-nasal ridge extending almost eye to eye.

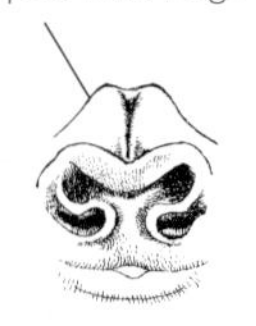

Greater Long-eared Bat *Nyctophilus timoriensis*

fa 39–50 mm; **hb** 50–75 mm; **t** 35–50 mm; **el** 25–30 mm; **wt** 11–20 (14.2) g

Body thickset and robust; head relatively large, muzzle broad; only long-eared bat with outer canine width >5.6 mm. Post-nasal ridge low and broad with indistinct central vertical groove. Fur dark grey-brown, only slightly paler on underparts. **Distribution, habitat and status** Found across semi-arid s. Aust to s. Qld. Population in Tas (subspecies *sherrini*) is probably separate species; population in sw. WA (subspecies *major*) may be a third species. Inhabits wet sclerophyll forest in Tas and sw. WA, and a range of dry woodland and shrubland communities in arid and semi-arid regions. **Behaviour** Roosts mostly in tree hollows. Forages low amongst the canopy and shrub layers gleaning arthropods from foliage and ground; also takes insects in flight. **Similar species** See under Gould's Long-eared Bat. Northern Long-eared Bat difficult to distinguish but more lightly built and outer canine width <5.6 mm.

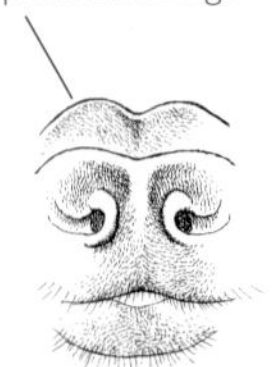

Lesser Long-eared Bat
Gould's Long-eared Bat
ears
folded
Greater Long-eared Bat

PLATE 64

Grassland Melomys *Melomys burtoni*

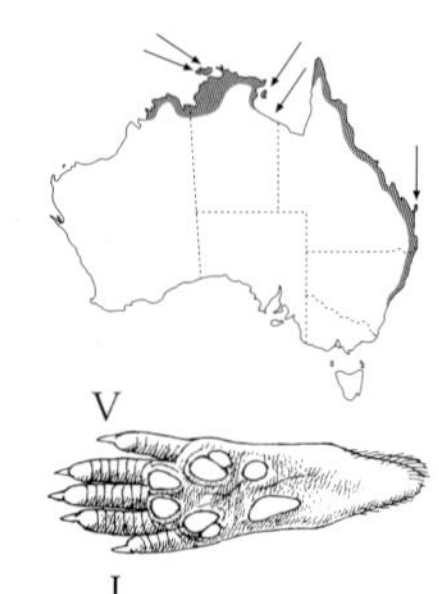

hb 90–170 mm; **t** 90–170 mm; **hf** 23–25 mm; **el** 16–18 mm; **wt** 30–120 (55) g; **teats** in 2 = 4
Fur of upperparts warm grey-brown to ginger above, grading through buff or pale orange on the flanks and legs to whitish or cream underparts. Ears short, rounded, pale grey. Feet dark grey or buff. Tail thin, grey-brown above, pinkish below. Animals from e. coast smaller than those from n. Aust. **Distribution, habitat and status** Occurs in 2 discrete areas: coastal e. Aust from about Gosford in NSW to top of Cape York; and n. Kimberley in WA and Top End, NT, e. to Sir Edward Pellew Is. In e. locally abundant in grasslands, sedgelands, canefields, edges of rainforest, sclerophyll forest and woodland with grassy understorey. WA and NT populations inhabit grassland, monsoon forest, riparian vegetation and mangroves. **Behaviour** Nocturnal, mostly terrestrial but an adept climber. Eats grass, berries, cultivated fruit, sugarcane, insects such as grasshoppers. Breeds mostly during dry season. **Similar species** See under Fawn-footed and Cape York Melomys.

Cape York Melomys *Melomys capensis*

hb 120–160 mm; **t** 120–170 mm; **hf** 25–27 mm; **el** 16–18 mm; **wt** 45–115 (70) g; **teats** in 2 = 4
Upperparts uniformly chestnut or pale grey-brown grading to buff flanks, occasionally small whitish spots on back. Underparts white, including feet, cheeks and lower jaw. Tail long, thin, naked, pale pinkish-grey below; pale brown above; scales in mosaic pattern, not overlapping. Ears brown. Juveniles uniformly grey. **Distribution, habitat and status** Confined to ne. Cape York, including the McIlwraith and Iron Ranges. Inhabits rainforest, monsoon forest and sclerophyll woodland. **Behaviour** Nocturnal, arboreal, terrestrial. Builds nest of leaves in tree hollow or cavity in building. Vegetarian, eats mostly fruit, seeds, leaves. Breeds year-round, several litters of 2 young may be raised per year. Can be confiding around human habitation, often living in buildings. **Similar species** Fawn-footed Melomys very similar but separated by gap of about 300 km between Cooktown area and McIlwraith Range. Grassland Melomys in the same area is smaller, grey or brownish fur (not white) on feet.

Fawn-footed Melomys *Melomys cervinipes*

hb 100–200 mm; **t** 115–200 mm; **hf** 23–30 mm; **el** 16–22 mm; **wt** 50–110 (80) g; **teats** in 2 = 4
Fur colour highly variable, greyish tan to bright orange-brown above and whitish or buff below; greyer on face. Immatures often uniformly grey. Pelage short, dense, without obvious guard hairs. Tail long, slender, naked; brownish-grey or black; scales in mosaic pattern, not overlapping. Ears short, round, dark grey. **Distribution, habitat and status** Along e. coast from far ne. NSW to about Cooktown, Qld. Common in rainforest, particularly where vines are abundant; in s. also inhabits wet sclerophyll forest, coastal woodlands, mangrove forests. **Behaviour** Nocturnal, arboreal, terrestrial, agile climber. Builds spherical nest of leaves in canopy. Vegetarian; eats leaves, shoots, fruits, mostly obtained in the canopy. Breeding season extended; births peak in spring–summer in s., midyear in n. Several litters, usually of 2 young, may be born per year. Young cling to mother's teats, dragged around while she forages until weaned at about 3 weeks. **Similar species** Cape York Melomys may be almost indistinguishable, but not known to overlap in range. Grassland Melomys very similar in colour, found in more open grassy habitat, builds nests close to ground or in tree hollow.

Bramble Cay Melomys *Melomys rubicola*

hb 140–160 mm; **t** 145–180 mm; **wt** approx 100 g; **teats** in 2 = 4
The only mammal endemic to a coral cay on the Great Barrier Reef. Differentiated from Cape York Melomys by protein differences; only morphological difference is that the tail feels rough due to the elevated scales. **Distribution, habitat and status** Confined to Bramble Cay, a tiny vegetated coral island at n. end of Great Barrier Reef in Torres Strait. Forages in herbfield and strandline vegetation, nests in burrows. Population only a few hundred individuals, classified as critically endangered because entire island is threatened by erosion.

Grassland Melomys
Cape York Melomys
Fawn-footed Melomys
Bramble Cay Melomys

PLATE 65

Black-footed Tree Rat *Mesembriomys gouldi*

hb 260–315 mm; **t** 310–410 mm; **hf** 66–72 mm; **el** 42–46 mm; **wt** 600–880 (630) g; **teats** in 2 = 4
Large, striking arboreal rat. Upperparts shaggy, grizzled straw and black. Black hairs dominant on rump, straw on flanks and crown. Underparts white or pale grey. Some individuals more grey, particularly on flanks. Tail long, completely furred, black except for distinctive white terminal tuft. Ears black, long, rounded. Hindfeet black, sometimes mottled white. Animals from Melville and Bathurst Is darker. **Distribution, habitat and status** Separate populations in n. Kimberley, Top End of NT, and far n. Qld between Ravenshoe and Iron Range. Occupies moist areas with well-developed shrub layer within tropical woodlands and open-forest. Locally common in Arnhem Land and Melville I., uncommon to rare elsewhere. **Behaviour** Nocturnal; arboreal but often forages on ground, shelters in tree hollow or nest built in dense foliage. Eats fruit, seeds, insects, green tips, nectar. Pandanus fruits are particularly favoured. Breeds year-round, peak of births in late dry season. Usually 2 young are weaned. **Similar species** Golden-backed Tree Rat is much smaller, has grey and white tail, white upper surface to hind feet, warm buffy brown upperparts. Giant White-tailed Rat has naked white and black tail, inhabits rainforest.

Golden-backed Tree Rat *Mesembriomys macrurus*

hb 190–270 mm; **t** 290–350 mm; **hf** 48–52 mm; **el** 24–27 mm; **wt** 240–330 (280) g; **teats** in 2 = 4
Upperparts grizzled and mottled grey-yellow with broad band of golden-rufous from crown to base of tail. Underparts whitish. Ears long, blackish. Tail long, thin, basal third grey then white with distinct terminal tuft of long white hairs. **Distribution, habitat and status** Now restricted to nw. Kimberley, including several islands in Buccaneer Arch. Last record in NT in 1969, but may persist on Wessel I. Has declined in pastoral country. Inhabits tropical woodlands and adjacent vine thickets, rainforest in rugged valleys; also recorded foraging on beaches. Most records are from high rainfall areas (> 600 mm). Vulnerable. **Behaviour** Shy and difficult to observe. Arboreal but spends considerable time foraging on ground, shelters in nest in tree hollow or in dense cover such as pandanus foliage. Eats flowers, fruits, insects, shoots, leaves. Probably breeds in most months. **Similar species** See under Black-footed Tree Rat.

Giant White-tailed Rat *Uromys caudimaculatus*

Giant Mosaic-tailed Rat

hb 275–380 mm; **t** 325–360 mm; **hf** 60–80 mm; **el** 30–40 mm; **wt** 500–900 (550) g; **teats** in 2 = 4
Fur coarse with dark guard hairs. Upperparts mid to dark grey-brown, greyer on flanks and legs. Underparts, including feet, creamy white. Muzzle long, seminaked; whiskers long and black; ears short and round. Feet long and strongly clawed. Tail long, naked, black basally with variable length of white, blotched with dark grey. **Distribution, habitat and status** NE. Qld from Mt Elliot, s. of Townsville to e. Cape York, including Hinchinbrook I. Common in rainforest and wet sclerophyll forest at all altitudes; also found in melaleuca forest, mangroves, semi-urban areas. **Behaviour** Mostly solitary, strong agile climber; omnivorous, eats fruit, seeds, shoots, fungi, insects, crabs, small vertebrates, birds eggs. Buries seeds in cache for future use. Breeds during wet season (Oct–Jan). Confiding, regularly found in camping grounds and around houses. **Similar species** Other similar sized rodents (tree rats and Water Rat) have furred tails.

Masked White-tailed Rat *Uromys hadrourus*

Thornton Peak Uromys

hb 170–185 mm; **t** 185–195 mm; **hf** 37–38 mm; **el** 20–25 mm; **wt** 140–205 (190) g; **teats** in 2 = 4
Fur of upperparts fawn or mid-brown with rufous tinge on crown, neck and below ears; underparts whitish. Distinct dark eye-ring (absent in juveniles). Tail long, narrow, naked, grey then white for last few cm. **Distribution, habitat and status** Restricted to 3 discrete areas of upland rainforest in ne. Qld: Thornton Peak and the McDowall Range n. of Daintree; Carbine Tableland w. of Mossman; Lammins Hill area, Atherton Tableland. Rarely enters traps so status and distribution difficult to estimate. **Behaviour** Poorly known. Eats nuts of rainforest trees and ground insects, including beetles. Breeding occurs in early wet season (Nov–Jan). **Similar species** Melomys are smaller and lack white tail tip. Immature Giant White-tailed Rats have dark mottling on the white tail, larger head length (>43 mm) and hindfoot (>40 mm).

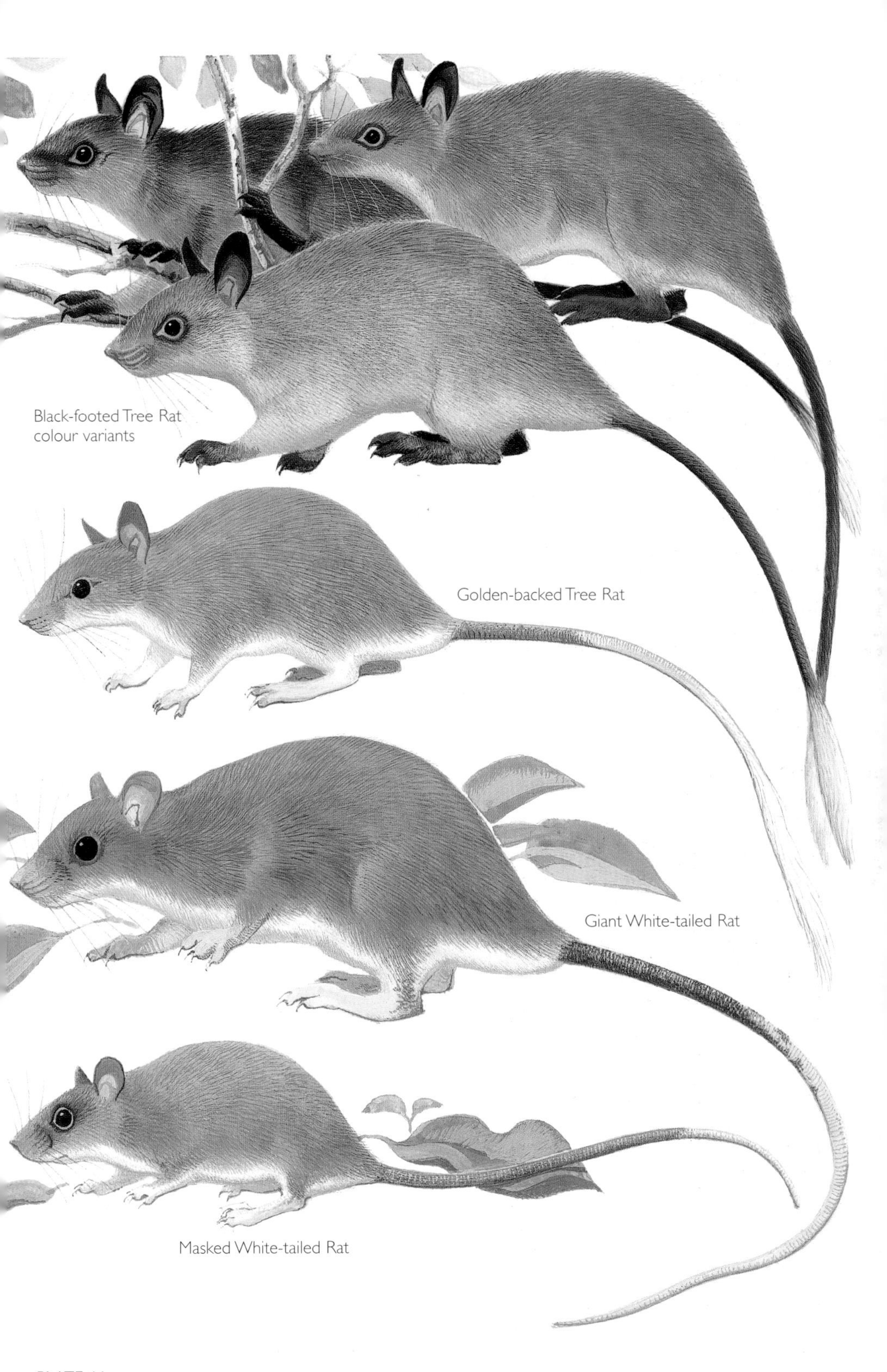
Black-footed Tree Rat
colour variants
Golden-backed Tree Rat
Giant White-tailed Rat
Masked White-tailed Rat

PLATE 66

Brush-tailed Rabbit Rat *Conilurus penicillatus*

Brush-tailed Tree Rat

hb 150–220 mm; **t** 180–230 mm; **hf** 40–45 mm; **el** 25–30 mm; **wt** 110–190 g; **teats** in 2 = 4
Broad head, short muzzle, large eyes and long naked ears give distinct head profile. Upperparts grizzled brown and straw, pencilled with black. Rufous wash to forehead, crown and hindneck, sometimes extending over much of back. Some individuals duller and more grey. Underparts whitish or cream with grey chest patch. Animals from Melville and Bathurst Is (subspecies *melibius*) have darker creamy brown underparts. Tail long, thin, blackish for basal half then dense tuft of long hair that may be black or white. **Distribution, habitat and status** Patchy within 130 km of coast in n. Kimberley and nw. Top End, NT. Formerly e. to about Burketown, Qld. Also Melville and Bathurst Is, Groote Eylandt, Sir Edward Pellew Group, and Bentinck I. in Wellesley Group. Inhabits moist areas with dense grassy understorey within coastal she-oak woodlands, sclerophyll forest, and pandanus thickets. Has declined in Kakadu NP. **Behaviour** Mostly nocturnal, semi-arboreal, shelters in tree hollow or in leaf nest in dense foliage. Breeds through late wet and early dry (Mar–Oct). Forages on ground or in trees, eats grasses, herbs, seeds. Bounds with tail held high and flicked from side to side. **Similar species** Black-footed Tree Rat >600 g in weight, has longer muzzle, black feet and ears. Golden-backed Tree Rat has entirely whitish tail and clear golden dorsal band.

White-footed Rabbit Rat *Conilurus albipes*

hb 230–260 mm; **t** 220–240 mm; **hf** 45–55 mm; **el** 25–30 mm; **wt** about 200 g; **teats** in 2 = 4
Large, robust build with broad head and short muzzle, ears long, narrow. Upperparts pale grey-brown; underparts whitish, including upper surface of feet. Tail furred and bicoloured: dark brown above, whitish below with a terminal tuft of blackish hairs. **Distribution, habitat and status** Extinct. Formerly inhabited temperate eucalypt woodlands of se. Aust from Darling Downs in se. Qld to sw. Vic. Last recorded in 1862. **Behaviour** Very little known. Semi-arboreal, nocturnal; sheltered in tree hollows and settlers' huts where sometimes a pest in food stores. **Similar species** Tail colour pattern diagnostic.

Water Rat *Hydromys chrysogaster*

hb 300–390 mm; **t** 230–320 mm; **hf** 55–76 mm; **el** 15–20 mm; **wt** 620–1200 (700) g; **teats** in 2 = 4
Highly adapted to aquatic life: streamlined shape; small eyes; nostrils and ears set high on head; tail thickened for use as rudder; hindfeet broad, partially webbed; fur very dense, shiny. Upperparts vary from grey-brown to rich golden brown to blackish, underparts cream to golden-orange, including lower half of head. Tail < hb, thick, densely furred, dark grey or blackish with broad white tip. Distinct pungent musky odour. Broad hindfeet leave distinctive footprints in mud. **Distribution, habitat and status** Widespread and common in much of coastal n. Aust, e. Aust including Tas and Furneaux Group, and sw. WA. Also Barrow and Dorre Is, WA. Inhabits great variety of aquatic environments including subalpine streams, slow inland rivers, lakes, farm dams, sheltered marine waters. Occurs in streams and estuaries in large cities. **Behaviour** Mostly nocturnal. Brings food to feeding platform to be eaten, leaving midden of inedible components. Mostly forages in water or adjacent vegetation; feeds on aquatic invertebrates, fish, frogs, small birds. Often seen swimming with only top of head and back visible. **Similar species** Sewer Rat often swims but is dull brown, has thin naked tail. Platypus has distinctive 'bill' and broad tail without white tip.

Water Mouse *Xeromys myoides*

False Water Rat

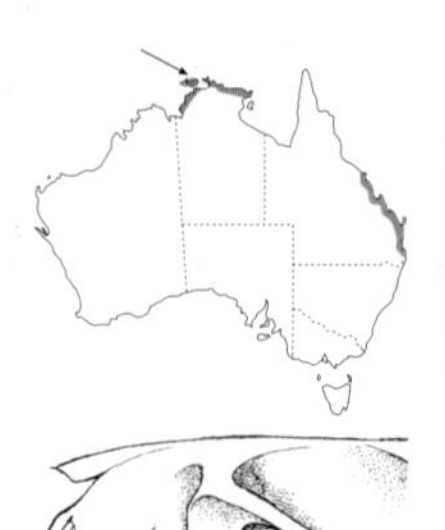

only 2 molars

hb 85–120 mm; **t** 75–90 mm; **hf** 23–26 mm; **el** 10–14 mm; **wt** 35–55 (42) g; **teats** in 2 = 4
Head flat with small eyes and ears; fur short, lustrous. Upperparts uniformly dark grey or grey-brown, sharply demarcated from clear white underparts, including chin and throat. Some individuals have small white flecks on back. Tail thin, dark grey, sparsely furred without obvious rings of scales. Pungent musty odour. **Distribution, habitat and status** Inhabits saline grassland, mangroves, margins of freshwater swamps, lakes close to foredunes in coastal Top End of NT including Melville I., and Qld coast from Cooloola to Proserpine, including Stradbroke and Bribie Is. Vulnerable. **Behaviour** Semi-aquatic, nocturnal, crepuscular. Eats marine and freshwater invertebrates. Constructs leaf nest in network of burrows in muddy bank or clay mound, with entrance in raised mound. **Similar species** None within habitat.

Water Rat, colour variants

Water Mouse

PLATE 67

Spinifex Hopping Mouse *Notomys alexis*

hb 95–115 mm; **t** 120–145 mm; **hf** 30–35 mm; **el** 22–26 mm; **wt** 27–45 (32) g; **teats** in 2 = 4
Upperparts fawn or chestnut, greyer on muzzle and between eye and ear; black guard hairs prominent over back and rump. Underparts, including lower jaw and cheeks, whitish. Tail long, thin, pale pink, darker above, sparse terminal tuft of silvery hairs. Both sexes have small throat pouch with central bare area and low fleshy ridge behind. **Distribution, habitat and status** Sandy deserts on dunes and swales covered with hummock grass, also loamy sands carrying mulga or melaleuca. Populations fluctuate greatly, depending on rainfall. **Behaviour** Nocturnal, gregarious. Groups shelter in deep burrow systems that have several vertical circular entrance holes without a spoil heap. Seen at night bounding across open areas on hindfeet with body and tail almost horizontal. Breeding is opportunistic when conditions are suitable. Omnivorous, eats seeds, most plant parts, arthropods. **Similar species** Dusky Hopping Mouse; both sexes have distinct throat pouch surrounded by fleshy ridge with stiff white hairs pointing inwards. Fawn Hopping Mouse inhabits gibber plains; female has no throat pouch or chest gland, male has bare, raised glandular area between forelegs. Mitchell's Hopping Mouse lacks throat pouch and chest gland in both sexes, has broad shiny white band from throat to chest. *Pseudomys* lack elongated hindfeet and legs, have post-interdigital pads on hindfeet, shorter ears and shorter, bicoloured tails without obvious tuft.

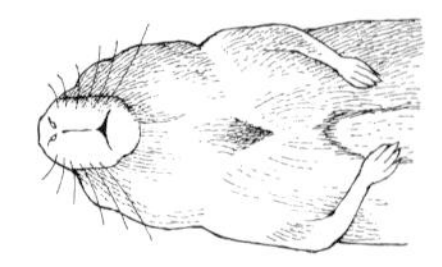

chest gland (after Watts & Aslin)

Fawn Hopping Mouse *Notomys cervinus*

hb 95–120 mm; **t** 120–160 mm; **hf** 32–37 mm; **el** 24–30 mm; **wt** 30–50 (35) g; **teats** in 2 = 4
Pale grey-yellow to pinkish fawn above; white below, including face below the large black eye. Tail bicoloured: brownish pink above, whitish below with terminal tuft of dark hairs. Short muzzle, broad head and very large ears give distinctive blunt profile. No throat pouch; males have raised flat area of bare glandular skin 2–3 mm across, between forelimbs. Has hallucal pad on sole of hindfoot. **Distribution, habitat and status** Patchily distributed on gibber plains and claypans in Lake Eyre basin of s. NT, ne. SA and sw. Qld. Vegetation mostly sparse low shrubland and herbfield. **Behaviour** Nocturnal, gregarious. Shelters in simple, shallow burrow system in hard gibber substrate. Eats seeds, green shoots, some insects; can survive without drinking. Probably breeds whenever conditions suitable. **Similar species** See under Spinifex Hopping Mouse. Dusky Hopping Mouse has distinctive throat pouch.

chest gland

Dusky Hopping Mouse *Notomys fuscus*

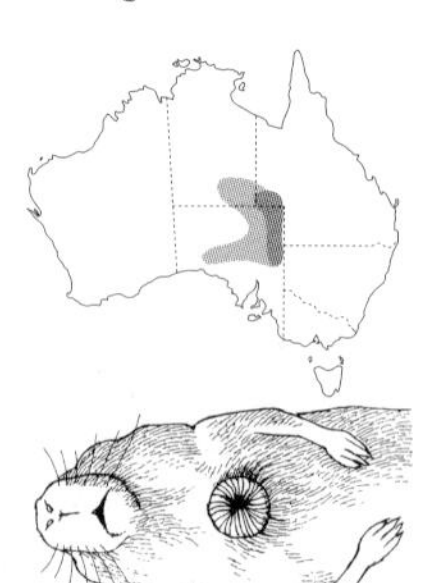

hb 80–115 mm; **t** 120–155 mm; **hf** 34–40 mm; **el** 24–28 mm; **wt** 30–50 (35) g; **teats** in 2 = 4
Both sexes have well-developed throat pouch (a forward-opening skin pocket) encircled by a fleshy lip covered in coarse, inward-pointing white hairs. Upperparts pale orange, sometimes tinged grey particularly about head. Underparts white. Only 3 pads on sole of hf—lacks hallucal pad. **Distribution, habitat and status** Now confined to Strzelecki Desert of ne. SA and sw. Qld; early records from Ooldea, SA and central Aust. Inhabits sand dunes carrying canegrass and ephemeral herbaceous species. Sparse, vulnerable. **Behaviour** Nocturnal, gregarious; shelters in deep burrow system with vertical shafts some 3 cm across and 1.5 m deep. Eats mostly seeds, also green shoots and insects. **Similar species** See under Spinifex and Fawn Hopping Mice.

chest gland

Mitchell's Hopping Mouse *Notomys mitchelli*

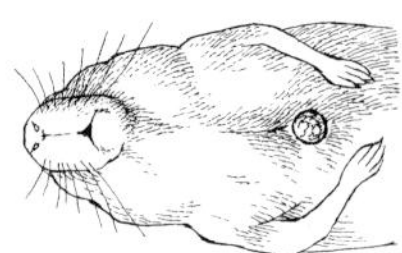

hb 100–125 mm; **t** 140–155 mm; **hf** 34–40 mm; **el** 24–28 mm; **wt** 40–60 (52) g; **teats** in 2 = 4
Both sexes lack throat pouch and chest gland, but have a broad tract of shiny white fur on throat and chest. Upperparts somewhat grizzled tawny-olive or rufous-grey, greyer on head and more orange on flanks. Underparts clear white or pale grey. Tail bicoloured—pinkish brown above with pale brown terminal tuft, pale grey below. Four pads on hf, including small hallucal pad. **Distribution, habitat and status** Inhabits mallee shrubland and heath growing on deep sand across s. Aust from sw. WA to Annuello, nw. Vic, formerly further e. to jn. of Murray and Murrumbidgee Rivers. **Behaviour** Similar to other hopping-mice but nests in logs as well as deep burrows; diet more varied—seeds, roots, green shoots, fungi, insects. Most births in late winter or spring but can occur in any month if conditions right. **Similar species** Slightly larger and duller than Spinifex or Fawn Hopping Mice. Complete lack of throat pouch and chest gland in both sexes is diagnostic.

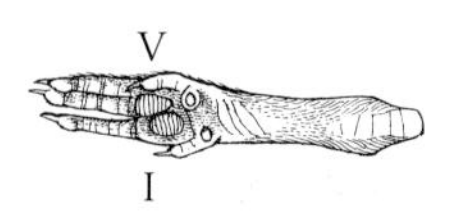

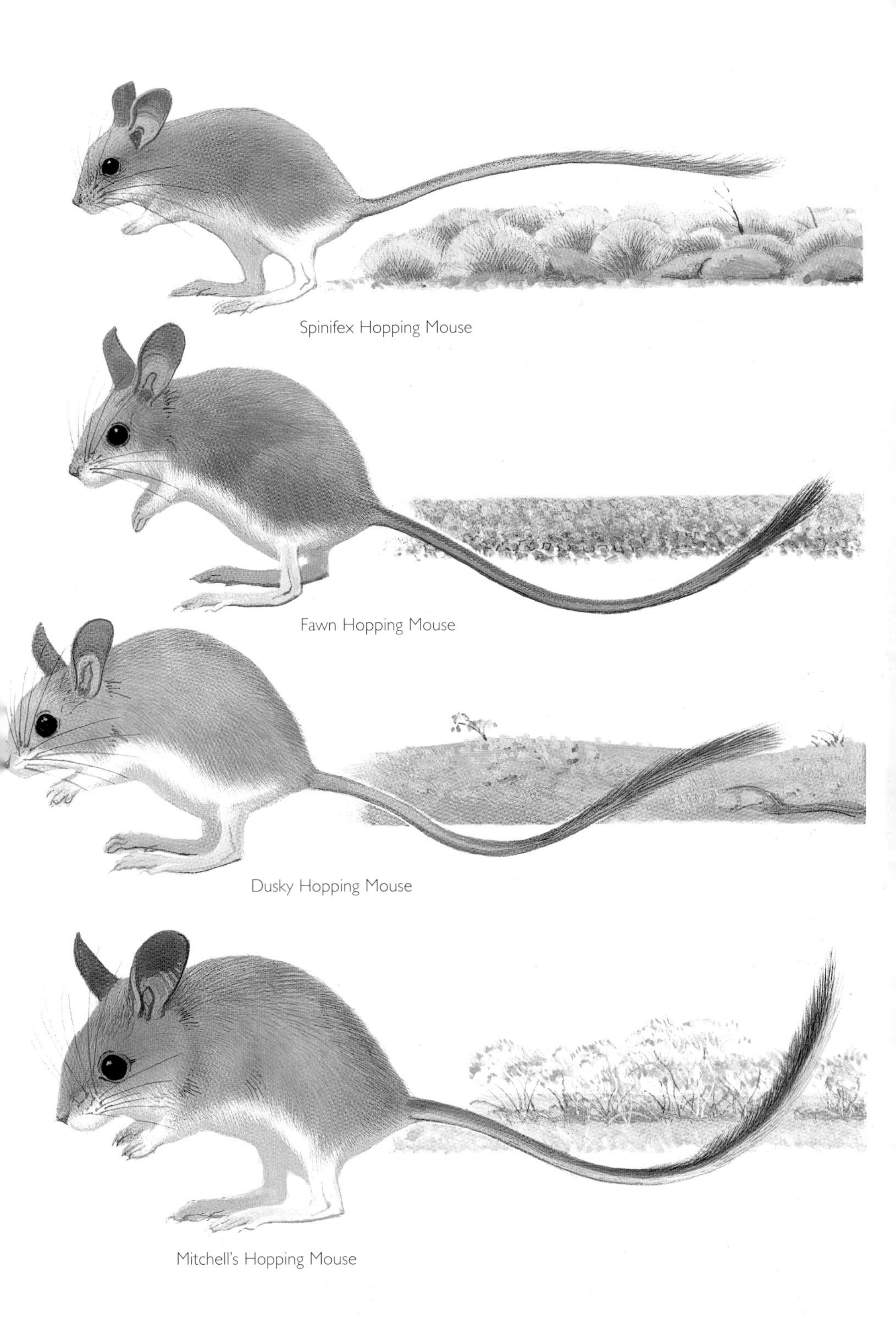
Spinifex Hopping Mouse
Fawn Hopping Mouse
Dusky Hopping Mouse
Mitchell's Hopping Mouse

PLATE 68

Northern Hopping Mouse *Notomys aquilo*

hb 100–115 mm; **t** 140–175 mm; **hf** 34–40 mm; **el** 18–22 mm; **wt** 25–50 (39) g; **teats** in 2 = 4
Upperparts pale brown; underparts white, extending onto sides of muzzle. Feet long, narrow, flesh coloured. Tail long, sandy brown; not obviously bicoloured. Ears short for a hopping mouse. Both sexes have a small throat pouch. **Distribution, habitat and status** Restricted to dune systems and sandsheets in ne. Arnhem Land, inland to Maningrida, and Groote Eylandt. Possibly also in Nassau R. area of Cape York. Inhabits a range of vegetation types including acacia scrub, heath, grassland. On Groote Eylandt favours sites where vine thickets occur immediately behind coastal dunes. Vulnerable. **Behaviour** Rarely trapped but presence readily detected by tracks of the entire hind feet, side-by-side, at spacings of 20–60 cm. Constructs burrow systems in deep sand and plugs the entrances after passing through. A spoil heap up to 40 cm in diameter is often found 1–2 m away from the active burrow entrance. On Groote Eylandt, births of 1–5 young occur mostly in late wet season (March–May) but can occur from Feb to Sept. Little known elsewhere. **Similar species** None within distribution.

Long-tailed Hopping Mouse *Notomys longicaudatus*

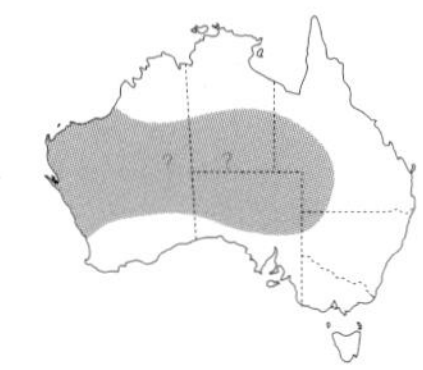

hb 110–160 mm; **t** 150–205 mm; **hf** 40–45 mm; **el** 24–28 mm; **wt** approx. 100 g
Slightly smaller than Short-tailed Hopping Mouse but t and hf proportionately longer. Males have oval chest gland. Upperparts pale sandy brown with numerous fine black hairs; underparts whitish. Tail pale brown above, whitish below; terminal third carries a tuft of short dark hairs, often tipped white. **Distribution, habitat and status** Presumed extinct, not recorded since 1901. Known as a living animal from widely divergent locations in sw. WA, s. NT and w. NSW, and from owl-pellet material from widespread localities across arid Aust. Apparently preferred clay soils rather than sand.

Short-tailed Hopping Mouse *Notomys amplus*

hb 145 mm; **t** 155–160 mm; **hf** 40–43 mm; **el** 35 mm; **wt** approx. 100 g
The largest hopping mouse (twice the weight of extant species). Ears long, but tail proportionately short, not much longer than hb. Upperparts brown, underparts whitish. Tail bicoloured, pale brown above, whitish below, and unlike all other hopping-mice, terminal tuft whitish not dark. Females at least have distinct throat gland. **Distribution, habitat and status** Presumed extinct; known from only 2 modern specimens, both females, collected at Charlotte Waters in se. NT in 1895. Skeletal remains also found in owl pellets in Flinders Ranges, SA.

Big-eared Hopping Mouse *Notomys macrotis*

hb 118 mm; **t** approx. 140 mm; **hf** 40 mm; **el** 26 mm; **wt** approx. 55 g
Similar to Fawn Hopping Mouse but hindfoot longer, skull larger and more robust with wider anterior palatal foramina. Despite name, ears not proportionately long for a hopping mouse. No glandular areas on chest or throat. Like Fawn Hopping Mouse has shallow groove down front surface of incisors. Upperparts grey-brown, underparts whitish. **Distribution, habitat and status** Extinct. Known from only 2 incomplete specimens collected near Moore R., 100 km n. of Perth, before 1844.

Darling Downs Hopping Mouse *Notomys mordax*

Known only from a single skull apparently collected on the Darling Downs, se. Qld, in the 1840s. The skull is closest to Mitchell's Hopping Mouse but has larger molars, a supplementary cusp at the front of M^1, broader incisors and longer anterior palatal foramina.

Broad-cheeked Hopping Mouse *Notomys* sp.

Great Hopping Mouse

Unknown as a living animal. Skulls of a large robust hopping mouse have been found in owl pellets, apparently of recent origin, in the n. Flinders Ranges, SA. Nothing is known of its external morphology or biology.

Northern Hopping Mouse

Long-tailed Hopping Mouse

Short-tailed Hopping Mouse

Big-eared Hopping Mouse

PLATE 69

Common Rock Rat *Zyzomys argurus*

hb 74–124 mm; **t** 90–122 mm; **hf** 18–24 mm; **el** 16–20 mm; **wt** 35–75 g; **teats** in 2 = 4
The smallest and most widespread rock rat. Upperparts golden brown or grey-brown, pencilled with black. Underparts white or pale grey. Upper surface of feet pinkish white. Tail > hb when not damaged; tapers gradually from swollen, naked base to narrow well-haired tip; pinkish grey, darker on top and towards tip; rings of scales obvious. Ears olive-brown. **Distribution, habitat and status** Four separate populations: Pilbara, WA; Kimberley, Top End and w. Gulf of Carpentaria coast; Tully Range, sw. of Winton, Qld; e. coast of Qld from s. of Cooktown to Mt Inkerman, near Ayr. Typically associated with fractured rock outcrops or scree slopes, usually within vegetation with monsoon forest elements. Locally common in w. and n. of range, highly localised in e. **Behaviour** Nocturnal, shelters in deep rock crevice during daylight; omnivorous, eats seeds, plant stem, some insects and fungi. Feeding ledges have accumulations of chewed seed coats. Births can occur throughout year but peak in late wet season (Mar–May). **Similar species** Kimberley, Carpentarian and Arnhem Land Rock Rats are larger (wt >60 g), tails not >hb, and more densely furred.

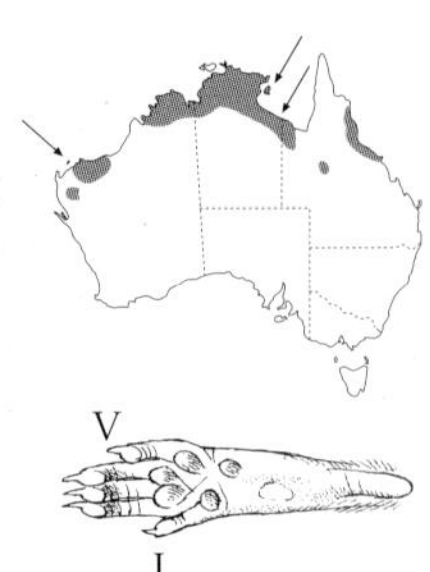

Arnhem Land Rock Rat *Zyzomys maini*

hb 100–135 mm; **t** 116–130 mm; **hf** 245–290 mm; **el** 17–20 mm; **wt** 70–180 g; **teats** in 2 = 4
Upperparts pale brown slightly grizzled with black, washed with pinkish buff on cheek. Underparts white, including upper surfaces of feet. Tail about = hb; thickened at base then tapering to narrow tip; moderately furred with bristly hairs and a sparse terminal tuft; pinkish brown above, white below. Ears have short pale brown hairs on outside. **Distribution, habitat and status** Confined to w. Arnhem Land plateau and sandstone outliers. Occupies boulder screes along gullies with monsoon forest and vine thickets. Can be locally abundant. **Behaviour** Nocturnal, shelters in rock crevices. Eats seeds and fruits of rainforest trees, and grass seeds. Collects hard fruits and carries them to a safe cranny before gnawing through the hard coat—leaving accumulations of seed coats on sheltered ledges. **Similar species** See under Common Rock Rat; Carpentarian Rock Rat has tail clearly <hb and short black hairs on outside of ears.

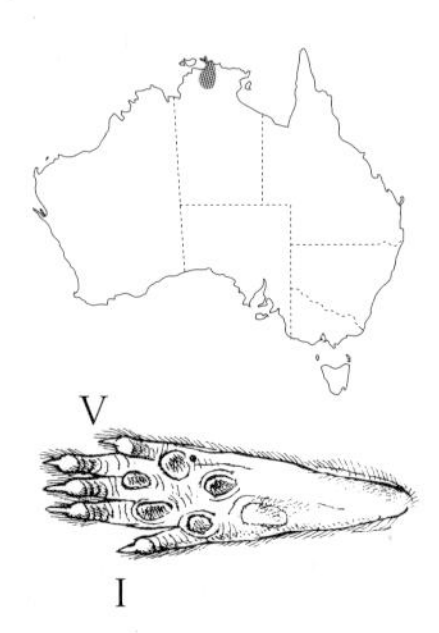

Carpentarian Rock Rat *Zyzomys palatalis*

hb 130–200 mm; **t** 100–150 mm; **hf** 27–28 mm; **el** 18–19 mm; **wt** 110–140 g; **teats** in 2 = 4
Upperparts grey-brown or buffy brown, pencilled with black; underparts whitish with patches of buff. Tail < hb, usually thickened at base, moderately haired with longer and darker hairs towards tip, olive-brown above, whitish below. Ears olive with short black hairs on outside. **Distribution, habitat and status** Endangered. Known from only 4 sandstone gorges in the Gulf of Carpentaria hinterland near the NT–Qld border. Occurs in dry rainforest patches growing on rocky screes in deep gorges. **Behaviour** Poorly known; probably similar to Arnhem Land Rock Rat. **Similar species** See under Common Rock Rat. Arnhem Land Rock Rat has slightly longer tail relative to hb and pale brown hairs on outside of ears.

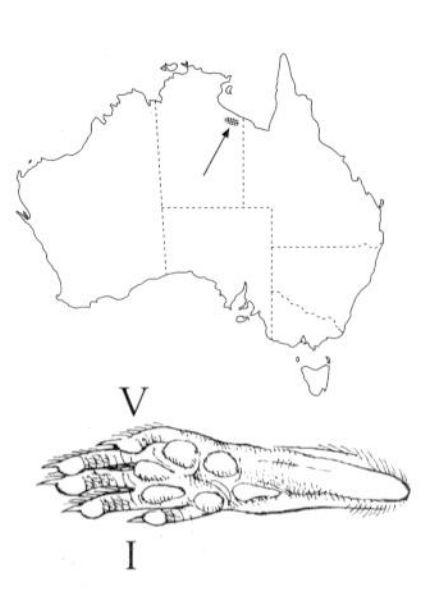

Kimberley Rock Rat *Zyzomys woodwardi*

hb 105–170 mm; **t** 95–135 mm; **hf** 25–30 mm; **el** 17–22 mm; **wt** 80–210 g; **teats** in 2 = 4
Upperparts cinnamon flecked with dark brown hairs, especially on forehead. Underparts whitish. Tail < hb; thick at base tapering smoothly, but often broken; moderately haired but scales visible, olive-brown above, whitish below. **Distribution, habitat and status** Patchily distributed in high-rainfall areas of n. Kimberley e. to Cambridge Gulf, and on numerous offshore islands. Occupies rugged boulder stacks and rock screes in a range of vegetation types, from monsoon forest to pandanus thickets and open woodland with hummock grasses. **Behaviour** Mostly nocturnal. Feeds on seeds from rainforest trees and grasses. Births can occur in any month. **Similar species** Overlaps in distribution only with Common Rock Rat, which is clearly smaller and has tail > hb.

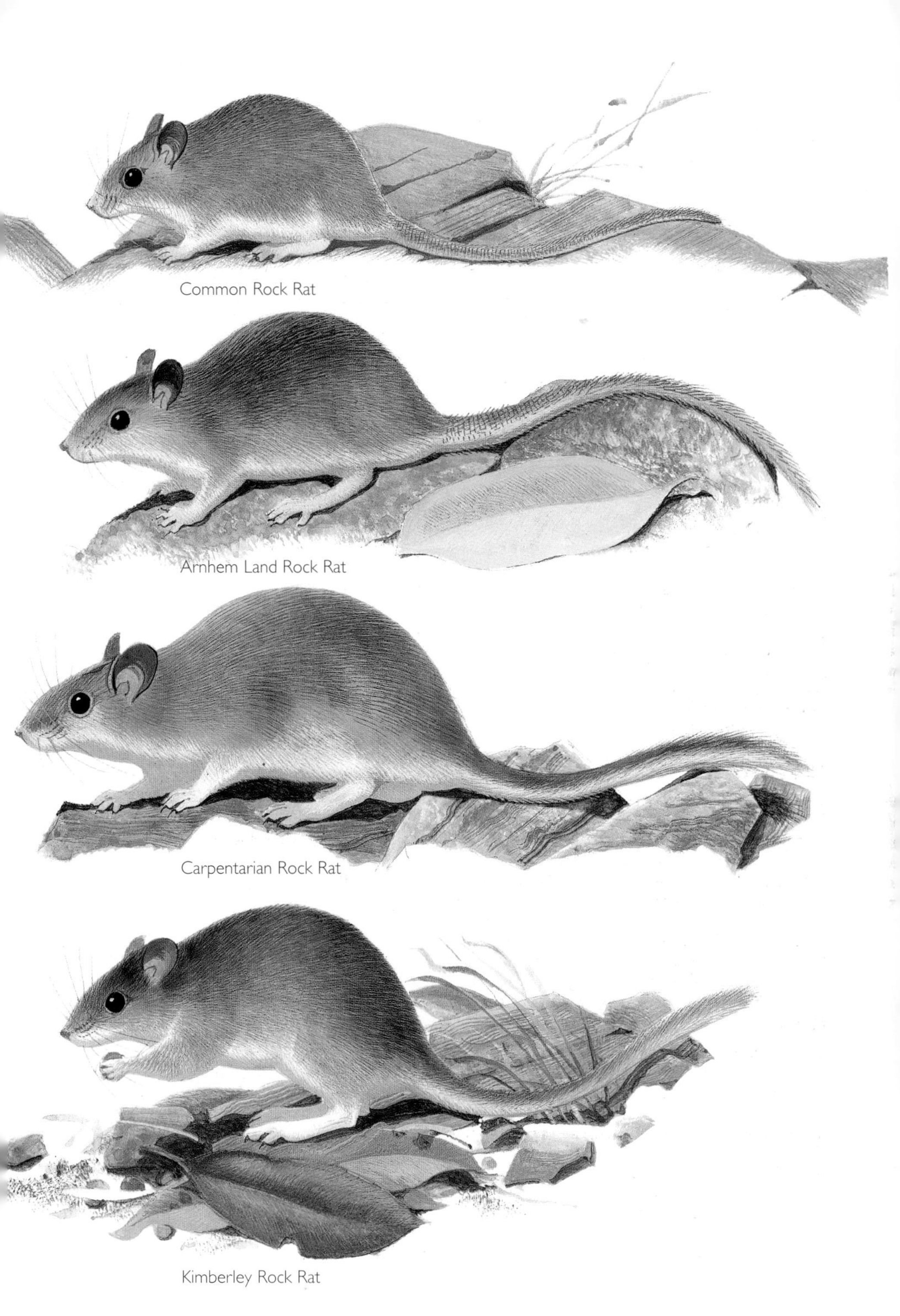
Common Rock Rat
Arnhem Land Rock Rat
Carpentarian Rock Rat
Kimberley Rock Rat

PLATE 70

Central Rock Rat *Zyzomys pedunculatus*

hb 110–124 mm; **t** 116–128 mm; **hf** 25–28 mm; **el** 20–23 mm; **wt** 50–80 (60) g; **teats** in 2 = 4
Fur thick and coarse, yellowish brown above, buff on flanks. Underparts cream or pale buff, including upper surfaces of feet. Tail about = hb, thick, well-furred so that scales not obvious, pale brown and distinctly tufted at tip. Distinct 'Roman nose' profile. **Distribution, habitat and status** Critically endangered. Confined to central Aust. Thought to be extinct until rediscovered in West MacDonnell Ranges in 1996. Early records from James Range, Davenport Range, Napperby Hills and The Granites, Tanami Desert. **Behaviour** Unknown. **Similar species** None within range.

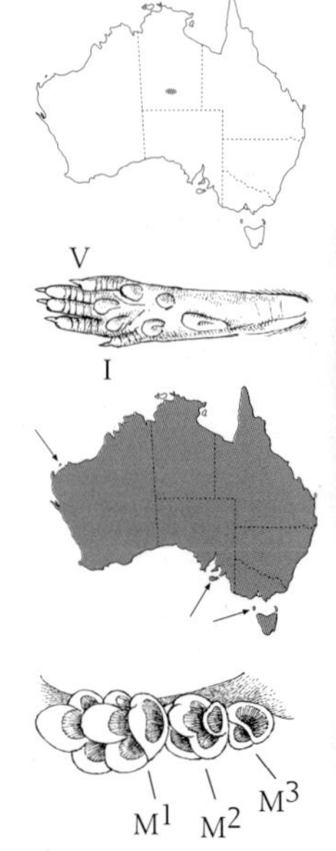

House Mouse *Mus musculus*

hb 60–90 mm; **t** 75–90 mm; **hf** 14–18 mm; **el** 10–15 mm; **wt** 10–25 g; **teats** pe 1, th 2, in 2 = 10
Colour variable; upperparts yellow-brown, grey-brown or mid grey; underparts pale grey-brown, pale yellow or whitish. Tail = or slightly > hb, virtually naked with clear rings of scales and sparse short hairs, pinkish brown. Ears large, rounded. Distinctive musty odour. Aggressive when handled. I^1 has occlusal notch on inner surface. **Distribution, habitat and status** Introduced, probably by Europeans, now throughout the continent, Tas and numerous islands. Common in buildings, cropland and open pasture, also recently burnt forest, scrub, heath; well adapted to arid country. In n. Aust does not occur away from human habitation. Populations fluctuate enormously according to climate and food availability. **Behaviour** Mostly nocturnal, shelters in grass nest in shallow burrow or beneath solid cover. Omnivorous, eats seeds, green shoots, fungi, insects and almost all human food. Breeds opportunistically, but mostly Oct–Apr. Females can produce up to 9 litters of 4–6 young per year. **Similar species** Small *Pseudomys* are superficially similar but lack musty odour, lack notch on inner surface of I^1, tail usually paler underneath, females have 2 pairs of teats not 5.

M1 M2 M3

M^1 longer than M^2 and M^3 combined

Desert Short-tailed Mouse *Leggadina forresti*

Forrest's Mouse

hb 70–100 mm; **t** 45–70 mm; **hf** 14–19 mm; **el** 11–14 mm; **wt** 15–25 g; **teats** in 2 = 4
Tail distinctly < hb. Thickset with broad, blunt muzzle. Upperparts lustrous pale yellow-brown or greyish fawn pencilled with darker hairs; small whitish patches behind ears. Sharp demarcation to white underparts including lower jaw, chin and feet. Tail short and thick (<70% of hb); grey above, pale grey below, sparsely-haired. Ears small, rounded, pinkish-grey. **Distribution, habitat and status** Widespread but sparse over much of arid inland Aust from e. WA to Rolleston in central Qld, and from Barkly Tableland in NT s. to Cockburn, SA. Habitats include riparian Coolabah forest, tussock grassland, stony saltbush plains, and spinifex-clad hillsides. **Behaviour** Nocturnal, shelters in grass nest in burrow. Omnivorous, eats seeds, green plant stems, arthropods. Breeds after rain, litter size 3 or 4. **Similar species** Tropical Short-tailed Mouse very similar but more tropical distribution, more greyish upperparts merge into whitish underparts, upper incisors point slightly forward rather than downward.

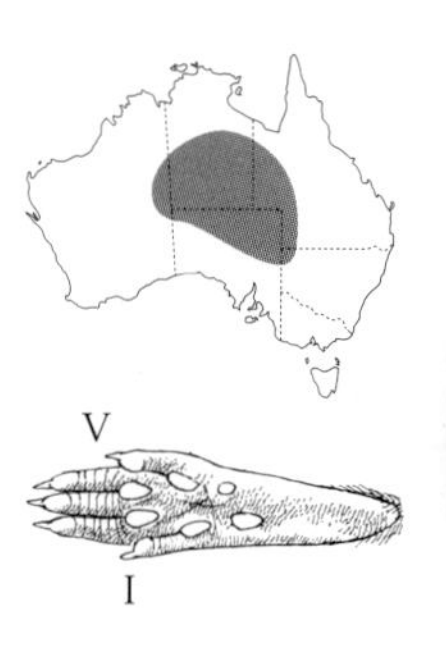

Tropical Short-tailed Mouse *Leggadina lakedownensis*

Lakeland Downs Mouse

hb 65–78 mm; **t** 40–50 mm; **hf** 14–16 mm; **el** 11–13 mm; **wt** 15–25 g; **teats** in 2 = 4
Similar to Desert Short-tailed Mouse but upperparts brindled grey-brown grading to whitish underparts and feet. Sometimes faint darker stripe between ears and pale eye-ring. Upper incisors angled slightly forward. **Distribution, habitat and status** Tropical coastal n. Aust from the Pilbara in WA to e. coast of Cape York between Princess Charlotte Bay and about Laura, and s. to Tennant Creek and The Granites, NT; isolated records w. of Paluma, Qld. Also Thevenard I. off Onslow, WA. Mostly in moist tussock grassland or tropical savannah but stony hummock grassland in Pilbara. **Behaviour** Little known; aggressive when handled. **Similar species** See under Desert Short-tailed Mouse.

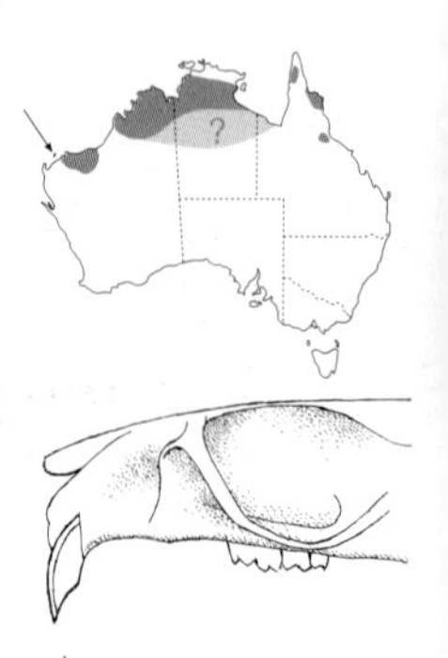

I^1 angled forward

Central Rock Rat

House Mouse

Desert Short-tailed Mouse

Tropical Short-tailed Mouse

PLATE 71

Delicate Mouse *Pseudomys delicatulus*

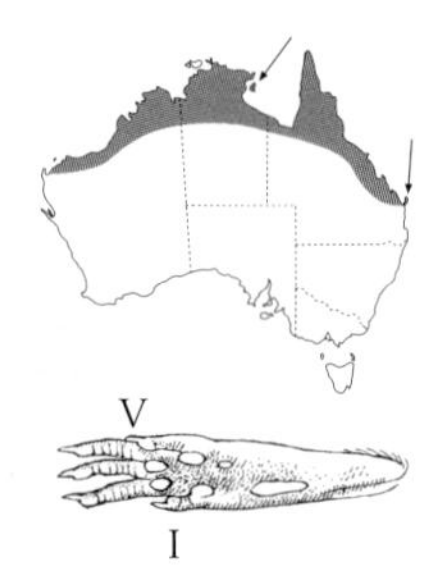

hb 50–62 mm; **t** 61–84 mm; **hf** 15–17 mm; **el** 11–13 mm; **wt** 6–12 (8) g; **teats** in 2 = 4; greatest skull length 20.5 mm.

Very small and delicately built. Upperparts yellow-brown to buffy-grey, warmer on face and flanks. Narrow dark brown ring around large, bulging eye. Underparts white or cream, including cheeks and lower muzzle. Feet and nose pink. Tail thin, slightly > hb, pale brown above, whitish below. Sole of hindfoot has small granules and hairs between and around posthallucal pads. **Distribution, habitat and status** Locally abundant across tropical coastal and subcoastal Aust from Pilbara coast in WA to Fraser I., Qld; s. to Purnululu NP, in WA and Newcastle Waters, NT; inland to Emerald, Qld. Found in open country with friable soils and sparse ground cover, such as sand dunes, sparse grassland or spinifex. **Behaviour** Nocturnal, terrestrial. Eats mostly seeds, fruit, plant stems, some insects. Shelters in simple burrow, or hollow log. Can breed in any month if conditions allow and produce a litter of 2–3 young every 5 weeks. **Similar species** Kimberley Pebble-mound Mouse is slightly larger, more brightly coloured, has larger greatest skull length (>22 mm). Sandy Inland Mouse is slightly larger with ear >12 mm and occurs in more arid areas, both it and Pilbara Pebble-mound Mouse lack granules and hairs between the posthallucal pads.

Kimberley Pebble-mound Mouse *Pseudomys laborifax*

hb 58–75 mm; **t** 63–85 mm; **hf** 16–17 mm; **el** 12 mm; **wt** 9–17 g; **teats** in 2 = 4; greatest skull length 22.2–23.5 mm.

Upperparts orange-brown, pencilled with black guard hairs, grading to cinnamon-buff on flanks and face. Sharp demarcation to white underparts including sides of mouth and feet. Tail > hb; thin, grey-olive above, whitish below. **Distribution, habitat and status** Widespread in Kimberley from moist nw. to arid Ord R. region, including Purnululu NP and Victoria R. district, NT. Occupies wide range of habitats including rocky plateaux and valley floors with sand or clay-loam soils. Vegetation mostly open woodland with dense grass cover. **Behaviour** Births probably mostly in dry season (Aug–Nov). **Similar species** Delicate Mouse is smaller (usually < 10 g), duller coloured, shorter skull. Sandy Inland Mouse can probably only be distinguished by skull measurements but not known to overlap in distribution. Western Chestnut Mouse is much larger.

Western Chestnut Mouse *Pseudomys nanus*

hb 90–140 mm; **t** 80–120 mm; **hf** 23–27 mm; **el** 14–16 mm; **wt** 25–50 (35) g; **teats** in 2 = 4

Large and stoutly built; short limbs and ears; tail clearly < hb. Upperparts orange-fawn heavily pencilled with long darker hairs; distinct pale chestnut eye-ring. Underparts white grading into rufous flanks. Tail dark brown above, whitish below, scales clearly visible. **Voice** Frequently gives high-pitched whistling call. **Distribution, habitat and status** Common in tropical n. Aust from about Port Hedland in WA to Barkly Tableland in nw. Qld, s. to The Granites, Tanami Desert, NT. Also South-west I., Sir Edward Pellew Group and subspecies *ferculinus* on Barrow I., WA. Inhabits woodlands with a shrubby or dense tussock grass understorey. Often along watercourses but also far from water at times. **Behaviour** Confiding and readily observed; mostly nocturnal. Eats mainly grass stems. Breeding opportunistic, populations can increase rapidly when conditions permit. **Similar species** Desert Mouse is less 'neat': long dark guard hairs give spiky appearance, has pale grey-brown underparts, tail not clearly bicoloured.

Calaby's Pebble-mound Mouse *Pseudomys calabyi*

Kakadu Pebble-mound Mouse

hb 75–95 mm; **t** 70–95 mm; **hf** 17–19 mm; **el** 13–14 mm; **wt** 15–24 g; **teats** in 2 = 4

Head long, flat; upperparts and face grey-brown, greyest on crown, grading to sandy flanks and rump, then sharp demarcation to white underparts including chin, sides of mouth and tops of feet. Eyes large; tail = hb, pinkish brown with black hairs along upper surface. **Distribution, habitat and status** Known only from headwaters of South Alligator and Mary Rivers, in s. of Kakadu NP, and from Litchfield NP, but may be more widespread. Occupies gravelly slopes with tall grass understorey. **Behaviour** Little known. Shelters in burrow system with the entrances surrounded and blocked by a small mound of pebbles. Eats mainly grass seeds. **Similar species** Delicate Mouse is smaller, has small granules and hairs between the post-hallucal pads.

Delicate Mouse

Kimberley Pebble-mound Mouse

Western Chestnut Mouse

Calaby's Pebble-mound Mouse

PLATE 72

Desert Mouse *Pseudomys desertor*

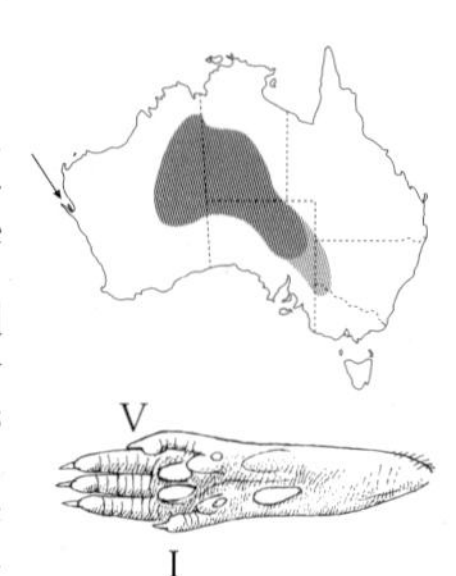

hb 70–105 mm; **t** 70–105 mm; **hf** 20–22 mm; **el** 11–14 mm; **wt** 15–30 (24) g; **teats** in 2 = 4
Fur long, with obvious guard hairs giving a spiny appearance. Ears short, hf long. Upperparts uniformly warm mid-brown pencilled with darker guard hairs, grading to pale grey underparts including upper lip and chin. Prominent buff-orange eye-ring. Tail = hb, brown above, whitish or cream below, scales readily visible. **Distribution, habitat and status** Widely distributed and locally common in arid central Aust, n. to Purnululu NP, WA, s. to Warburton area, WA and Roxby Downs, SA. Relict populations on Bernier Is., WA and Telowie Gorge, Flinders Ranges, SA. Usually associated with sandplains carrying mature hummock grass, or samphire, nitrebush and sedges in Lake Eyre region. Less common on scree slopes in the ranges with hummock grass. **Behaviour** Partly diurnal; shelters in shallow burrow, usually alone; eats mostly green stems and seeds; breeding is opportunistic when conditions suitable. **Similar species** Western Chestnut Mouse may overlap in the n., but is larger (> 25 g), neater, less obvious orange eye-ring, whiter underparts, more clearly bicoloured tail.

Bolam's Mouse *Pseudomys bolami*

hb 57–77 mm; **t** 89–96 mm; **hf** 18–20 mm; **el** 15–18 mm; **wt** 10–21 g; **teats** in 2 = 4
Upperparts olive-brown to yellow-brown pencilled with black guard hairs. Sharp transition to whitish underparts, feet, throat and sides of mouth. Tail clearly > hb, well furred, brownish above, paler below. Ears long and rounded. **Distribution, habitat and status** Across semi-arid and arid southern Aust from Woolgangie and Norseman, WA e. to the Murray Mallee of SA and Nanya Stn., sw. NSW. Usually found in lower parts of the landscape in floodout country with calcareous soils and chenopod vegetation. Also mallee shrubland with well-developed shrub layer. **Behaviour** Little known. Nocturnal; omnivorous, eating seeds, green shoots and some insects. **Similar species** Sandy Inland Mouse has smaller ears (<16 mm long and <11 mm wide), shorter hind feet (<18.5 mm), warmer brown dorsal pelage and tends to occur on sandy rather than clay-based soils.

Sandy Inland Mouse *Pseudomys hermannsbergensis*

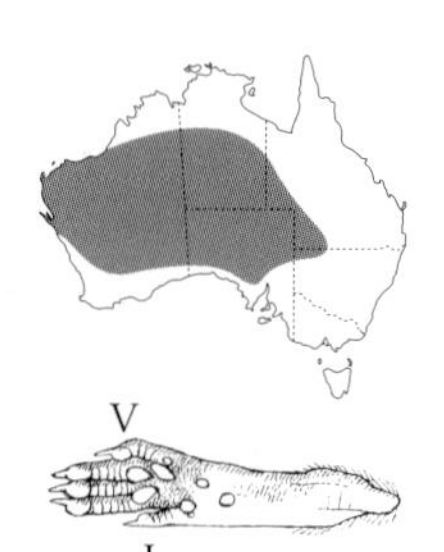

hb 62–77 mm; **t** 73–90 mm; **hf** 16–18 mm; **el** 13–15 mm; **wt** 9–17 (13) g; **teats** in 2 = 4
Slender and lightly-built; smooth haired. Upperparts pale ginger to grey-brown pencilled with short blackish guard hairs, flanks more rufous. Underparts white, including sides of mouth, chin and upper surfaces of feet. Tail > hb, sparsely-haired, pinkish brown above, slightly paler below. **Distribution, habitat and status** Common across much of arid Aust from the WA coast to near Hungerford, Qld; generally rare e. of Stuart Highway. Inhabits a wide variety of vegetation, mostly fairly open and on friable soil, including dune swales and loamy flats. **Behaviour** Nocturnal, gregarious, groups shelter in complex burrow system. Omnivorous, eats seeds, tubers, green shoots, arthropods. Breeding is largely opportunistic following rain, populations fluctuate widely. **Similar species** Central Pebble-mound Mouse difficult to distinguish but ratio of hf to el is 1.45 vs 1.25 for Sandy Inland Mouse. House Mouse has smaller ears, notch on inner surface of I^1 and characteristic musty odour. See also under Bolam's Mouse and Pilbara Pebble-mound Mouse.

Central Pebble-mound Mouse *Pseudomys johnsoni*

hb 61–74 mm; **t** 76–95 mm; **hf** 17–18.5 mm; **el** 12–13 mm; **wt** 9–17 (12) g; **teats** in 2 = 4
Upperparts pale yellow-brown to rufous-brown pencilled with black guard hairs. White underparts strongly demarcated from upperparts. Tail > hb, brown above, whitish below. Ratio of hf to el 1.45. **Distribution, habitat and status** Recorded only from the Davenport and Murchison Ranges, se. of Tennant Creek in NT, and the Mittiebah Range near Alexandria Station, Barkly Tableland, NT. Occupies gravelly slopes with hummock grasses and grevilleas. **Behaviour** Lives in complex burrow systems with piles of excavated stones dumped near one entrance, which is later closed. Lactating females have been found in Aug–Sept. **Similar species** Sandy Inland Mouse very similar: ratio of hf to el 1.25 in Sandy Inland Mouse, 1.45 in Central Pebble-mound Mouse.

Desert Mouse

Bolam's Mouse

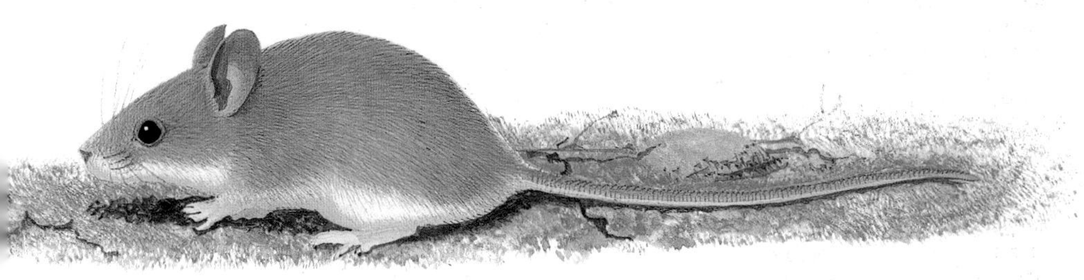
Sandy Inland Mouse

Central Pebble-mound Mouse

PLATE 73

Ash-grey Mouse *Pseudomys albocinereus*

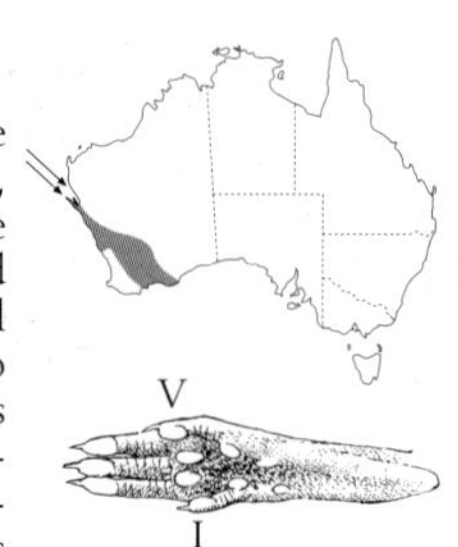

hb 70–100 mm; **t** 85–110 mm; **hf** 20–25 mm; **el** 17–19 mm; **wt** 15–40 g; **teats** in 2 = 4
Fur long, soft; upperparts mid-grey, sometimes mottled with fawn; extensive white areas on muzzle and below eyes; nose and feet pink. Tail slightly > hb, thin, sparsely haired, narrow dark brown line along upper surface near base, otherwise pale pink. In e. of distribution animals are larger (to 60 g) with darker fur and tail. Underside of hindfoot highly granulated, all interdigital and posthallucal pads smaller than terminal pads. **Distribution, habitat and status** Confined to sw. WA from Shark Bay to Israelite Bay, also Bernier and Dorre Is (subspecies *squalorum*), Dirk Hartog I. Widespread but not common in remnants of sand-plain heath and tall shrubland with tussock-grass understorey. In e. of distribution also in mallee shrubland. **Behaviour** Gentle and docile; nocturnal; shelters both in deep burrow systems and in surface nests in hollow log or deep litter. Omnivorous, eats green plant material, seeds, arthropods. Capable of climbing low shrubs when foraging. In w. of range births occur in spring, normally only 1 litter per year; in e. more opportunistic. **Similar species** Western Mouse overlaps in w., has proportionately longer tail that is grey not pink, buff-toned fur, interdigital pads larger than terminal pads, and underside of hf not highly granulated. House Mouse has smaller ears, notch on inner surface of I^1, and characteristic musty odour.

Western Mouse *Pseudomys occidentalis*

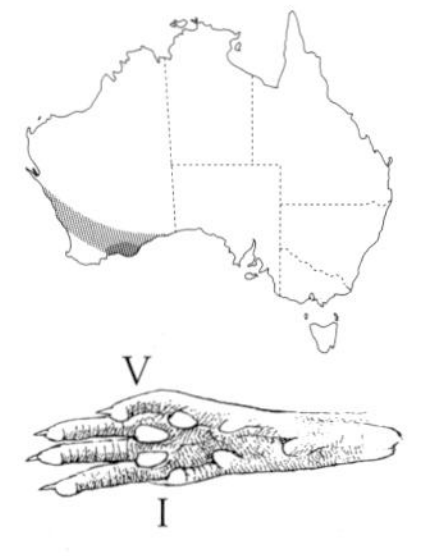

hb 88–110 mm; **t** 120–140 mm; **hf** 24–28 mm; **el** 18–20 mm; **wt** 30–55 g; **teats** in 2 = 4
Distinct 'Roman' nose and long tail. Upperparts dark grey with buff tinges and dark guard hairs; underparts, including feet, white or pale grey. Tail bicoloured: pale grey with narrow dark brown line along upper surface. Hindfoot long for a *Pseudomys*; interdigital pads larger than terminal pads and not highly granulated. **Distribution, habitat and status** Formerly in diagonal band across sw. WA and s. Nullarbor Plain; now confined to Ravensthorpe Range, Fitzgerald R. NP and several smaller reserves in the southern wheatbelt or s. of there. Prefers long-unburnt dense shrublands growing on gravelly soil. **Behaviour** Timid; nocturnal; gregarious; shelters during day in burrow. Eats mostly seeds, plant stems, flowers, fruit, some arthropods; readily climbs low shrubs when foraging. **Similar species** See under Ash-grey Mouse and Djoongari.

Djoongari *Pseudomys fieldi*

Shark Bay Mouse, Alice Springs Mouse

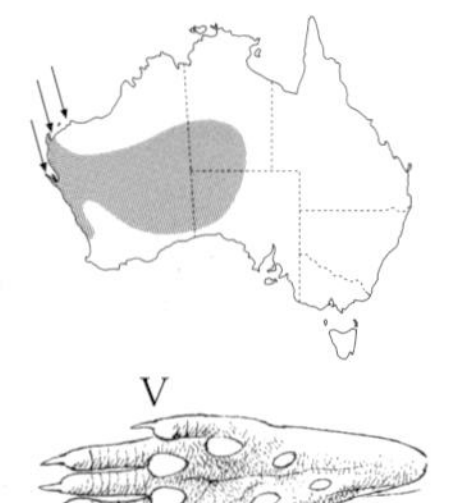

hb 90–115 mm; **t** 115–125 mm; **hf** 26–27 mm; **el** 19 mm; **wt** 30–50 (45) g; **teats** in 2 = 4
Large, shaggy-haired. Upperparts pale yellow-fawn grizzled with dark brown guard hairs, grading through buff flanks to whitish underparts and feet. Ears greyish. Tail slightly > hb, well furred, bicoloured, grey above and white below with a dark terminal tuft. **Distribution, habitat and status** Formerly found over most of the sw. quadrant of Aust except the WA wheatbelt. Now confined to Bernier I. and recently introduced to Doole I. in Exmouth Gulf and Trimouille I. in Montebello Group. On Bernier is most common on dunes covered in spinifex, less abundant in heath on island's plateau. Vulnerable. **Behaviour** Studied only on Bernier I. Constructs nests in a variety of surface cover, including piles of beachcast seagrass, and in shallow burrows. Mostly herbivorous, eats green plant material, flowers, some arthropods. Young born May–Nov. **Similar species** Western Mouse more grey, less shaggy, proportionately longer tail.

Pilbara Pebble-mound Mouse *Pseudomys chapmani*

Western Pebble-mound Mouse

hb 52–68 mm; **t** 63–94 mm; **hf** 15–17 mm; **el** 9–12 mm; **wt** 8–17 g; **teats** in 2 = 4
Tiny with long, flat head, small eyes, narrow muzzle, short ears. Upperparts buff-brown, greyer on crown and back. Underparts whitish. Tops of feet buff, posthallucal pads almost equal in size and slightly larger than terminal pads. Tail clearly > hb and uniformly pale pinkish-brown. **Distribution, habitat and status** Confined to central and e. Pilbara in WA, including Karijini NP. Formerly on Burrup Pen. and s. to Murchison district. Found on stony hillsides with hummock grassland. **Behaviour** Shelters in complex burrow systems built beneath mounds of pebbles collected from the surface. **Similar species** Delicate Mouse has small granules and hairs between the posthallucal pads. Sandy Inland Mouse has posthallucal pads half the size of terminal pads, feet and ears longer.

Ash-grey Mouse
Western Mouse
Djoongari
Pilbara Pebble-mound Mouse

PLATE 74

Silky Mouse *Pseudomys apodemoides*

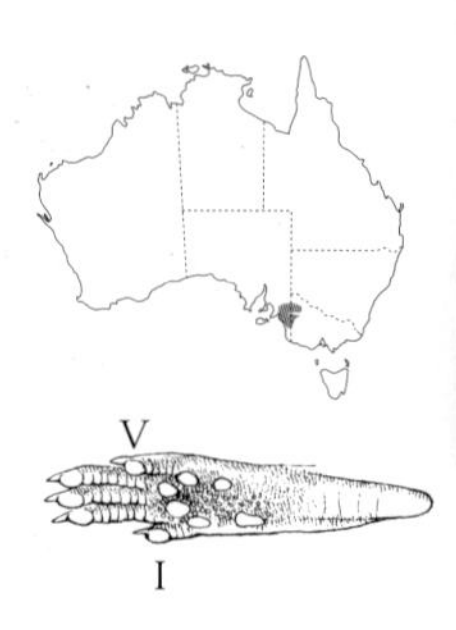

hb 68–80 mm; **t** 90–105 mm; **hf** 20–23 mm; **el** 16–18 mm; **wt** 15–22 g; **teats** in 2 = 4
Fur soft and dense; upperparts smoky grey mottled with buff and interspersed with dark guard hairs; ears pinkish grey; underparts, including sides of muzzle and lips, white. Tail > hb; pale pink with white hairs, brownish on upper surface; when disturbed carried horizontally or arched over back. **Distribution, habitat and status** Confined to the southern Murray Mallee from about Keith, se. SA to the e. edges of the Big and Little Deserts, Vic and s. to near Dergholm, Vic. Locally common in low semi-arid heath growing on deep sands. Prefers heath patches 3–10 years after a fire. **Behaviour** Shelters in deep burrow system with several entrances concealed beneath dense vegetation, often Desert Banksia. Omnivorous, eats seeds, fruits, flowers, nectar, fungi, arthropods. Breeds opportunistically when conditions are favourable. **Similar species** House Mouse brownish not grey and white, small brown ears, notched inner surface of I^1, distinct musty odour. Mitchell's Hopping Mouse yellow-brown, distinctly longer hind legs and feet (hf >34 mm), longer tufted tail, very long ears (>24 mm).

Blue-grey Mouse *Pseudomys glaucus*

hb 95 mm; **t** 100 mm; **wt** 25–30 g
Robust build; dense, fine fur. Upperparts pale blue-grey, underparts white. Tail slightly > hb, white haired. **Distribution, habitat and status** Presumed extinct. Known from only 3 specimens: 2 from Qld, one from Cryon, near Walgett in n. NSW. Last recorded 1956. Habitat and behaviour unknown.

Smoky Mouse *Pseudomys fumeus*

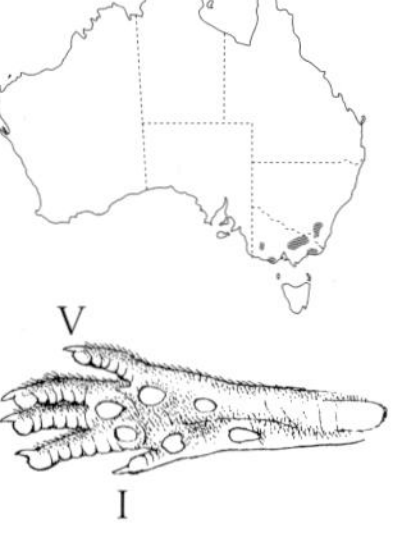

East of Melbourne: **hb** 95–118 (107) mm; **t** 105–130 (116) mm; **hf** 25–28 (26) mm; **el** 18–21 (20) mm; **wt** 26–46 (35) g. Grampians: **hb** 112–136 (122) mm; **t** 115–150 (132) mm; **hf** 26–29 (28) mm; **el** 19–22 (21) mm; **wt** 50–86 (65) g; **teats** in 2 = 4
Specimens from e. of Melbourne are soft mid-grey above, pale smoke grey below. (Those from the Grampians in w. Vic are larger and darker, slate grey above grading to greyish white below.) The muzzle is darker, there is a narrow blackish eye-ring and black guard hairs over upperparts. Feet pale pink with white hairs. Ears long, pinkish grey. Tail clearly > hb, pale pink with brownish stripe along top. **Distribution, habitat and status** Population fragmented: Grampians, w. Vic; Otway Range (possibly extinct there); GDR between upper Yarra R. ne. of Melbourne and Bulls Head, ACT; and coastal e. Gippsland between Marlo and Wingan Inlet. Sparse and patchy in dry sclerophyll forest on ridges with heath and tussock-grass understorey, coastal heath and subalpine heath. Endangered. **Behaviour** Shelters communally in surface nest. Eats seeds, fruit, fungi, arthropods. Breeds in late summer (Dec–Feb). When agitated elevates tail and slowly undulates it. **Similar species** None within range.

Heath Mouse *Pseudomys shortridgei*

Heath Rat, Dayang (WA)

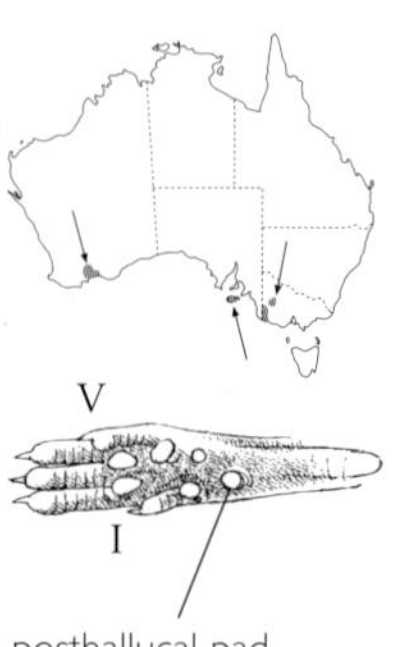

posthallucal pad rounded

hb 95–120 mm; **t** 85–100 mm; **hf** 25–27 mm; **el** 14–16 mm; **wt** 55–90 (70) g; **teats** in 2 = 4
Thickset with dense fur, broad face, short muzzle, short rounded ears. Upperparts uniformly warm brown, tipped buff; numerous dark guard hairs. Underparts pale grey, upper surface of feet covered with long grey hairs. Tail < hb, well haired, dark brown above, whitish below. **Distribution, habitat and status** Widely separated populations; one in lowland heath and heathy sclerophyll forest in sw. Vic between Nelson, Dergholm and Mt Clay, and in the Grampians; another of unknown status on Kangaroo I., SA; and in sw. WA in Ravensthorpe Range, Fitzgerald R. NP and Dragon Rocks NR where prefers long-unburnt tall heath and scrubby mallee. Vulnerable. **Behaviour** Docile, confiding when handled; partly diurnal; shelters in surface nest or shallow burrow; eats seeds and flowers when available, otherwise grass stems and fungi. Births occur Nov–Jan and 1–2 litters of 2–3 young may be raised. **Similar species** Bush Rat has unicoloured pinkish brown tail with obvious rings of scales, pink feet, posthallucal pad is elongated not round. Swamp Rat and Broad-toothed Rat have uniformly dark scaly tails, dark soles, brown fur on upper surface of feet.

Silky Mouse
Blue-grey Mouse
western form
eastern form
Smoky Mouse
Heath Mouse

PLATE 75

Hastings River Mouse *Pseudomys oralis*

hb 125–165 mm; **t** 120–160 mm; **hf** 28–34 mm; **el** 16–22 mm; **wt** 55–100 (92) g; **teats** in 2 = 4
Fur long, soft; muzzle short with distinct snub-nosed appearance. Upperparts grizzled brownish-grey and buff grading to whitish or buff underparts; narrow dark eye-ring. Upper surface of feet whitish. Tail = hb, well haired, bicoloured: brown above, cream below. **Distribution, habitat and status** Very patchy distribution in the ranges of ne. NSW and se. Qld from Lamington, Qld to Barrington Tops, NSW. Found at mid to high altitudes (to 1200 m), mostly in damp, dense fern or sedge understorey along drainage lines, but also utilises drier areas with grassy or heathy ground cover. Endangered. **Behaviour** Docile when handled. Builds nest in hollow log or other low cavity. Eats mostly seed and leaf in summer, green stem in winter, with smaller amounts of fungi, flowers, arthropods. Breeds Aug–Mar; up to 2 litters of 2–3 young may be raised. **Similar species** Eastern Chestnut Mouse occurs at lower altitudes, lacks obviously bicoloured tail, has grey not white hairs on feet. Bush Rat has pinkish-brown feet: tail unicoloured, near-naked, scaly; females have 10 teats.

Eastern Chestnut Mouse *Pseudomys gracilicaudatus*

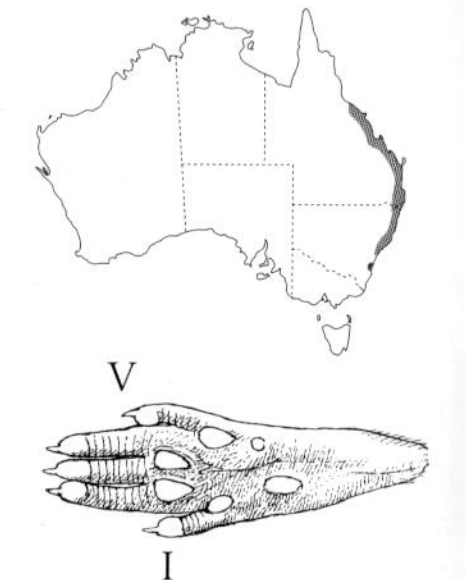

hb 100–145 mm; **t** 85–120 mm; **hf** 25–29 mm; **el** 15–18 mm; **wt** 45–115 (75) g; **teats** in 2 = 4
Large and compact with dense spiky fur. Upperparts uniformly warm brown, tipped buff, warmer on flanks. Underparts dark grey tipped whitish. Upper surface of feet covered with long grey hairs. Tail < hb, unicoloured dark brown, sparsely haired. **Distribution, habitat and status** Patchily distributed in coastal e. Aust, from Townsville in Qld to Brisbane Water NP, NSW; outlying population at Jervis Bay, NSW. Inhabits range of vegetation from grassy and heathy open-forests to heath and swampy depressions. Highest densities in heath reached 3–4 years after fire. Uncommon. **Behaviour** Mostly nocturnal, uses runways through dense sedge and grass ground cover. Omnivorous, eats seeds, fungi, green stem, arthropods. Breeds Sept–Mar; up to 3 litters may be reared. **Similar species** Bush Rat has pinkish-brown feet, unicoloured, near-naked, scaly tail, females have 10 teats. Swamp Rat is darker, particularly tail and soles of hindfoot. Hastings River Mouse has clearly bicoloured tail.

Pilliga Mouse *Pseudomys pilligaensis*

hb 73–80 mm; **t** 70–80 mm; **hf** 18–19 mm; **el** 12–13 mm; **wt** 10–14 g; **teats** in 2 = 4
Small, upperparts grey-brown, greyer on head and back, grading through russet flanks to white underparts. Upper surface of feet pale pink with white hairs. Tail slightly < or = hb, pale pink with distinct narrow brown line along top and small dark tuft. **Distribution, habitat and status** Known only from Pilliga Scrub between Baradine and Narrabri, ne. NSW. Occurs in mixed woodlands of cypress-pine, eucalypt and she-oak with sparse shrub and ground cover, but often close to creeks with dense stands of callistemon and tea-tree. **Behaviour** Little known, nocturnal, breeds spring and summer. **Similar species** New Holland Mouse is larger, hf > 20 mm, t > 80 mm and not known to overlap. House Mouse has shorter ears, unicoloured pale brown tail, notch on inner surface of I^1, distinctive odour.

Eastern Pebble-mound Mouse *Pseudomys patrius*

hb 56–78 mm; **t** 63–80 mm; **hf** 18 mm; **el** 12 mm; **wt** 12–17 (15) g; **teats** in 2 = 4
Small; head long and flat, incisors broad. Upperparts yellow-brown, heavily pencilled with black guard hairs, grading through rich buff flanks to whitish underparts, including sides of mouth so that below the eye and ear there is an abrupt change from buff to white. Upper surface of feet pale pink with short white hairs. Tail sparsely haired, uniformly pale pinkish-brown, rings of scales clearly visible. **Distribution, habitat and status** Patchy and poorly known. Found in broad semicircle in e. Qld from Paluma and Charters Towers w. to Burra Range and Clermont and se. to Springsure and Gympie. Inhabits dry rocky ridges carrying grassy woodlands. Vulnerable. **Behaviour** Unknown. Constructs mounds of pebbles into which several vertical burrows disappear. **Similar species** Delicate Mouse has shorter muzzle, narrower incisors, shorter head (<20.5 mm).

Hastings River Mouse

Eastern Chestnut Mouse

Pilliga Mouse

Eastern Pebble-mound Mouse

PLATE 76

New Holland Mouse *Pseudomys novaehollandiae*

M^1 not as long as M^2 and M^3 combined

hb 65–90 mm; **t** 80–105 mm; **hf** 20–22 mm; **el** 15–18 (12 in Tas) mm; **wt** 15–25 (18) g; **teats** in 2 = 4
Upperparts uniformly grizzled grey-brown, tipped pale brown, grading through buff flanks to pale grey underparts. Upper surface of feet white. Tail > hb, bicoloured: pale brown above, whitish below, darker at tip. **Distribution, habitat and status** Patchy in coastal e. Aust from Evans Head in NSW to Anglesea, Vic; also Flinders I., n. and e. Tas. In dry coastal heath or heathy sclerophyll forest where understorey <10 years old. Also further inland (to 100 km) and higher (to 600 m) in ne. NSW and se. Qld in dry sclerophyll forest often with little ground or shrub cover (may be separate taxon). Endangered in Vic, locally common in NSW. Type specimens are from Gwydir R., NSW, in very different habitat to that now occupied. **Behaviour** Nocturnal, gregarious, shelters in burrow systems up to several metres long. Omnivorous, eats seeds, green stems, fungi, arthropods. Breeds through spring and early summer, several litters of 2–6 young may be raised. **Similar species** House Mouse is smaller, has unicoloured pale brown tail which is not > hb, shorter ears (except in Tas), notch on inner surface of I^1, distinctive odour.

Plains Mouse *Pseudomys australis*

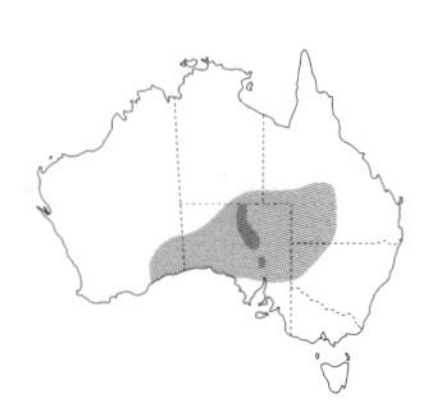

hb 100–140 mm; **t** 90–120 mm; **hf** 25–30 mm; **el** 20–25 mm; **wt** 40–75 g; **teats** in 2 = 4
Largest of the arid zone *Pseudomys* (>40 g). Stocky build, rounded muzzle, long ears. Upperparts lustrous sandy-grey or silver-grey, pencilled with darker guard hairs, grading through buff flanks to white or cream underparts. Tail usually < hb, clearly bicoloured, dark brown above, whitish below, paler towards tip. **Distribution, habitat and status** Formerly widespread across arid southern Aust. Now known only from a 600 km north–south band w. of Lake Eyre, with an outlying population w. of Lake Torrens. Inhabits open country primarily on clay-based soils and gibber plains, spreading to adjacent dunes during population eruptions. Breeds opportunistically following rain; populations fluctuate widely. Vulnerable. **Behaviour** Partly diurnal, gregarious, constructs complex systems of shallow burrows connected by runways. Eats mostly seeds, some green plant stems, arthropods. When alarmed may stand upright on hindfeet and squeal loudly.

Gould's Mouse *Pseudomys gouldi*

hb 100–120 mm; **t** 90–100 mm; **hf** 25–27 mm; **el** 16–18 mm; **wt** approx. 50 g; **teats** in 2 = 4
Similar to Djoongari but tail < hb, underparts pale grey. **Distribution, habitat and status** Extinct. Known only from upper Hunter R., Liverpool Plains and lower Darling R., NSW. Last collected in 1857. **Similar species** Plains Mouse very similar, lacks accessory cusp on M^1, has longer muzzle.

Long-tailed Mouse *Pseudomys higginsi*

hb 115–145 mm: **t** 145–195 mm: **hf** 30–35 mm; **el** 21–25 mm; **wt** 50–90 g; **teats** in 2 = 4
The largest *Pseudomys*. Upperparts uniformly sooty grey to grey-brown with small buff flecks, grading to pale grey underparts. Skin of muzzle and eye-ring darker. Upper surface of feet whitish, soles dark brown; Ears large, rounded, pinkish-grey. Tail long (1.5 × hb), clearly bicoloured: sooty grey above, cream below. **Distribution, habitat and status** Endemic to Tas where widespread in wet sclerophyll forest, rainforest, alpine heath and boulder screes, up to 1600 m, where population densities are highest. **Behaviour** Mostly nocturnal; terrestrial; omnivorous, eats ferns, fruits, seeds, fungi, grass, moss, arthropods; shelters in surface nest in hollow log or other cranny; breeds Oct–Mar, usual litter size is 3. **Similar species** Long, bicoloured tail diagnostic.

New Holland Mouse

Plains Mouse

Gould's Mouse

Long-tailed Mouse

PLATE 77

Broad-toothed Rat *Mastacomys fuscus*

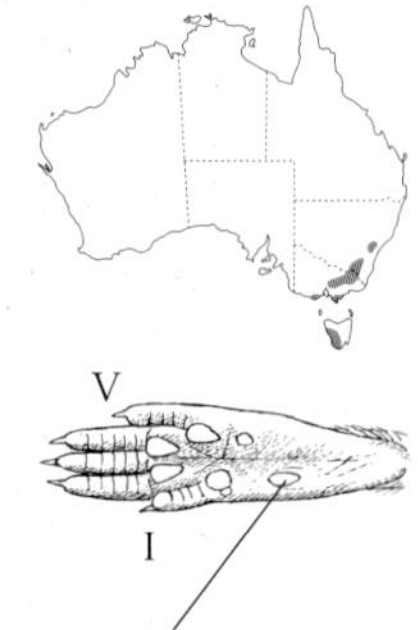

rounded posthallucal pad

hb 145–175 mm; **t** 100–130 mm; **hf** 30–35 mm; **el** 20–23 mm; **wt** 95–145 g; **teats** in 2 = 4
Rotund build; broad, short head; long, fine dense fur; hunched posture. Upperparts dusky brown heavily flecked with yellow and rufous, grading to buff-grey underparts. Feet dusky brown above and below; ears short, broad, rounded. Tail short, uniformly brown, only slightly paler below, sparsely haired, rings of scales visible. Incisors broad; molars large, broad, not graduated in size (M^3 as long as M^1). Young furred at birth. Scats distinctive: large, fibrous, green. **Distribution, habitat and status** Patchy; inhabits alpine sedges and heath in Aust Alps and at Barrington Tops, NSW; also wet sedge and grass patches in forest in Eastern Highlands, South Gippsland Highlands and Otway Range, Vic (down to sea level); and buttongrass sedgeland up to 1000 m in w. Tas. **Behaviour** Nocturnal, shy, builds nests in dense cover, constructs runways through dense ground vegetation and beneath snow. Vegetarian, eats mostly stems of sedges and grasses, some seeds and moss sporangia. Litters of 2–3 born in summer (Nov–Mar). **Similar species** True rats (genus *Rattus*) have elongate not rounded posthallucal pad, give birth to naked, helpless young, females have 8 or 10 teats. Swamp Rat very similar, darker hindfeet, narrower muzzle, narrower incisors, more nervous and elongated posture. Bush Rat has longer, pinkish tail and hindfeet; longer muzzle. Heath Rat not known to overlap, tail clearly bicoloured, hindfeet covered with grey hairs.

Swamp Rat *Rattus lutreolus*

Velvet-furred Rat (Tas)

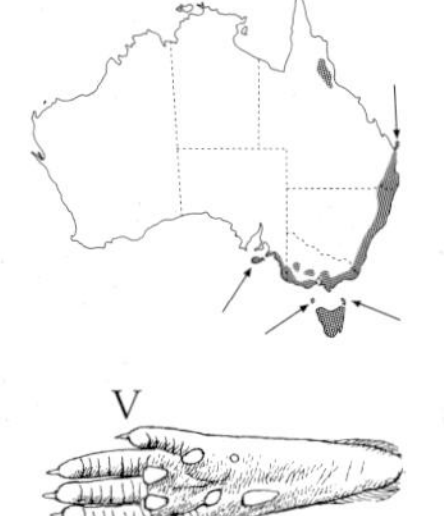

hb 120–200 mm; **t** 80–145 mm; **hf** 27–35 mm; **el** 15–20 mm; **wt** 55–160 (122) g; **teats** pe 1, th 2, in 2 = 10; in Tas th 2, in 2 = 8
Fur of upperparts dense, long; colour varies from reddish-brown through chocolate to blackish; underparts buff-grey. Hindfeet dark brown above, soles blackish; ears short, rounded, dark grey; eyes small, not obvious. Tail clearly < hb; naked, uniformly dark grey or blackish with obvious rings of scales. **Distribution, habitat and status** Lowland se. Aust from Fraser I. in Qld to Mt Lofty Ranges and Kangaroo I., SA, and much of Tas (subspecies *velutinus*), including Furneaux Group and King I. In ne. Qld subspecies *lacus* occurs in isolated patches of high-altitude rainforest between Atherton Tableland and Paluma. Found in wet, dense vegetation including heath, fern thickets, sedgeland, dune scrub, rank grassland including ungrazed pasture. In Tas found in most forest types up to 1600 m. Locally abundant. **Behaviour** Partly diurnal, nocturnal, terrestrial. Forms runways through dense vegetation and extensive shallow burrow systems. Vegetarian, eats mostly stems of grass and sedges. **Similar species** See under Broad-toothed Rat. Bush Rat has pinkish feet, pink-brown tail is about = hb. Heath Mouse has clearly bicoloured tail, feet paler above and below.

Bush Rat *Rattus fuscipes*

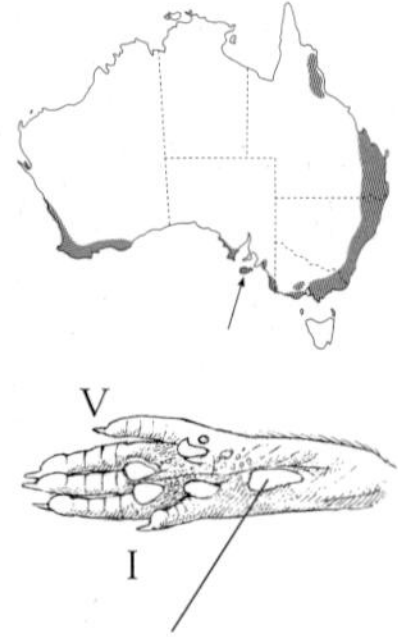

elongated posthallucal pad

hb 100–205 mm; **t** 100–195 mm; **hf** 30–40 mm; **el** 18–25 mm; **wt** 50–225 (125) g; **teats** pe 1, th 2, in 2 = 10; in n. Qld th 2, in 2 = 8.
Highly variable in body size and colouration. Upperparts grey-brown to warm reddish-brown, blending through rufous flanks to grey or cream underparts. Subspecies *coracius* tends to be more reddish-brown. Tail approx = hb, virtually naked, pinkish-brown or grey with obvious rings of overlapping scales. Feet pinkish-brown (darker in subspecies *coracius*) with short pale hairs. Eyes large, prominent. **Distribution, habitat and status** Widespread and common in coastal e. and s. Aust from Rockhampton in Qld to Timboon, Vic (subspecies *assimilis*); from e. of Portland in Vic to Eyre Pen. and Kangaroo I., SA (subspecies *greyii*); sw. WA from Jurien Bay to Israelite Bay (subspecies *fuscipes*); lowland and montane rainforest between Townsville and Cooktown, ne. Qld (subspecies *coracius*). Inhabits most moist vegetation with dense ground cover from sea level to sub-alps. Subspecies *greyii* and *fuscipes* can live in drier habitats. **Behaviour** Nocturnal terrestrial, shy; omnivorous, eats mostly plant tissue, also seeds, fungi, arthropods. Shelters and nests in shallow burrows. Three litters of 5 young can be produced in the 1-year life span. **Similar species** See under Swamp Rat. Cape York Rat overlaps on Atherton Tableland and is difficult to distinguish from subspecies *coracius*; has smaller eyes and narrower muzzle, faint dark eye-rings, female has 6 teats compared to 8. Black Rat has tail clearly > hb and longer ears. Brown Rat is larger, more thickset with thick, coarsely haired tail, commensal (Bush Rat never commensal).

Broad-toothed Rat

Swamp Rat

Bush Rat

PLATE 78

Pacific Rat *Rattus exulans*

Polynesian Rat

hb 80–135 mm; **t** 125–135 mm; **hf** 23–28 mm; **el** 16–19 mm; **wt** 35–100 g; **teats** th 2, in 2 = 8

Pale and lightly built. Fur short and lustrous; upperparts light brown to cinnamon, sharply demarcated from whitish underparts. Rufous wash to face. Upper surface of hindfeet whitish. Tail > hb, grey-brown, sparsely furred. **Distribution, habitat and status** Widespread through the Pacific islands but only 3 Australian records: from Maer I., e. of Torres Strait, and Adele and Sunday Is, off the Kimberley coast, WA. Presumably introduced by humans; current status unknown. **Behaviour** Nocturnal, omnivorous, partly arboreal.

Black Rat *Rattus rattus*

Ship Rat, Roof Rat

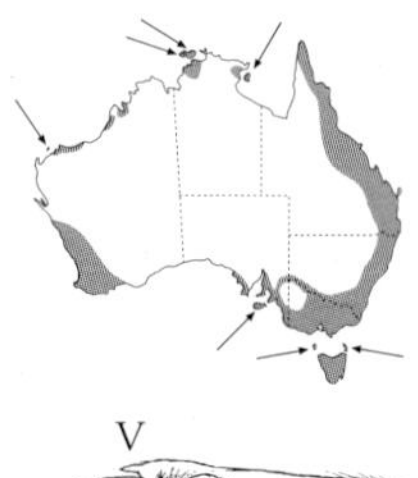

hb 165–220 mm; **t** 185–245 mm; **hf** 30–40 mm; **el** 24–27 mm; **wt** 95–300 g; **teats** pe 1 or 2, th 2, in 2 = 10 or 12

Body slender; fur sleek and shiny; ears reach past middle of eye when bent forward. Colour varies from blackish to creamy brown but mostly mid grey-brown above and creamy white or pale yellow below. Upper surface of feet whitish with black guard hairs. Tail clearly > hb, slender, sinuous, naked with obvious rings of overlapping scales. **Distribution, habitat and status** Introduced. Scattered and locally common in moist regions of Aust. Usually associated with human settlement and disturbed environments. **Behaviour** Crepuscular and nocturnal. Omnivorous, eats seeds, fruit, insects, green plant matter, carrion and human food scraps. Shares communal nest in burrow or roof cavity. Climbs expertly. Breeding is continuous in good conditions, 5–10 young per litter. **Similar species** No other true rat in Aust has tail obviously > hb. Brown Rat is more thickset with coarse brown fur, blunt muzzle, short ears that just reach eyes when bent forward, thick tail < hb. Bush Rat has shorter tail and ears.

V

I

Brown Rat *Rattus norvegicus*

Sewer Rat

hb 180–255 mm; **t** 150–210 mm; **hf** 35–45 mm; **el** 16–20 mm; **wt** 280–500 g; **teats** pe 2, th 2, in 2 = 12

Large and robust, muzzle thick and blunt, aggressive when cornered. Fur short, coarse, grizzled. Upperparts dark warm brown heavily grizzled with buff; darkest on crown and forehead, warmest on flanks. Underparts cream or yellow-grey. Completely blackish or whitish specimens also occur. Ears short, only just reach eye when bent forward. Tail ≤ hb, thick, tapering, grey-brown, naked, obvious overlapping rings of scales. **Distribution, habitat and status** Introduced. Cities and towns in wetter parts of Aust but rare in tropics. Also wet grassland along streams. Usually close to fresh or salt water. **Behaviour** Mostly nocturnal, terrestrial, gregarious. Nests in extensive burrow systems or cavities at ground level, does not climb readily. Capable swimmer. Truly omnivorous. Litters of 5–10 young can be produced at 5–6 week intervals. **Similar species** See under Black Rat. Water Rat has furred tail with distinct white tip.

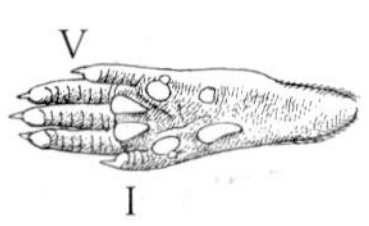

Long-haired Rat *Rattus villosisimus*

Plague Rat

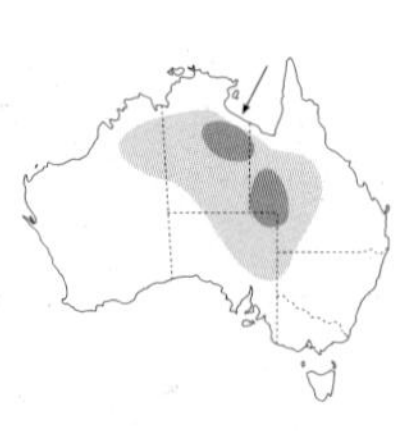

hb 130–225 mm; **t** 120–180 mm; **hf** 30–40 mm; **el** 16–22 mm; **wt** 60–280 (134) g; **teats** pe 2, th 2, in 2 = 12

Fur long, shaggy. Upperparts light buff to rich rufous heavily grizzled with dark grey; long black guard hairs (50 mm) on back. Underparts cream or pale grey. Tail < hb, dark grey with obvious rings of overlapping scales. Some individuals paler, often with white flash on forehead. **Distribution, habitat and status** Fluctuates dramatically in population size and distribution. Strongholds are Barkly Tableland and Channel Country of w. Qld. Following widespread rains can expand rapidly s. and w., as far as Woomera in SA, Uluru in NT, Halls Creek in WA and Mt Isa in Qld. At these times may occupy most habitats, then contracts back to moist patches along streams and bore overflows as conditions deteriorate. Isolated populations on Sir Edward Pellew Group, NT. In recent decades large populations have developed around Ord R. Scheme, WA. **Behaviour** Omnivorous, terrestrial, mostly nocturnal. Gregarious, shelters in extensive burrow systems. Breeding is opportunistic. When in plague proportions, readily enters buildings and camps seeking food.

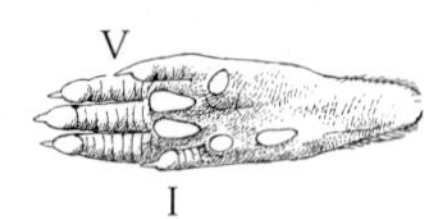

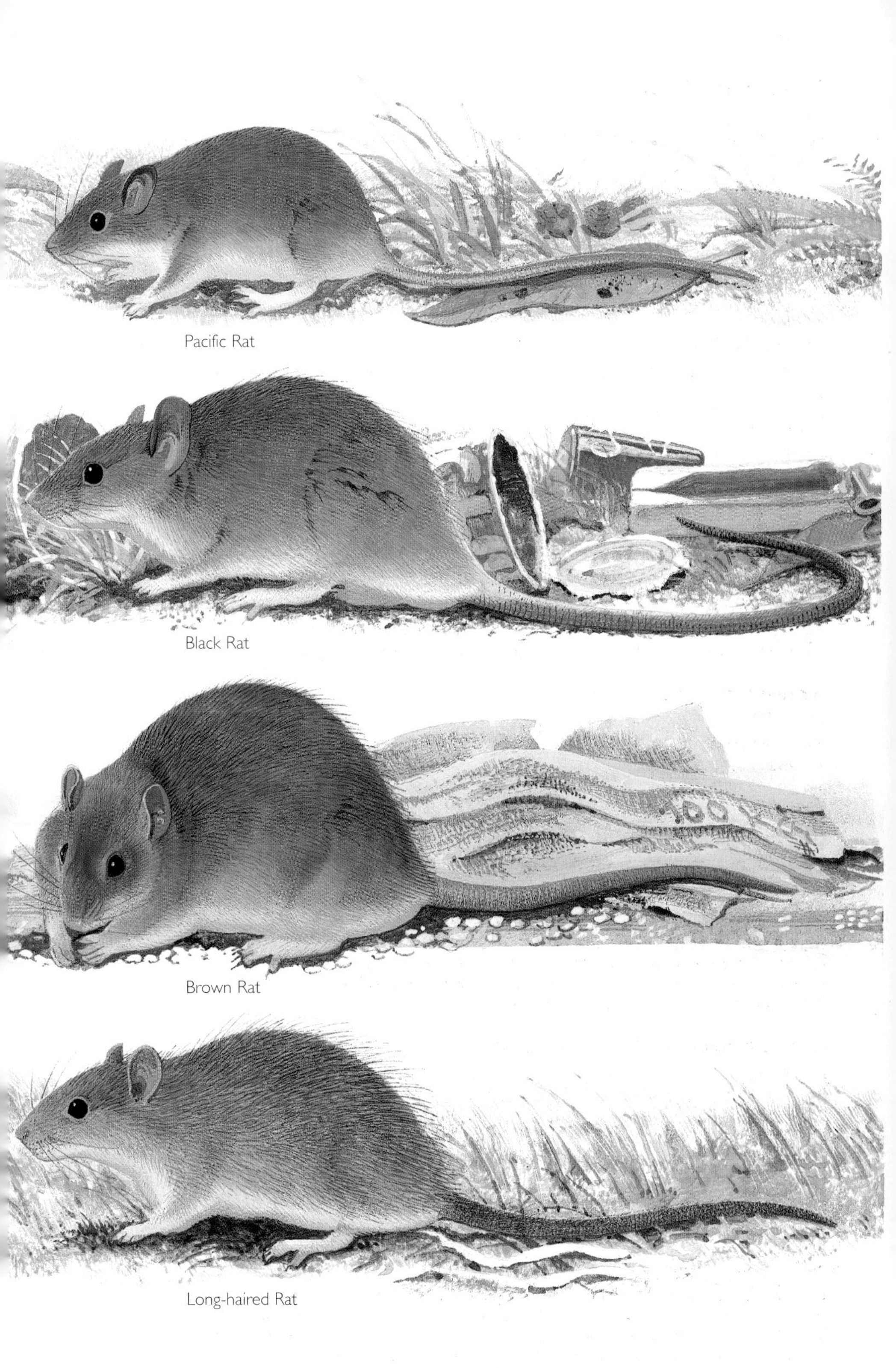

Pacific Rat

Black Rat

Brown Rat

Long-haired Rat

PLATE 79

Dusky Rat *Rattus colletti*

hb 120–200 mm; **t** 95–150 mm; **hf** 28–33 mm; **el** 17–21 mm; **wt** 60–215 g; **teats** pe 2, th 2, in 2 = 12

Fur coarse, spiny; muzzle long. Upperparts brown, grizzled grey and buff, grading to yellowish flanks and throat; rest of underparts pale grey-yellow. Tail clearly < hb, dark grey to black with sparse blackish hairs. Ears short, dark brown; upper surface of feet brown, soles blackish. **Voice** High-pitched whistling alarm call. **Distribution, habitat and status** Confined to flat coastal flood plains of nw. NT, inland to the Daly R. Inhabits dense grassland on swampy treeless plains that are flooded for long periods in wet season. Populations fluctuate widely in response to monsoonal rainfall. **Behaviour** During wet season retreats to higher ground and shelters in shallow burrows; as waters recede populations spread out over the vast plains, sheltering in cracks in clay soil. Mostly nocturnal, vegetarian, eats mostly corms of sedges and stems of grasses. Breeds from late wet season (Mar–Apr) to early dry (July). **Similar species** Pale Field Rat has paler fur, tail and feet. Long-haired Rat has paler fur and tail. Black Rat has t > hb.

Cape York Rat *Rattus leucopus*

hb 135–210 mm; **t** 140–210 mm; **hf** 33–40 mm; **el** 18–24 mm; **wt** 90–205 g; **teats** th 1, in 2 = 6

Body elongate, muzzle long and pointed. Indistinct dark area around eyes and muzzle. Upperparts grizzled blackish brown to golden brown, becoming yellowish on flanks and throat. Underparts whitish or yellow-grey. Upper surface of feet whitish, soles not strongly pigmented; tail = hb, slender, naked, brown with obvious rings of scales, n. subspecies (*leucopus*) has whitish mottling on tail and whitish underparts. **Distribution, habitat and status** Subspecies *cooktownensis* inhabits rainforest between Cooktown and Paluma, ne. Qld. Subspecies *leucopus* occurs in McIlwraith and Iron Ranges, and patches of rainforest to n. and w. Locally common. **Behaviour** Nocturnal, terrestrial, shy. Omnivorous, large component of insects in diet plus leaves, fruit, seeds. Breeds throughout year, litter size 2–5. **Similar species** The only rat in rainforest n. of about Cooktown. Bush Rat has more rounded muzzle and head, lacks dark eye-ring, has 8 teats.

Pale Field Rat *Rattus tunneyi*

hb 120–195 mm; **t** 80–150 mm; **hf** 25–35 mm; **el** 15–20 mm; **wt** 50–210 g; **teats** pe 1, th 2, in 2 = 10

Head broad, rounded. Upperparts 'toffee' coloured, merging to pale grey or cream underparts. Tail clearly < hb, naked, uniformly pink-brown. Ears short, pale pink-brown; upper surface of feet whitish; eyes large and bulging. **Distribution, habitat and status** On e. coast from Coen to Brisbane (subspecies *culmorum*) and n. coast from w. Kimberley to w. Gulf of Carpentaria (subspecies *tunneyi*). Inhabits wide range of environments, including damp grassland, woodland and monsoon forest with dense grass or sedge understoreys; also pasture and canefields in e. In Kakadu NP common on rocky slopes and often seen around campsites. Formerly much more widespread including along inland watercourses. **Behaviour** Nocturnal, terrestrial, vegetarian, eats grass stems, tubers, seeds. Shelters in shallow burrow systems. In NT breeds in dry season (Mar–Aug); in e. breeds mostly Sept–Nov. **Similar species** Dusky Rat has dark hind feet. Canefield Rat and Cape York Rat have darker fur. Long-haired Rat occurs in much drier habitats, has shaggy, greyer fur, 12 teats.

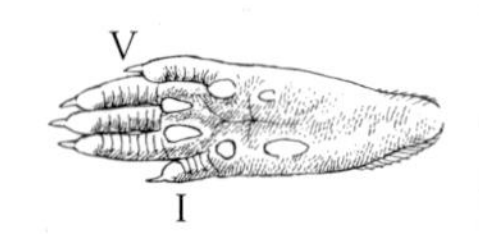

Canefield Rat *Rattus sordidus*

hb 125–190 mm; **t** 100–160 mm; **hf** 27–32 mm; **el** 16–22 mm; **wt** 60–250 g; **teats** pe 2, th 2, in 2 = 12

Small, dark; distinct 'Roman nose' profile; fur coarse, spiky. Upperparts grizzled dark golden brown to black with buff tips; underparts pale grey. Feet pinkish white above, pale brown soles. Tail < hb, dark grey to black with obvious rings of overlapping scales. **Distribution, habitat and status** E. coastal Qld from Cape York to sw. of Rockhampton; also Moreton and North Stradbroke Is. Isolated population on South West I., NT. Inhabits open areas with dense cover of grasses and sedges; also canefields. **Behaviour** Nocturnal, terrestrial, lives communally in burrow systems, often at great densities. Vegetarian, eats mostly grass or sedge stems, some seed and insect material. Breeds in any month but most births in Mar–May. **Similar species** Swamp Rat has dark brown soles. Pale Field Rat has paler fur, pinkish tail, 10 teats. Bush Rat has sleeker, greyer fur, pink-brown tail, 8 or 10 teats.

Dusky Rat

Cape York Rat

Pale Field Rat

Canefield Rat

PLATE 80

Prehensile-tailed Rat *Pogonomys mollipilosus*

hb 120–165 mm; **t** 160–205 mm; **hf** 22–27 mm; **el** 14–16 mm; **wt** 45–82 (60) g; **teats** th 1, in 2 = 6
Muzzle short, broad; top of head flat giving squirrel-like profile. Tail > hb, thin, naked, grey-brown often blotched whitish, last 2 cm of upper surface covered with tough skin without scales. Fur soft and dense. Upperparts uniformly mid grey-brown, narrow black eye-ring. Underparts white, including chin and upper jaw, sharply demarcated from upperparts. Ears small and rounded; feet pinkish. The only Aust non-*Rattus* rodent with thoracic teats. **Distribution, habitat and status** Restricted to 2 areas in ne. Qld: in low elevation monsoon forest at Iron Range; and between Shiptons Flat (s. of Cooktown) and Millaa Millaa, w. of Innisfail, in a range of rainforest types at all elevations. Often in ecotones. **Behaviour** Nocturnal, agile climber after fruit and leaves of rainforest trees. Shelters during day in a burrow. Uses tail as support by coiling it upwards around branches. **Similar species** Fawn-footed Melomys lacks clearly demarcated white underparts and has shorter tail. Long-tailed Pygmy-possum has blackish areas forward of eyes, has long ears and steep forehead, can have swollen tail.

tail detail

Greater Stick-nest Rat *Leporillus conditor*

hb 190–260 mm; **t** 148–180 mm; **hf** 42–48 mm; **el** 28–32 mm; **wt** 190–450 g; **teats** in 2 = 4
Size of small rabbit; head short, broad; fur soft, fine; ears broad, rounded. Upperparts uniformly grey-brown, tinged rufous particularly on flanks, merging with pale grey-buff underparts. Upper surface of hindfeet grey-brown on outside, whitish on inside. Tail < hb, dark brown above, grey below. Gentle, passive disposition. **Distribution, habitat and status** Formerly widespread in arid s. Aust from Nullarbor Plain to junction of Murray and Darling R., and n. to about 28°S. Declined drastically in late 1800s until found only on Franklin I. in Nuyts Arch., SA. Recently released onto several island and mainland sites in SA and WA: St Peters and Reevesby Is, and Venus Bay in SA, Salutation I. in WA. Usually in limestone country with dense low shrub cover. Endangered. **Behaviour** Mostly nocturnal; herbivorous, eats mainly leaves of succulent plants; gregarious, constructs large nests of interwoven sticks around base of shrub, under rock overhang or in shallow cave. Nests may be 2 m across and 1 m high with a grass nest in the centre. Births of 1–4 young peak in March-April but can occur in any month. **Similar species** Lesser Stick-nest Rat has more slender build, longer, more clearly bicoloured tail with distinct white end, white upper surface to feet with distinct demarcation to grey legs.

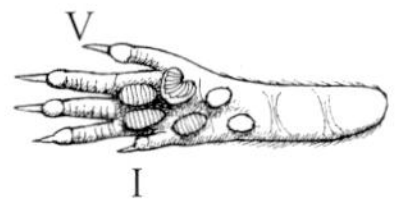

Lesser Stick-nest Rat *Leporillus apicalis*

hb 170–200 mm; **t** 220–240 mm; **hf** 41–44 mm; **el** 27–33 mm; **wt** approx. 150 g; **teats** in 2 = 4
Upperparts mid grey-brown merging to pale grey flanks and grey-white underparts including chin and lips. Tail > hb, well-furred, clearly bicoloured: dark brown above, whitish below with distinct white tuft. **Distribution, habitat and status** Presumed extinct; last collected in 1933. Formerly widespread in arid s. and central Aust from North West Cape in WA to central Aust and nw. Vic. Occupied shrubby habitats in rocky ranges, breakaway country, mulga flats, mallee. **Behaviour** Little-known. Nocturnal. Vegetarian, ate fleshy stems and leaves, particularly of chenopods, some seeds; gregarious, sheltered communally in tree hollow or in stick nest as described for Greater Stick-nest Rat. **Similar species** See under Greater Stick-nest Rat.

Five-lined Palm Squirrel *Funambulus pennanti*

hb 140–160 mm; **t** 110–120 mm; **hf** 34–36 mm; **el** 15–17 mm; **wt** approx 135 g
Small, alert, quick darting movements, bushy tail and bold red-brown and white body stripes. Upperparts grizzled yellow-grey with 5 white and 4 red-brown stripes. Underparts whitish. Ears short, grey; tail < hb, grizzled grey-rufous basally, pale rufous and dark brown bars becoming more intense towards tip. Feet pale grey-brown. **Distribution, habitat and status** Introduced. Feral population exists in grounds of Perth Zoo and surrounding suburbs. Another occurred around Taronga Zoo and Mosman, Sydney but may be extinct. **Behaviour** Climbs trees and buildings with great agility and darting movements. Builds nests of coarse vegetation in trees. Eats fruit, seeds, buds, insects and human food scraps. In Perth births occur Aug–May, with peaks in Oct, April.

Prehensile-tailed Rat
Greater Stick-nest Rat
Lesser Stick-nest Rat
Five-lined Palm Squirrel

PLATE 81

Red Fox *Vulpes vulpes*

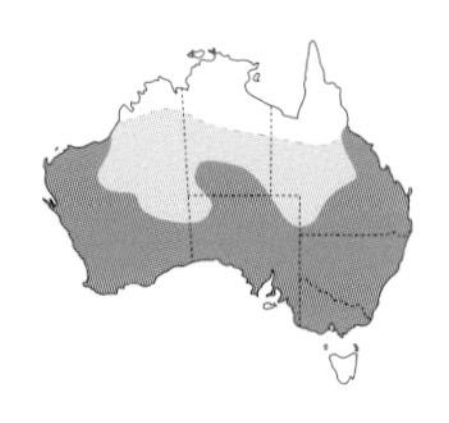

hb 600–740 mm; **t** 360–450 mm; **sh** 350–400 mm; **el** 80–92 mm; **wt** 3.5–8 kg
Slender build, long thin limbs, long bushy tail, narrow pointed muzzle, erect pointed ears. Grizzled reddish-brown or sandy above, whitish below, including chin and muzzle; dark diagonal line from muzzle to front of eye. Back of ears, lower legs and feet blackish; tail long, bushy, held semi-horizontally, tip often whitish. Distinctive scent deposited in urine. Eye-shine white or pinkish. **Voice** Many vocalisations including high-pitched bark, and screams during mating period. **Distribution, habitat and status** Deliberately introduced in 1860s. Now widespread and common s. of Tropic of Capricorn and n. in WA to edge of the Kimberley. Absent from Tas, islands of Bass Strait, French and Kangaroo I. Found in most habitats from wet forest to desert and suburbs. Responsible for local extinction of many populations of small to medium-sized mammals. **Behaviour** Nocturnal; diurnal where undisturbed. Dens in thick cover, burrow or other cavity. Opportunistic omnivore but predominantly carnivorous, taking the most readily available live prey, supplemented by carrion, fruit, berries, insects. A single litter of 4–5 cubs is born in late winter or spring, reaching independence in late summer. **Similar species** Narrow muzzle, erect, black-backed ears, tail are diagnostic.

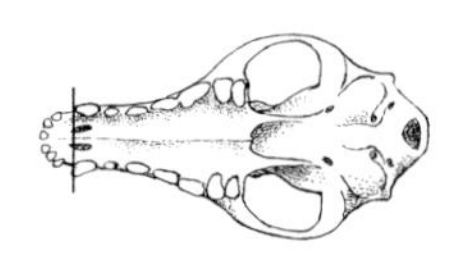

position of palatine vacuity relative to C^1

Dingo *Canis lupus dingo*

hb 860–1000 mm; **t** 260–360 mm; **sh** 440–620 mm; **el** 95–105 mm; **wt** 12–24 kg
Primitive dog derived from Gray Wolf. Probably introduced to Aust by Indonesian seafarers about 4000 years ago. Coat usually red-ginger or sandy-yellow, also black, whitish or black with tan patches on cheeks, ears and legs. Most have white feet, chest, cheeks, lips and tail tip. Other coat colours or patterns indicate hybrids with Domestic Dog. Ears erect and pointed, tail bushy. **Voice** Howling used for pack cohesion, throughout the night during breeding season, seldom barks. **Distribution, habitat and status** Formerly throughout continent but never in Tas. Now absent from densely settled parts of se. and sw. A 9600 km fence extending from e. edge of Nullarbor Plain to NSW border and in big loop through central Qld keeps Dingoes out of much agricultural land. Endangered due to hybridisation with Domestic Dog, and persecution. **Behaviour** Opportunistic carnivore, eats mostly mammals, overwhelmingly rabbits and macropods, also reptiles, arthropods, carrion. Forms packs when hunting large prey but often forages alone for small prey or carrion. Breeds only once per year, during winter. Dominant pair usually the only successful breeders; other pack members help feed litter of 1–10 pups. **Similar species** Field identification from similar-looking Domestic Dogs difficult, perhaps impossible. Dark specks in the white markings, or dark midline, indicate hybrids. Female Dingoes on heat only once per year.

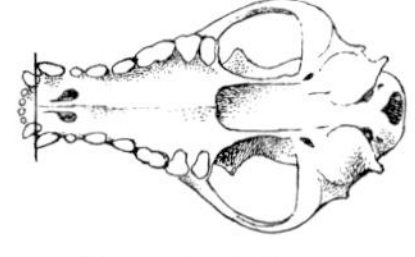

position of palatine vacuity relative to C^1

House Cat *Felis catus*

hb 380–600 mm; **t** 230–330 mm; **el** 53–62 mm; **wt** 2.5–6.5 kg
The only cat in Aust. Feral House Cats have thin, lithe body, long short-haired, pointed tail. Fur usually short, often tabby (narrow dark vertical stripes on lighter background), black, ginger, or rarely tortoiseshell (orange–tabby or orange–black). Often has white feet and underparts, and white flecks or patches in upperparts. Eye-shine bright greenish white. Feral cats generally not much larger than domestic ones: few exceed 5 kg. **Distribution, habitat and status** Introduced, was probably widespread before European settlement. Throughout, in all habitats from rainforest to desert and cities. **Behaviour** Exhibits a variety of lifestyles, from urban strays depending on human refuse or goodwill for food to truly feral animals surviving entirely by hunting. Eats range of small to medium-sized vertebrates, also large invertebrates including spiders, centipedes, crickets. Where present, rabbits are major prey, also reptiles in arid and semi-arid regions. Active day and night, often crepuscular, mostly terrestrial but agile climber. Two litters may be produced per year, usually in spring and late summer–autumn. Litters born and reared in lair in rabbit burrow, hollow log or dense thicket.

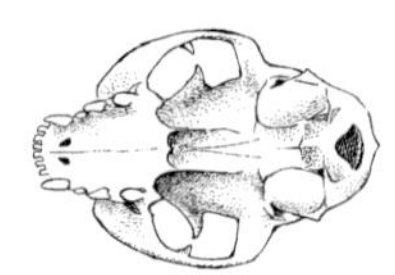

PLATE 82

European Rabbit *Oryctolagus cuniculus*

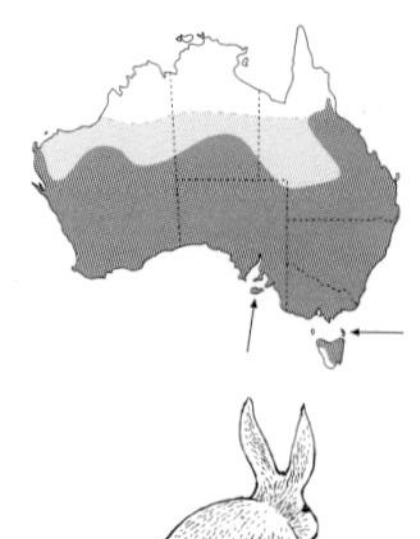

hb 350–450 mm; **t** 40–65 mm; **hf** 75–95 mm; **el** 60–70 mm; **wt** 1–2.4 kg
Upperparts yellowish grey-brown, pencilled with black, cream eye-ring; underparts pale grey to whitish. Ears long, but when laid forward do not reach past nostrils; pinkish inside, grey outside with narrow black rim. Tail short, brown above, white below (held vertical when pursued so white is visible). Blackish individuals occur, especially in Tas. **Distribution, habitat and status** Introduced. Throughout Aust, except for monsoonal tropics. Has established permanent populations around reliable water in deserts; in temperate and subtropical regions absent only from extensive, undisturbed wet forest or heath. **Behaviour** Gregarious, lives in warrens, nocturnal and crepuscular, diurnal if undisturbed. Grazes on the most nutritious forage available, especially green growing tips. In droughts can survive on dry grass, tubers, bark. Breeding seasonal, but season varies across range. Young born naked in fur-lined nest within warren. Each adult female usually produces 11–25 young per year but most do not survive. Thumps ground with hindfeet as warning signal. **Similar species** See under Brown Hare.

rear view

Brown Hare *Lepus capensis*

European Hare

hb 480–650 mm; **t** 80–110 mm; **hf** 125–150 mm; **el** 85–105 mm; **wt** 2.5–6.5 kg
Large; long ears and hindlegs; sits upright. Mottled black and fawn on crown and back, tawny or greyish on cheeks, shoulders, flanks and rump. Whitish stripe from chin to nose, through eye to base of ear. Ears have pale edges and black tips, extend 3 cm beyond nostrils when laid forward. Underparts white. Tail white with black stripe on upper surface. **Distribution, habitat and status** Introduced. Has spread through se. Aust from Eyre Pen. in SA to central Qld and Tas. Mostly in farmland or adjacent scrub; also grassy woodland and alpine grassland. **Behaviour** Solitary, nocturnal and crepuscular. Spends day crouched in 'form' (open nest) in long grass or other cover above ground. Sits tight until intruder very close, then dashes for safety. Eats grass, cereal and vegetable crops, bark. Spectacular courtship and aggressive behaviour involving chasing, leaps and boxing. Breeding begins soon after shortest day (late June) and extends to summer. Young, usually 2, born in the 'form', fully furred with eyes open. **Similar species** European Rabbit has shorter ears without black tips; shorter hind legs; shorter tail with brown upper surface; more uniformly grey pelage; and young born naked and blind in underground nest.

rear view

Goat *Capra hircus*

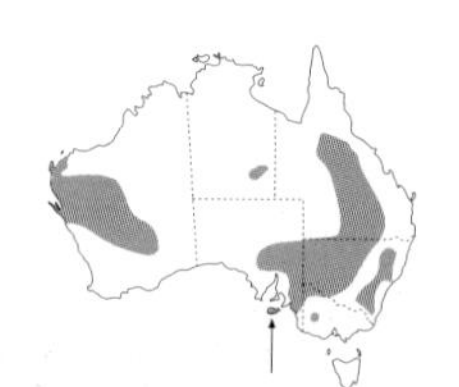

hb 1.1–1.6 m; **t** 12–17 cm; **sh** 60–70 cm; **wt** 20–80 kg
Feral Goats usually short-haired, may be shaggy on hindquarters of both sexes, and on neck and shoulders of male. Male has beard extending as mane down throat to chest. Fur white, black, pale brown or dark brown, or almost any combination of these. Tail short with long hairs, may be held erect over rump. Ears long, narrow, project horizontally from side of head below horns. Both sexes have horns; very variable but usually curve upwards and backwards then spiral; longer and broader in male, may touch at base. Male has distinct pungent smell. **Voice** 'meh meh' like Sheep but higher pitched. **Distribution, habitat and status** Introduced. Widespread but patchy outside monsoonal tropics. Absent from the wettest and driest regions, most established in sheep-grazing country where Dingoes are few, particularly semi-arid shrubland and woodland with rocky outcrops. Requires surface water in hot weather; can survive on nutrient-poor herbage. **Behaviour** Diurnal, gregarious. Browser with catholic taste. In good years females may produce 2 litters of 2 young; breeding stops during drought.

PLATE 83

Horse *Equus caballus*

Brumby

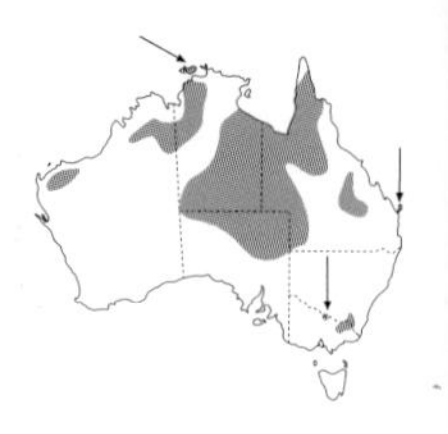

hb to 2.5 m; **t** to 80 cm; **sh** to 1.6 m

In Aust feral horses are derived from a variety of domestic breeds, including Arab, thoroughbred and draught horses. They are generally smaller, more heavily boned and strongly muscled than domestic horses. Colour brown to black and white, often with white flash on muzzle and feet. Large size, long head, flowing mane with forelock, and long-haired tail are diagnostic. **Distribution, habitat and status** Introduced. Scattered populations across much of Aust, particularly in w. Qld, central Aust, central WA, the Aust Alps and Barmah–Moira Forest on Murray R. Prefers grassland and shrub steppe habitats, needs access to water and will dig in dry streambeds to obtain it; absent from densely settled areas and true deserts. Environmentally damaging where abundant. **Behaviour** Gregarious; dominant stallion defends harem of mares and their offspring; young males form bachelor groups. Specialised grazer but will browse when necessary. Most births in spring, early summer. **Similar species** Donkey has long ears, short upstanding mane, base of tail is short-haired.

Donkey *Equus asinus*

hb to 2 m; **t** to 60 cm; **sh** to 1.4 m; **wt** 300–350 kg

Very long ears; mane of short upstanding hairs; hairs short at base of tail, longer towards tip, creating tufted appearance. Fur usually fawn-grey with narrow dark midline and dark stripe across shoulder perpendicular to midline. Underparts often whitish, also lower muzzle and eye-ring. Black and whitish individuals also occur. **Voice** Loud braying territorial calls. **Distribution, habitat and status** Introduced. Through much of arid central Aust, the Kimberley and Top End. Can survive without surface water. Prefers tropical savannah and arid hill country. Locally abundant; a serious pest. **Behaviour** Diurnal grazer and browser, survives on much coarser vegetation than Horse. Females may produce one young per year under favourable conditions. **Similar species** See under Horse.

One-humped Camel *Camelus dromedarius*

Dromedary

hb 2.0–3.3 m; **t** to 70 cm; **sh** 1.7–2.2 m; **wt** to 900 kg

Feral descendants of domestic Camels brought to Aust as beasts of burden in second half of 19th century. Unmistakable; very large body, long, slender legs with large two-toed feet, single hump. Orange-brown to sandy-brown fur. **Distribution, habitat and status** Introduced. Aust is now only country with wild camel populations. Widespread in arid regions of central and w. Aust; small satellite population n. of Cloncurry in Qld. Mostly in sandplain and arid shrubland habitats. Total population may approach 250 000. **Behaviour** Highly mobile, rarely stays in one place long, can easily travel 50 km per day. Browses on shrubs and forbs, prefers succulent species. Gregarious, cows live together in cow groups and are temporarily herded by a bull. Bulls live in bachelor groups or alone. Not territorial; aggregations of hundreds can occur. Mates in May–Sept, births 12–13 months later.

PLATE 84

Swamp Buffalo *Bubalus bubalis*

Water Buffalo

hb 2.5–3 m; **t** 70 cm; **sh** to 1.6 m; **wt** 450–1200 kg
Introduced in NT to Melville I. in 1828 and Port Essington on Cobourg Pen. in 1838; has since spread widely. Huge size; wide, flat, backswept horns, triangular in cross-section; short neck; head held horizontally; large splayed hooves. Hair sparse, coarse; skin uniformly dark grey; calves brownish. **Distribution, habitat and status** Confined to flood plains and adjacent slopes of Top End of NT, s. to Pine Creek and e. to Borroloola. Cause of serious environmental degradation and carrier of bovine tuberculosis, so being vigorously eradicated. **Behaviour** Nocturnal and crepuscular, spend heat of day in muddy wallow and rest in near-by wooded country. For much of year segregated into single-sex herds. Calves formed into creches watched over in turn by a roster of cows. In wet season grazes selectively on aquatic vegetation; less selective during dry, when forced to eat wider variety of terrestrial vegetation including some browse. **Similar species** Shape of horns readily distinguishes buffalo from wild cattle.

Bali Banteng *Bos javanicus*

Bali Cattle

hb 1.8–2.2 m; **t** 70 cm; **sh** 1.2–1.5 m; **wt** males to 550 kg; females to 400 kg
Domesticated form of Banteng, a wild ox of South-East Asia. Twenty introduced to Cobourg Pen. in NT from Bali in 1849. Like small, slender domestic cattle; uniformly reddish-fawn or sandy-brown with distinctive white oval patch on rump and white stockings from just above the hocks and heels. Old bulls are blackish and some animals have dark midline. Cows have small, curved, back-ward-pointing horns; bulls have longer, straighter horns with tips pointing out. **Distribution, habitat and status** Confined to Cobourg Pen., mostly in monsoon forest and adjacent coastal plains. Population of 2000–3500 animals is subject to controlled safari hunting. **Behaviour** Nocturnal and crepuscular; nervous and alert. Shelters by day in forest, feeds in open areas, including along coast. Grazes on sedges and sometimes seaweed, browses trees and shrubs. Herds comprise several females, their offspring and a single male; other males are solitary or in small groups. Mating peaks at start of wet season (Oct–Nov); most births June–Sept.

Pig *Sus scrofa*

hb 1.1–1.6 m; **t** 20–30 cm; **sh** 40–60 cm; **wt** 25–175 kg
Body stocky, powerful, especially forequarters; legs short, thin; head long, muz-zle tapers to mobile, cartilaginous snout disk, long triangular ears, small eye; tail straight and furred. Fur coarse, bristly, usually olive-brown or black but often piebald; old boars have thickened skin forming shield over shoulders and flanks, and long curved tusks (canine teeth). **Distribution, habitat and status** Introduced; serious pest. Widespread in e. Aust and Top End of NT, scattered populations in Kimberley and sw. WA. Needs access to water; prefers margins of wetlands and floodplain country, but also inhabits wet forest. **Behaviour** Mostly nocturnal, solitary. Opportunistic omnivore, eats almost any plant or animal material but prefers succulent green herbage and tubers. Wallows in mud and dust. Breeding can occur in any month and females may produce 2 litters of 4–6 young per year.

Swamp Buffalo

♀ ♂

Bali Banteng

Pig

PLATE 85

Sambar *Cervus unicolor*

hb 1.6–2.5 m; **t** 25–30 cm; **sh** 1.0–1.2 m; **wt** 110–240 kg
The most successful deer in Aust. Solidly built; uniformly dark pelage; large, broad, rounded ears; fur short and coarse, mature males have mane of longer fur around neck. Colour varies from pale brown to blackish, often tan on rump, particularly in females; underparts and buttocks paler and greyer; lower lip whitish. Tail long, well-furred, blackish above, pale grey below, raised over back when alarmed. Calves uniformly dark brown. Antlers heavy, 3-tined; on main beam, outer (front) tine longer than inner. Old antlers shed Oct–Nov, fully replaced by April. **Distribution, habitat and status** Introduced to Vic in 1860s. Now widespread in forests of Eastern Highlands from near Melbourne to ACT; range still expanding. Isolated populations at Mt Cole, w. Vic and Sunday I., Vic. Also introduced to Cobourg Pen. in NT but status there unknown. **Behaviour** Nocturnal and crepuscular; wary and alert; if startled explodes from cover with loud coughing bark. Browses on variety of shrubs and coarse plant material. Stags solitary, hinds and young form small groups of 2–5. Stags thrash shrubs and small trees with antlers and wallow in muddy pools. Hinds attracted to stag by his roaring and scent. Births year-round. **Similar species** Rusa is smaller and daintier, has pale upper and lower lips, rear or inner tine of antlers is easily longest.

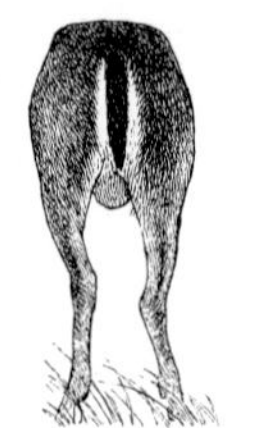

Rusa *Cervus timoriensis*

hb 1.5–1.8 m; **t** 18–25 cm; **sh** 0.8–1.1 m; **wt** 50–120 kg
Stags uniformly dark grey-brown, warmer reddish-brown after summer moult; hinds more reddish and paler. Underparts cream or pale grey. Some males have distinctive whitish throat patch separating grey-brown chin and lower neck. Lips pale grey. Calves not spotted, tan above, whitish below. Tail relatively long and narrow, brown, tipped blackish. Ears < 1/2 head length, somewhat pointed. Antlers 3–tined, with innermost (rear) tine easily longest and usually nearly vertical. Cast in Dec–Jan and regrown by June. **Voice** Alarm bark; stags give shrill roar, mostly at night during rut (June–Aug in NSW, Sept–Oct in NT). **Distribution, habitat and status** Introduced to numerous sites in e. Aust but has survived only in Royal NP s. of Sydney, Possession and Prince of Wales Is in Torres Strait, and Groote Eylandt, NT. **Behaviour** Wary, nocturnal and crepuscular, agile, flees with head low and neck outstretched. Gregarious, forms single-sex herds outside rut. During rut males may adorn antlers with vegetation to enhance their display. Breeding can be year-round but most births in Apr–May. Grazer, preferring lush grass but will browse coarse material when necessary. **Similar species** Sambar heavier build; short, round ears; carries tail erect when alarmed; on main beam of antler the outer (front) tine is longest. Red Deer has clear pale rump patch bordered at sides with black extending down thighs, and complex antlers.

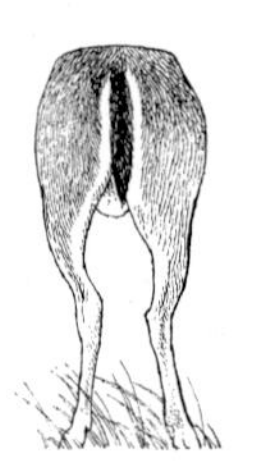

Red Deer *Cervus elaphus*

hb 1.7–2.5 m; **t** 12–15 cm; **sh** 0.9–1.3 m; **wt** 90–160 kg
The largest deer in Aust; long, tapering muzzle; long, pointed ears > 1/2 head length. Fur uniformly reddish brown except for cream-buff rump patch bordered on outside by blackish stripes from rump to thighs. Head, throat and underparts greyish, paler eye patch and lips. Stags have mane of longer brown hair and may have blackish belly. Tail short, pale reddish brown above, whitish below. Calves have scattered white spots on neck, back and rump, fading after about 2 months. Antlers have 5–8 tines on each side, the uppermost pointing upwards in cluster; 2 lowest tines point forward from close together on lower stem, then large gap to third tine. **Voice** Alarm call a gruff bark. During rut stags roar with deep guttural bellow reminiscent of cattle, mostly in morning and evening. **Distribution, habitat and status** Introduced. Populations persist in Grampians NP in w. Vic, and in headwaters of Mary and Brisbane Rivers inland from Sunshine Coast, Qld. Locally common. **Behaviour** Opportunistic herbivores, graze and browse on foliage and bark of wide range of plants, also lichen and fungi. Gregarious in small single-sex groups for much of the year. Both sexes wallow. During late Mar–Apr stags disperse and establish separate rutting areas on which they attempt to gather a number of hinds. Single young born Nov–Dec. **Similar species** Sambar is darker with simpler antlers (max. 3 tines on each beam), long blackish tail, lacks black-bordered cream rump.

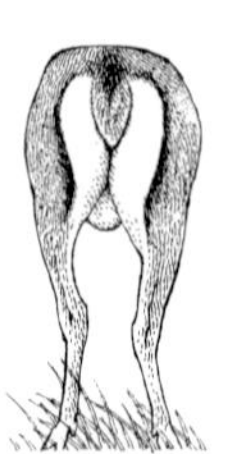

PLATE 86

Fallow Deer *Dama dama*

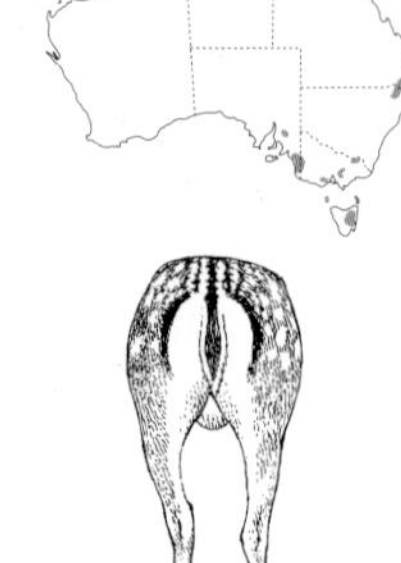

hb 1.4–1.7 m; **t** 20–24 cm; **sh** 75–100 cm; **wt** 35–90 kg
Mid-sized, long-legged deer with many colour variants. Most common summer coat is reddish brown above with large white spots, cream flanks and underparts, black midline from crown to rump, whitish rump bordered above and to side by black crescent, and accentuated by black tail. Winter coat is grey-fawn with indistinct spots. Some individuals blackish brown without spots, or entirely whitish. Ears long, pointed. Tail curled over rump when alarmed, displaying whitish underside. Antlers multi-tined, upper half flattened and webbed in older individuals; shed in Oct–Nov, regrown by March. **Voice** Mostly silent; short bark alarm call given at 10–20 s intervals; during rut males produce deep, loud belch-like sound called groaning. **Distribution, habitat and status** Introduced. Widely scattered small populations in all states except WA and NT. Confined to fringes of cleared land and plantations. Locally common in central Tas. **Behaviour** Gregarious, nocturnal or crepuscular, diurnal if undisturbed, fast and agile, jumps fences with ease. Eats mostly soft grasses and herbs, also browses on coarser material when necessary. Mates Mar–Apr, births Nov–Dec. Rut involves much noisy fighting between males, and thrashing of vegetation with antlers. **Similar species** Only deer in Aust with curved black edging above whitish rump patch and flattened webbed antlers. Black phase animals much darker than any other deer in Aust.

Chital *Axis axis*

Spotted Deer

hb 1.3–1.9 m; **t** 25 cm; **sh** 80–100 cm; **wt** 50–80 kg (rarely to 100 kg)
Both sexes and all ages have light chestnut upperparts with lines of clear white spots from hindneck to rump and onto thighs. Head, foreneck and legs fawn, often with clear white chin and throat. Broad, dark midline from lower neck to tip of long, pointed tail, bordered by 1 or 2 lines of white spots. Underparts and inside of legs whitish. Antlers tall (to 80 cm), slender, 3-tined. When alarmed tail raised to reveal white underside. **Distribution, habitat and status** Introduced; only feral population remaining is centred around Maryvale Creek, 170 km w. of Townsville, Qld. Inhabits grassy dry open-forest, Requires access to permanent water. **Behaviour** Gregarious, herds of up to 50 not uncommon; nocturnal and crepuscular; grazes on grass, also browses foliage and fruit; births occur in all months.

Hog Deer *Axis porcinus*

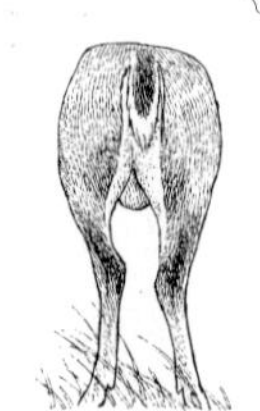

hb 1.0–1.3 m; **t** 12–14 cm; **sh** 60–80 cm; **wt** 25–45 kg
Smallest deer in Aust; size of large sheep and solidly built. Legs short, rump higher than shoulder. Upperparts reddish-brown or sandy, greyer on legs; older animals have buff foreneck and head with pale border to black rhinarium. Many individuals have darker midline from nape to rump with, in summer coat, a row of buff spots along either side. Underparts dark grey-brown. Tail short, brown above tipped white, white below; raised to display white hairs when alarmed. Ears broad and rounded. Calves distinctly spotted. Antlers short (20–30 cm), simple, typically 3-tined but can have more. Brow tines form an acute angle with main beam; inner top tines are short and angled backwards and inwards. **Voice** Sharp alarm bark and shrill whistling call when startled. **Distribution, habitat and status** Introduced. Confined to coastal plains and islands in South Gippsland, Vic, from Cape Liptrap in w. to Lake Tyers in e. Found in damp coastal woodland, scrub, heath, dune swales, edges of cleared land. Locally common. **Behaviour** Partly diurnal where infrequently disturbed, otherwise nocturnal; solitary in loose associations; eats primarily grasses, sedges and herbaceous plants, some browse. Births in most months, peak July–Sept.

PLATE 87

Australian Fur Seal *Arctocephalus pusillus doriferus*

tl males 2.0–2.3 m; females 1.3–1.7 m; **wt** males 220–360 kg; females 40–112 kg
Adult males entirely grey-brown with darker, thickened neck, chest and shoulders, and mane of coarse hair. Pelage darkens with age, except for mane which becomes paler. Adult females silvery grey above, yellowish throat and chest, and brown belly. Pups blackish with grey underparts. Head large, broad, forehead low. When sitting, neck held close to vertical with head at right angles. When moving slowly on land each flipper moves alternately, resulting in a 'walking' gait and side-to-side swaying with the head moving in a circular motion. **Distribution, habitat and status** Coast, continental slope and shelf waters of Vic, Tas and NSW, up to 180 km offshore; comes ashore on rocky islands and rock platforms. Breeds on 9 islands in Bass Strait, formerly bred off se. NSW and s. Tas. Haul-out sites extend from Kangaroo I. in SA to s. Tas and near Jervis Bay, se. NSW. Population 40 000–60 000. **Behaviour** Eats pelagic and mid-water fish and cephalopods; can dive to at least 200 m. Pregnant females return to colonies in late Oct – early Nov, give birth to single pup within a week; pup is suckled until about Aug. Eight days after birth, female mates with dominant male. Dominant males depart colony by mid Jan, then many bachelor males move into colony. Pups gather in groups while mothers feed at sea. Pregnant females feed intensively at sea in early spring before returning to give birth. **Similar species** New Zealand Fur Seal is more thickset, has narrower more pointed muzzle with steeper forehead; foreflippers very triangular. When sitting, neck slopes forward, head held at >90° to foreneck. On land, foreflippers move together in bounding gait with little sideways swaying of body, head moves up and down. Vocalisations generally higher-pitched, more whining. Pups small, docile, very short muzzle.

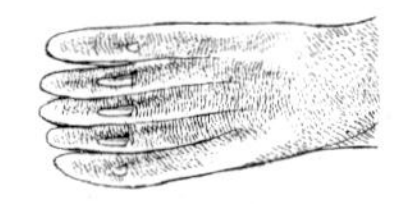

nails of hind flipper

New Zealand Fur Seal *Arctocephalus forsteri*

tl males 1.6–2.2 m; females 1.0–1.5 m; **wt** males 120–180 kg; females 35–50 kg
Males dark grey to brown above with pale grey or tan muzzle, paler below. Females more variable, metallic grey after moult (Feb–Mar) gradually becoming browner through year; paler underparts but brown belly. Pelage has grizzling of white, giving a silvery sheen when dry. Whiskers long and whitish; ears dark tan. Juveniles rich dark brown to burgundy with cream moustache. Pups dark brown, pale blaze on forehead. **Distribution, habitat and status** Breeds on islands off s. Aust, from Flinders I., off Cape Leeuwin, e. to Kangaroo I., plus Maatsuyker I., s. Tas, and a very few among Vic colonies of Australian Fur Seal. Formerly also in Furneaux Group. Population 35 000 and rising. Breeds and hauls out on areas of jumbled boulders rather than open rock platforms. Non-breeding animals range from s. WA waters, including Perth, e. and n. to Fraser I., Qld. **Behaviour** Feeds mainly at night on cephalopods and pelagic fish, also lobsters and penguins. Breeding cycle as for Australian Fur Seal but one month later. **Similar species** See under Australian Fur Seal.

Sub-antarctic Fur Seal *Arctocephalus tropicalis*

tl males 1.5–2.0 m; females 1.0–1.4 m; **wt** males 95–160 kg; females 30–50 kg
Compact body with short, thick neck; muzzle short, flat, deep; whiskers white, very long, often reaching past ears; ear pinnae not prominent, short, black, naked; foreflippers short and broad. Adult males heavily built with barrel chest, short neck, thick mane, distinctive dark 'mohawk' crest forming a point forward of eyes. Upperparts vary from dark orange-brown to charcoal, contrasting with pale yellow chest and pale brown belly. Sides of muzzle and face to above eyes and behind ears pale cream or sandy. Subadult males paler, more grey upperparts and only partial crest. Females paler with less contrast, more cream on face, lack crest. **Distribution, habitat and status** Breeds on many sub-antarctic islands; subadults wander widely and are being recorded with increasing frequency, mostly July–Oct, along s. coast as far n. as Kalbarri in WA and Urunga, ne. NSW. **Similar species** Distinguished by short, blunt muzzle, pale face, chin and throat contrasting with dark upperparts, and mohawk crest of males. Antarctic Fur Seal has similar head profile, but males lack crest; has uniformly silver-grey to brown fur with little contrast between back and chest; longer, narrow foreflippers.

♀
breeding
♂
Australian Fur Seal
breeding
♂
♀
New Zealand Fur Seal
♀
breeding
♂
Sub-antarctic Fur Seal

PLATE 88

Australian Sea-lion *Neophoca cinerea*

tl males 2.0–2.5 m; females 1.3–1.8 m; **wt** males to 350 kg; females 60–115 kg
Head and muzzle flat-topped, muzzle rounded, not pointed; toes of hindflippers unequal in length. Adult males have massive head and forequarters with mane of longer, rough hair over crown and neck; chocolate brown above, older males have golden-white crown and mane giving distinctly bicoloured head; paler brown throat and underparts. Females and subadult males silvery grey or fawn above, yellow or cream below and on face including broad eye-ring. Maturing males progressively develop brown blotches over creamy underparts. Pups dark brown with paler crown. **Distribution, habitat and status** Breeds on 66 islands around the w. and s. coast between Houtman Abrolhos (off Geraldton, WA) and The Pages, e. of Kangaroo I., SA. About 40% of 11 000–13 000 total population breeds at 3 largest colonies, all e. of Port Lincoln, SA; other colonies small. Non-breeding range extends from Shark Bay in WA to s. Tas and mid-coast NSW, but rare outside core breeding range. Prefers sheltered rocky bays and reefs for birthing, and sandy beaches for loafing. **Behaviour** Unique among seals in non-seasonal, non-annual, asynchronous breeding. Births extend over 5–7 months at a given colony, but birthing months vary between colonies. Interval between births is 17–18 months. Forages mainly in continental shelf waters at depths <150 m and mostly 20–30 km offshore. Eats cephalopods, benthic fish, lobsters, seabirds. **Similar species** Distinguished from fur seals by paler colouration, deeper, rounded muzzle, unequal length of toes on hindflippers.

Leopard Seal *Hydrurga leptonyx*

tl males 2.0–2.8 cm; females 2.0–3.0 m; **wt** males 200–320 kg; females 225–380 kg
Strangely reptilian appearance imparted by long, slender body; massive flat head; thick, round muzzle with nostrils set on top; very wide gape; long foreflippers (30% of tl); and distinct neck. Upperparts dark blue-grey or pale grey; darker on head and hindneck, often with cream upper lip and eye patch. Sharp demarcation to silvery underparts variably spotted with grey on throat and lower flanks. **Distribution, habitat and status** Throughout Southern Ocean. Regular visitor to southern Australia, mostly July–Nov and mostly as subadults, frequently in poor condition. Also vagrant to Qld and NSW waters, including Heron and Lord Howe Is. **Behaviour** Breeds on Antarctic pack ice. Very pelagic, solitary. Eats a variety of marine animals, including krill, penguins and other seabirds, fish, cephalopods and young seals. **Similar species** Size, shape of head and length of foreflippers diagnostic.

Southern Elephant Seal *Mirounga leonina*

tl males 3.0–5.0 m; females to 2.6 m; **wt** males 2.0–3.8 tonnes; females 200–400 kg
The largest seal; extreme sexual dimorphism in size and nostril conformation. Adult males unmistakable, huge size with erectile proboscis that hangs over face when relaxed, and neck shield of thick (to 40 mm), heavily creased and scarred skin. Females and young males have robust body, particularly head and neck, blunt muzzle, large eyes, bulbous nostrils giving a pug-like appearance. Pelage is uniformly brown, yellow-brown or grey, only slightly paler below; older males develop paler face and proboscis. **Distribution, habitat and status** Circumpolar in Southern Ocean, mostly n. of pack ice and at sub-antarctic islands. Breeds on Macquarie and Heard Is; former breeding colonies in Bass Strait on King, Hunter and New Year Is were hunted to extinction in early 1800s. Occasional vagrants haul out to moult during summer and autumn on sandy beaches in s. Aust, as far n. as Sydney. Births have occurred on beaches in Tas, Vic and SA during spring and early summer. Most regularly used site is Maatsuyker I., Tas. Vulnerable. **Behaviour** Eats mostly cephalopods and fish; can dive to 1200 m and remain submerged for over 1 hour. Annual moult necessitates spending about 3 weeks ashore, usually Nov–Mar, when it is most likely to be seen in Aust. **Similar species** All other southern true seals have spotted or streaked colour patterns.

PLATE 89

Antarctic Fur Seal *Arctocephalus gazella*

tl males 1.7–2.0 m; females 1.0–1.4 m; **wt** males 130–200 kg; females 30–40 kg
Muzzle short, moderately pointed; nose does not extend far past mouth; forehead steep. Whiskers white, very long (35–50 cm in bulls). Adult females and subadults of both sexes medium grey-brown above, paler below, palest, often whitish, on chest, throat, muzzle and face below eye. Usually a pale blaze on flanks extending towards hindflippers. Foreflippers darker than back. Adult males dark grey-brown to charcoal with paler guard hairs giving distinct silvery sheen to mane when dry; muzzle and face entirely grey. **Distribution, habitat and status** Confined to seas and islands around Antarctic Convergence, including Heard and Macquarie Is. Possible vagrant to Australia: one unconfirmed report from Kangaroo I., SA. **Similar species** Sub-antarctic Fur Seal has distinct yellow tones to chest and face contrasting boldly with upperparts. Australian and New Zealand Fur Seals have longer muzzle, nose extends well beyond mouth.

Crab-eater Seal *Lobodon carcinophagus*

tl to 2.6 cm; wt 200–350 kg
Slender, lithe body; head and muzzle moderately long and narrow relative to body size; top of head and muzzle flat with slight forehead; nostrils on top of muzzle near tip; foreflippers long and oar-shaped. Pelage silvery grey or pale yellow-brown, paler below with patches of dark brown blotches and rings, particularly on flanks. After moult coat is bright and shiny but fades steadily to overall creamy-white in summer. **Distribution, habitat and status** Circumpolar and abundant in Southern Ocean, breeds on pack ice. Stragglers occasionally reach s. Aust: 20 records from WA, SA, Tas, Vic and NSW. **Behaviour** Eats mostly krill sieved from the water at night using highly modified interlocking post-canine teeth; also some fish and cephalopods. **Similar species** Leopard Seal has massive head, long foreflippers. Weddell Seal has more rotund body, smaller head, distinctly spotted pelage.

Weddell Seal *Leptonychotes weddelli*

tl 2.1–3.3 m; **wt** 300–550 kg
Head appears small on long plump body; head flat without discernible forehead; muzzle short, blunt; mouth upturned at corners; eyes large, close-set. Foreflippers very short. Upper canines and incisors project forward, used to maintain openings in fast ice. Upperparts bluish-black after moult, fading to brownish-grey; sides of muzzle and patch over eye pale yellow or fawn. Underparts grey, streaked white; flanks heavily and variably spotted and blotched with silvery grey. **Distribution, habitat and status** Circumpolar in Southern Ocean between fast ice and Antarctic Convergence. Several records from Macquarie and Heard Is, one from Aust: Encounter Bay, SA in 1913. **Behaviour** When disturbed usually rolls onto side and raises a foreflipper. **Similar species** Crab-eater Seal is more slender, has longer foreflippers, proportionately larger head, noticeable forehead. Ross Seal has thick head and neck, very short muzzle, long brownish stripes along throat and flanks.

Ross Seal *Ommatophoca rossii*

tl males 1.7–2.1 m; females 1.9–2.3 m; **wt** males 130–200 kg; females 160–225 kg
The smallest southern true seal. Broad head, thick neck, short round muzzle. Uniformly dark grey or chestnut above, silver-white below with spotted sides. Distinctive parallel streaks of chestnut or brown along throat from eyes and mouth to foreflippers, and less obviously along flanks. **Distribution, habitat and status** Inhabits pack ice around Antarctica. One record from Heard Is and one from mainland Aust: Beachport in SA, Jan 1978. **Behaviour** When disturbed raises head almost vertically. **Similar species** Weddell Seal is larger, has small narrow head with long muzzle, lacks long streaks down throat and flanks.

PLATE 90

Irrawaddy Dolphin *Orcaella brevirostris*

Snubfin Dolphin

tl to 2.7 m; **wt** 90–135 kg

Small slow-moving dolphin; no beak; bulbous head with large melon, depression and creases at the 'neck', mouthline straight but flexible, eyes relatively large and bulging. Body thickset, tapering rapidly to tailstock. Crescent blowhole, offset to left of midline. Flippers long, broad with distinctly curved trailing edge; fin low (4.4% of tl), and well rounded; flukes broad with central notch. Subtly tri-coloured: bluish-grey cape, pale grey or brown-grey flanks, whitish underparts from flippers to anal region. **Distribution, habitat and status** Restricted to coastal and estuarine waters, mostly <20 m deep, in n. Aust from Onslow, WA to Brisbane R., se. Qld. Prefers muddy and brackish waters, will travel long distances up tropical river systems. **Behaviour** Inconspicuous, little activity above water, rarely bow-rides. Occurs in small groups of 2–5, often feeds by working into tidal flow taking fish, squid and prawns from the water column. **Similar species** Dugong lacks dorsal fin.

Indo-Pacific Humpback Dolphin *Sousa chinensis*

tl to 2.6 m; **wt** males to 250 kg, females to 170 kg

Robust dolphin with long, clearly defined, cylindrical beak, steep forehead with slight melon. In adults fin sits on ridge in middle of back; fin swept back at acute angle with rounded tip. Flippers short, rounded; flukes with distinct central notch. Body deep, adults develop keels above and below tailstock. Skin uniformly pale, light grey above, whitish or pale pink below. Young animals darker grey. **Distribution, habitat and status** Confined to coastal tropical and warm temperate waters, mostly <20 m deep, estuaries, tidal rivers, channels through mangroves. On open coasts is often in surf zone. In Aust recorded from about Exmouth in WA around n. coast to Coffs Harbour area, NSW. **Behaviour** Gregarious in small groups of 2–4, occasionally up to 25; eats mostly fish, some squid and crustaceans. Groups follow prawn trawlers but not known to bow-ride. When surfacing, exposes beak and head, then arches back strongly; flukes often lifted clear when diving. **Similar species** Pale colour, long beak that is exposed when surfacing, and ridged back of adults are diagnostic.

Rough-toothed Dolphin *Steno bredanensis*

tl to 2.6 m; **wt** to 155 kg

Characteristic beak and head shape: long, slender, cone-shaped beak merging into low forehead; curved mouthline extends to below the large eye. Long flippers with rounded tip set well back on body; fin tall, strongly back-curved. Skin mostly greyish; darker cape from top of head expanding to include most of flanks behind fin. Belly may be blotched pinkish white; lips and lower jaw white or pale pink. **Distribution, habitat and status** Tropical and subtropical waters of Pacific, Indian and Atlantic, usually in deep water beyond continental shelf. In Aust known only from a few records in northern waters. **Behaviour** Gregarious, usually in groups of 10–20, sometimes up to several hundred. When travelling fast, 'porpoises' with only head and back briefly visible. Rides bow-waves. Eats pelagic fish, cephalopods; stays submerged for longer than other dolphins: up to 15 min. **Similar species** Other long-snouted dolphins have a distinct crease at base of steep, sharply defined forehead, and lack white lips and lower jaw (except sometimes Pantropical Spotted Dolphin).

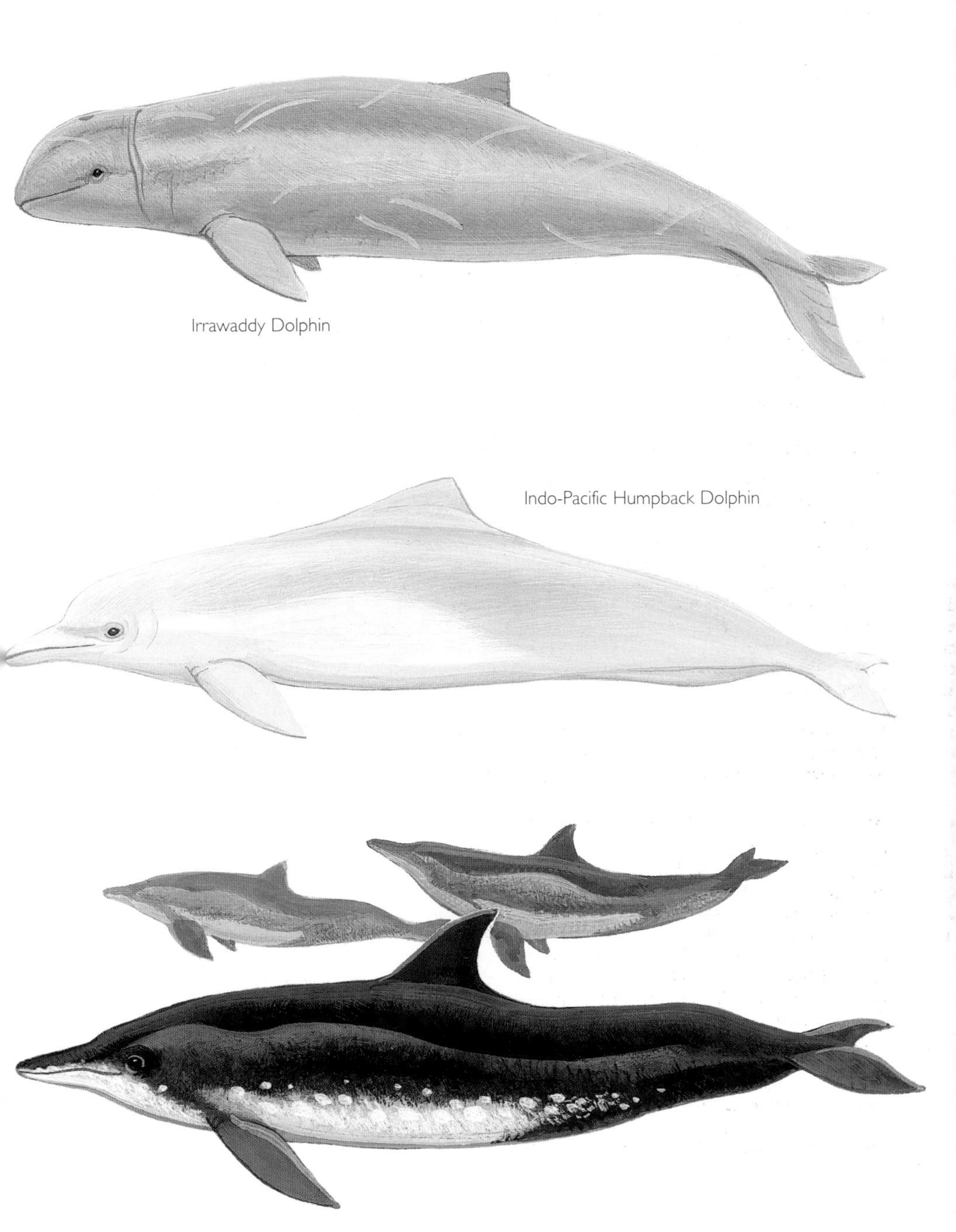
Irrawaddy Dolphin
Indo-Pacific Humpback Dolphin
Rough-toothed Dolphin

PLATE 91

Dusky Dolphin *Lagenorhynchus obscurus*

tl 1.6–2.1 m; **wt** 40–90 kg
Short, stocky; beak short but clearly demarcated from forehead; fin tall, not markedly hooked, dark grey with paler trailing edge; flippers strongly curved, grey with darker margins, darker than adjacent body; flukes blackish, narrow, pointed. Upperparts blue-black with dark diagonal band across flanks from below fin towards vent and along tailstock. Whitish or grey diagonal stripe from forward of fin onto tailstock. Flanks and sides of tailstock grey; underparts pale grey, chin white. Tips of beak and lower jaw dark grey continuing to eye and onto insertion of flipper. **Distribution, habitat and status** Throughout southern oceans, mostly s. of 26°S. Most sightings in cold inshore waters of continental shelf; rare in Aust waters: records from off Kangaroo I., Bass Strait and e. Tas, all Oct–Apr. **Behaviour** Gregarious in groups of up to 20; inquisitive and approachable; frequently bow-rides, leaps and somersaults; eats schooling fish and squid, especially Southern Anchovy. **Similar species** Common Dolphin has longer, more obvious beak, lacks diagonal flank blazes and has yellowish-white hourglass flank pattern.

Southern Rightwhale Dolphin *Lissodelphis peronii*

tl males to 2.9 m, females to 2.3 m; **wt** to 115 kg
Only dolphin in Southern Hemisphere that lacks dorsal fin. Striking black and white colour pattern: short white beak and forehead, rest of upperparts entirely black encompassing eyes to corners of mouth and near flipper inserts; very sharp boundary to clear white underparts. Flippers small, curved, pointed, may be edged blackish; flukes narrow, pointed, central notch. **Distribution, habitat and status** Circumglobal in s. hemisphere at mid to high latitudes (30–65°S) in deep water, often associated with cold currents and upwellings. Probably not uncommon but few sightings documented; recorded off s. and sw. Tas, sw. WA and in Great Australian Bight. **Behaviour** Highly gregarious, often in herds of 30–200, sometimes over 1000. Wary, usually does not remain near ships; travels fast in graceful, long, low-angle leaps. Feeds on pelagic fish. **Similar species** At a distance can be confused with porpoising penguins.

Risso's Dolphin *Grampus griseus*

tl to 4 (3) m; **wt** 250–500 kg
No beak, rounded blunt head with low melon and diagnostic longitudinal crease along midline from upper lip to blowhole. Mouthline slopes steeply upwards towards eye. Body robust forward of fin, tapering rapidly to narrow tailstock. Fin tall (to 50 cm), narrow, pointed (can be strongly hooked at tip), at midpoint of body; flippers long, curved, pointed; flukes broad, pointed, deeply notched. Colour when young uniformly dark grey, becoming paler with age except for fin, flippers and flukes. Mature animals are pale grey, whitish on head, throat and belly. Most mature individuals bear numerous pale scars crisscrossing body, particularly on back forward of fin. **Distribution, habitat and status** Found in low–medium latitudes in all oceans, usually well offshore over continental slope, but can be found close inshore, e.g. near Fraser I., Qld. In Aust recorded off s. and e. coasts. Probably common. **Behaviour** Gregarious in groups of 20 to several hundred. Feeds in pelagic waters, dives deep, leaps and 'porpoises', rarely bow-rides. **Similar species** Pilot Whales are much larger and blackish, have wide, curved fin. False Killer Whale and Melon-headed Whale are blackish.

Spectacled Porpoise *Australophocoena dioptrica*

tl 2.2 (1.65) m; **wt** to 80 (50) kg
The only porpoise known from Australian waters. Distinctive colour pattern with sharp boundary between black upperparts and white underparts, the white extending above the mouth and encircling the black eye-patch. Lips black; underside of flukes whitish. Narrow pale grey line from gape to flipper and along leading edge of flipper. Snout bluntly pointed with no beak. Fin broadly triangular with straight rear edge; taller with rounded tip in males. Flippers small and rounded. **Distribution, habitat and status** Circumpolar in sub-antarctic waters; only 2 records from Aust, both in SA, 1997. **Behaviour** Rarely observed and very little known. Inconspicuous, not known to leap or follow vessels. Occurs in 2s or 3s; eats squid and fish. **Similar species** Southern Right Whale Dolphin has similar colour pattern but lacks dorsal fin and has distinct beak.

Dusky Dolphin
Southern Rightwhale Dolphin
Risso's Dolphin
Spectacled Porpoise

PLATE 92

Fraser's Dolphin *Lagenodelphis hosei*

tl to 2.7 m; **wt** 130–190 kg
Small, robust, with short well-defined beak; flippers, fin and flukes appear rather small; fin triangular, sharply hooked at tip; adults have pronounced keels above and below tailstock. Medium grey above, pinkish white below with 2 parallel stripes from base of beak to near vent, the upper stripe is pale grey or cream, the lower blackish, becoming darker and wider with age. Throat and chin white; tip of lower jaw blackish; adults have blackish band from eye to flipper. **Distribution, habitat and status** Circumglobal in tropical and subtropical seas. Several records from Aust waters and strandings in WA, Qld, n. NSW and, exceptionally, at Geelong, Vic. Largely oceanic, in pods of 50–100, rarely thousands. Rarely bow-rides but often leaps clear of water. **Similar species** Spinner, Striped and Pantropical Spotted Dolphins have long beak, slender body, tall fin.

Pantropical Spotted Dolphin *Stenella attenuata*

tl to 2.4 m; **wt** to 120 kg
Slender, light build; long, narrow beak; tall, backswept fin; strongly convex pointed flippers; distinct keel on underside of tailstock. Colour variable; usually dark grey cape, narrow on head then sweeping low to flanks forward of fin; pale grey below cape with dark fin, flippers and flukes. With age skin becomes increasingly spotted and mottled, first on belly, then flanks and back. Facial pattern distinctive: dark eye-rings joined by narrow dark 'bridle' across base of beak; brownish line from flipper to lower jaw; beak blackish, edged white or pinkish. **Distribution, habitat and status** Widespread in tropical and subtropical waters, mostly oceanic but also on continental slope. Recorded off n., w. and e. coasts of Aust, s. to Augusta, WA and Sydney, NSW. **Behaviour** Highly gregarious in pods of up to 1000; larger herds always well offshore. Highly acrobatic, regularly bow-rides and leaps. **Similar species** The only heavily spotted dolphin in Aust waters, but Long-beaked Bottlenose Dolphin develops spotted belly. Spinner Dolphin has tall, upright fin, dark 'lips', and spins on long axis when leaping.

Spinner Dolphin *Stenella longirostris*

tl to 2.0 (1.8) m; **wt** to 95 kg
Small, lightly built; very long, narrow beak (to 19 cm); tall upright fin sometimes appears to lean forward; long, pointed flippers. Colour and fin shape variable; generally upper surfaces and appendages dark grey; broad pale grey or brownish patch on flanks from behind eye to flukes; belly paler still, often whitish. Upper mandible dark grey, lower pale grey but with dark 'lips'; dark eye-patch and 'bridle', dark line from eye to flipper. **Distribution, habitat and status** Throughout tropical and subtropical seas; in Aust recorded from WA n. of Bunbury, NT, Qld and ne. NSW. Primarily in deep waters. **Behaviour** Highly gregarious in herds of 5–200, sometimes 1000s. Supreme aerialist among cetaceans; leaps and spins on its long axis, often bow-rides for long periods. **Similar species** Pantropical Spotted Dolphin has backward curved fin, whitish 'lips', line from flipper insertion to lower jaw rather than to eye.

Striped Dolphin *Stenella coeruleoalba*

tl to 2.6 m; **wt** to 150 kg
Spindle-shaped body; tall backward-curving, moderately pointed fin; beak not elongated, forehead pronounced. Colour distinctive: upperparts blue-grey or brownish grey, including beak, fin, flippers and flukes; pale grey flanks; white or pinkish underparts with clear narrow black stripe from eye to vent, another from eye to flipper, and often a third thin line behind eye; may have upswept pale blaze from above eye towards fin, and corresponding dark blaze pointed forward and down from behind fin. **Distribution, habitat and status** Widespread in tropical to warm temperate waters around the globe. In Aust recorded from WA s. to Augusta, Qld and NSW s. to Illawarra coast. Usually well offshore. **Behaviour** Highly gregarious in herds of 100s or even 1000s. Conspicuous and active; frequent acrobatic leaps, including rotation of flukes in mid-leap; frequently bow-rides. Eats small fish, squid and prawns. **Similar species** Narrow dark stripes beginning at eye (not lower jaw), and flank blazes diagnostic.

Fraser's Dolphin
Pantropical Spotted Dolphin
Spinner Dolphin
Striped Dolphin

PLATE 93

Common Dolphin *Delphinus delphis*

tl 1.7–2.4 m; **wt** to 135 kg

Medium-sized; shortish beak; tall, pointed, slightly hooked fin; long, narrow flippers. Distinctive tricolour pattern: dark grey-brown back forming a 'V' shape below fin where the other colour elements intersect; white or pale grey belly and face; hourglass flank pattern comprising tan or pale yellow patch forward of fin and pale grey patch on tailstock. Distinct blackish stripe of varying width from flipper to lower jaw; dark eye-rings joined by narrow 'bridle'. Beak grey or blackish, often darkest at tip. An inshore form (subspecies *capensis*) is more slender, has longer beak, more extensive black on lower jaw and chin, less contrast in body colour. **Distribution, habitat and status** Worldwide distribution where water temperature is 10–28°C. Most common in deep water but also occurs near shore. **Behaviour** Gregarious, often in herds of >1000, but also in groups of <10. Highly active with much leaping, slapping and bow-riding. Opportunistic feeder on shoaling and mid-water fish and squid. **Similar species** Prominent hourglass flank pattern with cream, yellow or tan forward half is diagnostic. Striped Dolphin has stripe from eye, not lower jaw, to flipper.

Bottlenose Dolphin *Tursiops truncatus*

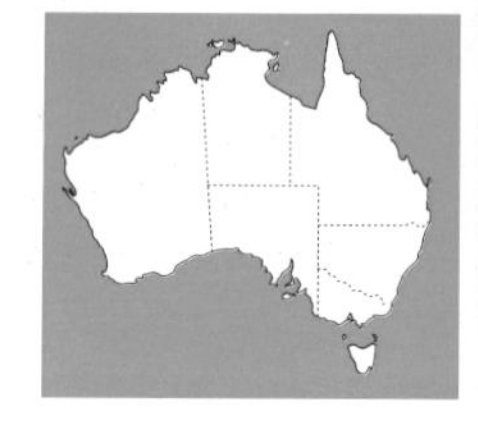

tl to 3.1 m; **wt** to 200 kg

Large, robust; uniformly mid-grey or brownish grey without obvious stripes or colour patches, paler below including lower mandible; distinct stubby beak (12 cm long) with obvious crease at base; steep forehead; mouthline curves upward giving characteristic amused expression; fin tall, backswept; flippers pointed. **Distribution, habitat and status** Throughout tropical and temperate oceans, inshore and offshore, in wide variety of habitats including bays, estuaries, harbours. In Aust most common around s. coast from about Hervey Bay in Qld to at least Albany, WA. Locally common. **Behaviour** Gregarious in small groups mostly <20; opportunistic feeder with catholic taste, but eats mostly fish and squid. Regularly bow-rides and 'porpoises'. **Similar species** Pantropical Spotted Dolphin can have uniformly grey upperparts but has much longer, narrower beak, bolder facial pattern of eye-rings and bridle, and spotted underparts. Rough-toothed Dolphin has long narrow beak which blends gently into forehead without a crease. See also Long-beaked Bottlenose and Common Dolphins.

Long-beaked Bottlenose Dolphin *Tursiops aduncus*

tl 1.8–2.1 m; **wt** to 120 kg

A warm-water inshore relative of the Bottlenose Dolphin, recently elevated to species status. Smaller, more slender, paler than that species, with longer beak (>15 cm) and larger flippers. Mature individuals have dark spots on whitish underparts. **Distribution, habitat and status** Mostly in warm, shallow, inshore waters, often <10 m deep but sometimes up to 10 km offshore. In Aust known from the n. and w. coasts from Perth in WA to Port Macquarie, NSW. Possibly also in Spencer Gulf, SA. The famous dolphins at Monkey Mia in Shark Bay, WA, are this species. **Behaviour** Little known; eats mostly fish from inshore reef and sandy benthic habitats.

Killer Whale *Orcinus orca*

Orca

tl to 9.5 m; **wt** to >8000 kg

By far the largest dolphin; unmistakable. Large, thickset, muscular body; blunt head. Tall fin (10–20% of tl), huge broad flippers. Body black with large white or yellow-white patch behind eye, also on underside from chin to vent with backward-pointing extensions up onto each flank behind the fin. Variable grey saddle just behind fin. Underside of flukes white or pale grey edged black. Blow visible in cold air—low and thick. **Distribution, habitat and status** Throughout all oceans; most common in cold waters and within 800 km of continents. In Aust recorded from all States but not NT; frequently recorded off Tas, Vic and SA, especially near seal colonies. **Behaviour** Opportunistic carnivore, eats fish, squid, seabirds, seals, other cetaceans including Minke, Blue and Sperm Whales. Hunts cooperatively when attacking large prey. Gregarious in small pods (mostly <10 in Aust waters, but up to 50). Inquisitive, breaching, spy-hopping and lob-tailing common, rarely bow-rides. **Similar species** Bold black and white pattern and tall fin are diagnostic.

Common Dolphin
Bottlenose Dolphin
Long-beaked Bottlenose Dolphin
♂
♀ fin
Killer Whale

PLATE 94

Long-flippered Pilot Whale *Globicephala melas*

tl males to 7 (6) m, females to 6 (5) m; **wt** males to 3000 kg, females to 1800 kg
Stocky; bulbous head, melon may overhang beak; mouthline slopes steeply up towards eye; distinctive low, rounded fin set $^{1}/_{3}$ of length along back, very long base, highly concave trailing edge; flippers long, slender, pointed (18–27% of tl) with distinctive 'elbow' in older animals; tailstock deep, strongly keeled; flukes deeply notched. Body grey-black or very dark brown; pale grey anchor-shaped patch on throat and chest; some individuals have whitish saddle behind fin and whitish diagonal streak behind eye. **Distribution, habitat and status** Temperate and subpolar waters of Southern Hemisphere. Mostly in deep water but will enter shallow water and strandings not uncommon. In Aust known from s. coasts n. to se. Qld. Strandings recorded in all States but not NT. **Behaviour** Highly gregarious in herds of 10s to 100s, often associates with other small cetaceans. Never bow-rides, moves at steady sedate pace; tends to ignore boats; entire herds sometimes lie motionless on surface (logging). **Similar species** Shape of head and fin is characteristic of *Globicephala*. Short-flippered Pilot Whale very difficult to distinguish at sea: flippers proportionately shorter, and 'crease' near blowhole gives impression of a neck. Little overlap in distribution.

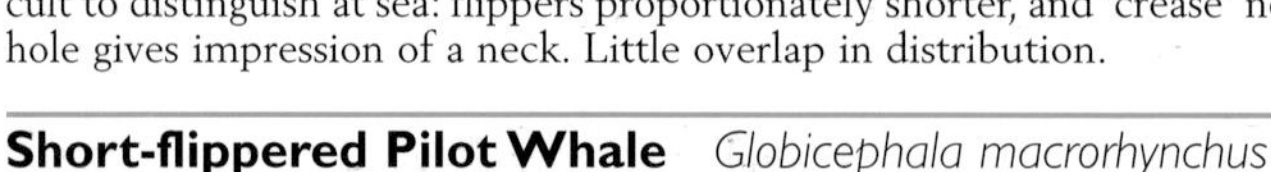

Short-flippered Pilot Whale *Globicephala macrorhynchus*

tl males to 6 (5.5) m, females to 5 (4.2) m; **wt** males to 2500 kg, females to 1500 kg
Very like Long-flippered Pilot Whale but flippers shorter (14–19% of tl), sickle-shaped without distinct elbow; head larger, melon rises from near blowhole giving impression of 'neck'. **Distribution, habitat and status** Tropical to temperate waters, s. to e. Vic. Usually in deep water. **Behaviour** Similar to Long-flippered Pilot Whale. **Similar species** See under Long-flippered Pilot Whale.

Melon-headed Whale *Peponocephala electra*

tl to 2.8 m, mostly 2.2–2.5; **wt** to 210 (160) kg
Small, slender, elongated; head slim, bluntly conical and slightly bulbous, giving a triangular plan; fin tall, backward-pointing, located midway along back, trailing edge often damaged; flippers medium length, smoothly curved, pointed. Colour charcoal or dark brown; darker mask on face and sometimes a darker cape, but this can be difficult to discern at sea. Variable pale grey area on throat and chest continuing as a midline to vent, where it widens; narrow whitish lips. **Distribution, habitat and status** Throughout tropical and subtropical seas, s. to about 35°S; predominantly oceanic in deep, warm waters (>25°C), rarely inshore. Recorded from WA, Qld and NSW waters. Nowhere common. **Behaviour** Highly gregarious, in herds of 50–500, rarely 1000s. Swims rapidly with low 'porpoising', will bow-ride. Eats mostly deepwater squid and small fish. **Similar species** Pygmy Killer Whale has more rounded head without taper, shorter fin with rounded tip, short blunt flippers; white 'goatee' patch on chin, and does not occur in very large herds. False Killer Whale also lacks tapered head, is much larger and more robustly built, flippers are widest at distinct 'elbow' before tapering to backward point.

Pygmy Killer Whale *Feresa attenuata*

tl to 2.7 m; **wt** to 225 kg
Small whale. Body robust forward of fin, slender behind; rounded head, mouthline angles steeply upwards to near eye; tall, curved fin with indentations on trailing edge; flippers gently curved, blunt. Head and back black, extending lowest onto flanks below fin; flanks grey from above eye to tailstock, narrowing below fin (this colour darkens rapidly after death); greyish anchor-shaped patch between flippers of variable intensity; large whitish vent patch. When breaching, entire head is exposed, revealing narrow white 'lips' that broaden markedly at tip of lower jaw, suggesting a white goatee. **Distribution, habitat and status** Inhabits deep tropical and warm temperate waters s. to 35°S. Few records from Aust waters but known from WA, NSW. **Behaviour** Little known and rarely seen; occurs in small groups, mostly about 10, rarely to 50; wary and difficult to approach; swims rather slowly, rarely bow-rides or leaps; eats fish and squid, may also prey on dolphins. **Similar species** Melon-headed Whale lacks chin patch, has longer pointed flippers, more tapered head profile. False Killer Whale is larger, more boisterous, often leaps and bow-rides, lacks white chin patch, has distinct 'elbow' in flipper.

Long-flippered Pilot Whale
Short-flippered Pilot Whale
Melon-headed Whale
Pygmy Killer Whale

PLATE 95

False Killer Whale *Pseudorca crassidens*

tl males to 6 (5.3) m, females to 5 (4.4) m; **wt** males to 2000 kg, females to 1200 kg
Body slender. Head long, slender; profile is flat-topped with rounded snout; lower jaw underslung; mouth long, reflecting long snout; from above, head tapers to a blunt snout; fin tall, back-curved; strongly keeled tailstock. Flippers set well forward, uniquely shaped: widest point at a distinct bent 'elbow' and flexing downward at tip. Colour is blackish with paler chest patch which can extend back towards navel. **Distribution, habitat and status** Tropical to temperate waters to about 45°S., usually in deep water. Numerous records in Aust waters, including mass strandings in most states. Uncommon. **Behaviour** Gregarious in herds of 20–50, sometimes aggregations of several hundred. Fast and acrobatic; leaps clear of water, bow-rides. Often swims with mouth open revealing robust conical teeth. Eats squid and large pelagic fish. **Similar species** Pilot whales have very bulbous head and broad, low, rounded fin. Pygmy Killer Whale and Melon-headed Whale are smaller, lack sharp bend in flipper, have white lips and grey patches on chest and throat.

Pygmy Sperm Whale *Kogia breviceps*

tl to 3.4 (3.0) m; **wt** to 400 (350) kg
Small, thickset body forward of fin, tapering quickly behind fin; head somewhat funnel-shaped with apex high above the tiny underslung mouth; fin tiny (<20 cm high), backswept, placed behind midpoint of back; flippers short and broad, placed well forward; tailstock strongly keeled right to fluke notch; flukes finely pointed; blowhole displaced to left of midline; snout-to-blowhole distance >10% of tl. Each lower jaw has 11–16 narrow, sharp teeth, none in upper jaws. Colour dark steel grey above, underparts pale grey tinged pink in life; whitish crescent between eye and flipper, can have whitish patch immediately forward of eye. **Distribution, habitat and status** Tropical to temperate seas, usually close to continental slope. Uncommonly seen. In Aust most records are from the se. but strandings recorded in all states. **Behaviour** Solitary or in small groups. Feeds on fish, squid, crabs. Slow, deliberate movements, rarely leaps or breaches, often rests at surface with body vertical and only back of head exposed; when startled may produce a rust-coloured defecation thought to provide cover while the whale dives; no visible blow. **Similar species** See under Dwarf Sperm Whale. Pygmy Killer Whale has long tapered head, no visible pale markings and tall, obviously backswept fin. Melon-headed Whale has long narrow flippers, tall fin, and uniformly curved head profile.

Dwarf Sperm Whale *Kogia simus*

tl to 2.7 (2.4) m; **wt** to 250 (160) kg
Similar to Pygmy Sperm Whale in form and colour pattern but the head is even more pointedly conic, the fin is taller (> 20 cm high), slightly backswept, situated close to midpoint of back; flukes less pointed. Each lower jaw has 7–12 narrow, curved, sharp teeth and up to 3 pairs of vestigial teeth may be present at tip of upper jaw. Snout to blowhole distance <10% of tl. Whitish crescent from behind eye to flipper. **Distribution, habitat and status** Recorded sporadically from tropical to temperate seas, mostly in deep water. Rarely recorded in Aust but strandings known from WA, SA, Tas, NSW. **Behaviour** Little-known, secretive. Swims slowly, submerges directly without a forward roll; often alone but also in small groups of <10; eats small deepwater fish, squid, cuttlefish. Like Pygmy Sperm Whale, stores large quantities of faeces in enlarged rectum and defecates when startled, perhaps to provide a camouflage cloud.

False Killer Whale

Pygmy Sperm Whale

Dwarf Sperm Whale

PLATE 96

Sperm Whale *Physeter macrocephalus*

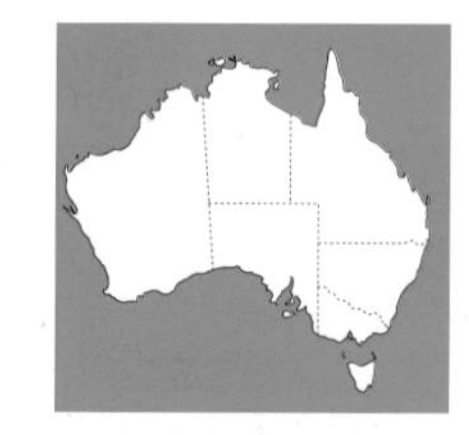

tl males to 18 (15–16) m, females to 12.5 (10–11.0) m; **wt** males to 57 (45) tonnes, females to 24 (20) tonnes
The largest toothed whale. Profile unmistakable. Huge square head occupying almost 1/3 of tl; blunt snout extends well beyond tiny lower jaw; fin reduced to broad triangular hump; several undulations along back towards flukes; flukes massive, broadly triangular with straight trailing edge and deep notch; flippers short, stubby; deep keel to underside of tailstock; blowhole at tip of head, slit-like, slightly raised, offset to left of midline. Skin dark purplish grey with patches of off-white on snout and underside of head; extent increases with age, almost entirely white animals are known. Blow is characteristic: angled forward, bushy, to 5 m high. Blows 5–6 times per minute and usually >20 times between dives. **Distribution, habitat and status** Worldwide in deep water off continental shelves. Concentrates where seabed rises steeply from great depth. In Aust recorded from all states; concentrations occur 20–30 nm off sw. WA, sw. of Kangaroo I. in SA, off s. and w. coasts of Tas, and off Stradbroke I., Qld. The most abundant of the great whales. **Behaviour** Gregarious in herds of about 20–30. Travels slowly when at surface, often almost motionless. Dives deep, lifting flukes high; often remains submerged for > 1 hour and may descend to 3000 m; eats deep-sea squid, fish and shrimps. **Similar species** Humpback Whale has white markings on underside of flukes and irregular, knobbed, concave trailing edge to flukes; usually blows only a few times before diving.

Humpback Whale *Megaptera novaeangliae*

tl to 18 (13.5) m; **wt** to 45 (25–35) tonnes
Stout body; broad rounded head with clumps of tubercles and a distinct lump beneath tip of lower jaw; flippers extremely long (to 5 m, 25–33% of tl) with knobby leading edge; fin low, broad-based, hooked; flukes large, deeply notched with fluted trailing edge and backswept tips; throat pleats extend well beyond flippers. Upperparts dark grey to bluish black; underside of flippers always white; throat, flanks and underside of flukes variably white or black, often in an individually recognisable pattern. Flukes often raised high out of water when diving. Blow a single vertical bushy plume to about 3 m. **Distribution, habitat and status** Worldwide. In Aust a winter migrant along w. and e. coasts. Usually travels singly or in small groups, aggregates in favoured warm, shallow breeding locations such as Hervey Bay and inner Great Barrier Reef on e. coast, near Rottnest I., off Shark Bay and Dampier Arch. on w. coast. Vulnerable. **Behaviour** The most energetic of the large whales, frequently breaches, slaps tail or flippers and spy-hops. Swims slowly; curious, will approach boats. **Similar species** Sperm Whales also raise flukes high when diving but have no white on undersides. Southern Right Whale flukes are all black, sharply pointed, lack knobby trailing edge.

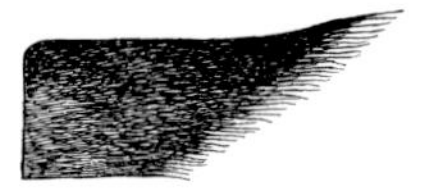

baleen plate

Southern Right Whale *Eubalaena australis*

tl 13.5–17.5 m; **wt** 40–80 tonnes
Distinctive bulky, rotund body tapering rapidly to narrow tailstock; very large head with obvious white callosities around mouth, above eyes and on top of head; strongly arched mouth to accommodate baleen plates almost 2 m long; lacks fin and throat pleats, flippers broadly fan-shaped; flukes all dark, pointed, smoothly concave trailing edge. Colour mostly blue-black or dark brown; white patches of variable size and shape on belly and sometimes on back. Blow V-shaped when seen from ahead or behind, left plume taller than right. **Distribution, habitat and status** Circumglobal in temperate and subpolar waters, mostly 30–60°S; feeds at high latitudes in summer and migrates n. in winter to birth in shallow bays around southern continents. In Aust occurs May–Oct around s. coast, mostly between Perth and Sydney, but occasionally n. to North West Cape and s. Qld waters, often very close to shore and readily observed. Aust population 600–800 in any one year. Vulnerable. **Behaviour** Does not feed in Aust waters. Swims slowly, blowing every 1–2 min., dives rarely exceed 15 min. Often breaches, lobtails, slaps flippers or rolls belly up. Births in late July–Sept, on average at 3 year intervals. Suckling mothers fast for over 4 months. **Similar species** Only other large whales to routinely show flukes are Sperm and Humpback. Sperm Whale flukes are dark, heavy, broadly triangular, straight trailing edge, deep notch; Humpbacks always have some white underneath, a serrated trailing edge and backswept tips.

baleen plate

PLATE 97

Pygmy Right Whale *Caperea marginata*

tl to 6.5 (5) m; **wt** to 4.5 tonnes

The smallest baleen whale. Head long, narrow with highly-arched jaw, cream baleen and white gums; fin triangular, backswept, set 2/3 of tl back from snout, not often exposed; flippers small, narrow, rounded tips. Colour dark grey or blue-grey above, underparts, including all of lower jaw whitish; pale diagonal stripe from behind blowhole to flanks. **Distribution, habitat and status** S. hemisphere between about 30 and 50°S. In Aust recorded as singles, mostly stranded, in sw. WA, SA, w. Vic, Tas, and se. NSW. **Behaviour** Feeds on plankton skimmed from the surface. Usually alone or in small loose group. Inconspicuous and slow-moving; when breathing, top of head and mouth exposed momentarily then sinks from view. Tailstock and flukes rarely exposed. **Similar species** Both Minke Whales are similar in size but lack strongly bowed jawline, arch tailstock and fin when diving, have distinct throat pleats.

baleen plate

Dwarf Minke Whale *Balaenoptera acutorostrata*

tl males to 7 (6) m, females to 8 (7) m; **wt** to 9 tonnes

The smallest rorqual. Body slender; rostrum flat, acutely pointed with a single prominent central ridge, mouthline straight; fin tall, backswept, placed in rear 1/3 of back; flippers pointed, about 12% of tl; flukes narrow, pointed, not strongly backswept. Throat pleats do not extend far beyond flipper insertion. Upperparts dark blue-grey; flippers have striking white patch or band close to their base. Darker pigment in the neck region extends down onto throat pleats, and 2 other V-shaped dark patches along body extend down onto flanks. Dark patch on underside of tailstock. **Distribution, habitat and status** Summers in sub-antarctic waters. General movement n. in winter as far as n. Qld, often found close inshore, will enter bays and estuaries. **Behaviour** When surfacing, blowholes and fin become visible simultaneously, flukes do not normally show but tailstock is arched high when submerging. Mostly alone or in pairs, readily approaches ships. Eats mostly krill and small shoaling fish, herding them into compact shoal before cruising through with mouth open. **Similar species** Antarctic Minke Whale lacks white patch on flippers, dark pigment on neck does not extend below level of eye. Bryde's Whale is larger, has less pointed rostrum with 3 longitudinal ridges, taller more erect fin, throat pleats extend past half way along underside; when surfacing, head and blow appear before fin.

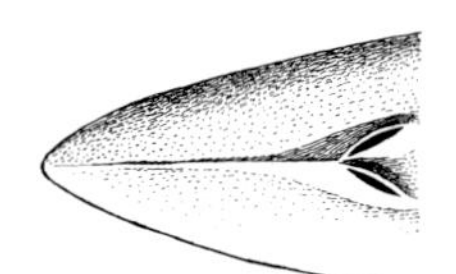

plan of rostrum

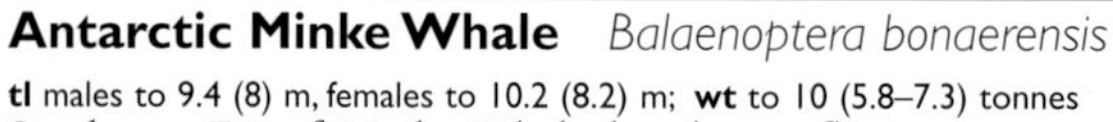

Antarctic Minke Whale *Balaenoptera bonaerensis*

tl males to 9.4 (8) m, females to 10.2 (8.2) m; **wt** to 10 (5.8–7.3) tonnes

Similar to Dwarf Minke Whale but larger, flippers are grey without distinct white patch, dark pigment of the back does not extend down onto flanks or throat, lacks dark patch on underside of tailstock. **Distribution, habitat and status** Summers in Antarctic waters, moves n. in winter but mostly remains s. of 35°S, rarely n. to Qld waters. Tends to remain offshore. **Behaviour** As for Dwarf Minke Whale. **Similar species** See under Dwarf Minke Whale.

Sei Whale *Balaenoptera borealis*

tl males to 17 (15) m, females to 19 (16) m; **wt** to 26 (12–15 tonnes)

Long, streamlined, fast. Rostrum sharply pointed, flat-topped, with single central ridge and noticeably arched lower jaw; fin tall (50 cm), erect, moderately hooked, about 2/3 tl from snout; flippers relatively small (9% of tl); tailstock laterally compressed, forming sharp ridge to notch between flukes. Throat pleats end at about midbody. Upperparts uniformly steel grey, whitish throat grading to grey underparts. Blow cone-shaped, to 3 m high. **Distribution, habitat and status** Worldwide, but avoids the highest latitudes; usually in deep water. Infrequent in Aust waters: records from WA, Qld, e. Great Australian Bight, w. Bass Strait, and s. of Tas. Vulnerable. **Behaviour** Slow sedate movement, submerges without a splash, does not arch tailstock or show flukes; when surfacing, blowholes and fin become visible simultaneously. Shy, never approaches ships; usually in small groups. Feeds by surface-lunging in swarms of crustaceans and small shoaling fish, rolls to one side with mouth open and one flipper raised out of water. **Similar species** Bryde's Whale has 3 ridges along rostrum, white throat may be visible, throat pleats extend half way along underside, arches fin and tailstock high when submerging but flukes do not emerge; when surfacing, head and blow appear before fin; often inquisitive. Fin Whale is longer, more slender; has asymmetrical colour pattern on head, fin meets back at acute angle.

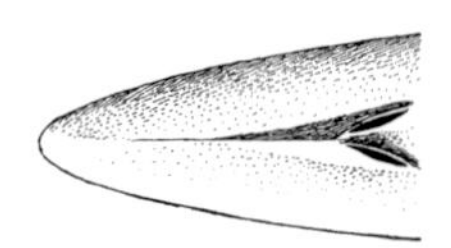

plan of rostrum

Pygmy Right Whale
Dwarf Minke Whale
Antarctic Minke Whale
Sei Whale

PLATE 98

Bryde's Whale *Balaenoptera edeni*

Tropical Whale

tl to 16 (14) m; **wt** to 24 tonnes
Similar in form and colour to Sei Whale but is shorter, more slender; has 3 ridges along rostrum; throat pleats extend to about the navel, more than half way along underside; fin less upright, more strongly hooked, can have ragged trailing edge; undersides may have yellowish or mauve tinge, underside of flukes dirty white. When surfacing, head and blow appear before fin. **Distribution, habitat and status** Circumglobal in tropical to warm temperate seas (water temperature >17°C). Can be inshore or offshore. Recorded from all Aust states but not NT; concentrations occur off central WA and near Great Barrier Reef. **Behaviour** Not migratory; feeds year-round on swarming crustaceans and small schooling fish. Energetic when feeding, often erratic, churns the water. Usually alone or in pairs but can form groups of up to 20; inquisitive, will approach boats. **Similar species** See under Sei Whale.

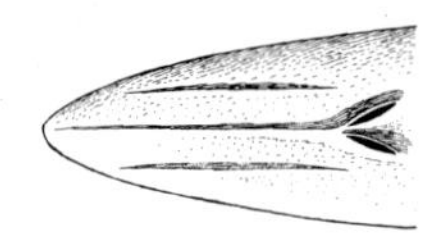

plan of rostrum

Fin Whale *Balaenoptera physalus*

tl males to 25 (21) m, females to 27 (22) m; **wt** to 90 tonnes
Narrow, elongated, laterally compressed body; rostrum pointed with distinct central ridge; fin tall (60 cm), forming acute angle to back (< 40° set 3/4 of way back from snout; flippers shortish (12% of tl), pointed; Throat pleats extend to midpoint of body; tailstock distinctly ridged. Upperparts dark grey or grey-brown, underparts whitish. Undersides of flippers and flukes have distinct white areas. Unique bicoloured head pattern: right-hand lower jaw and baleen plates white (sometimes extending onto head near blowhole), left side dark grey. May be pale chevrons behind head. Upon surfacing, head and blow appear shortly before fin; blow tall, slender, 4–6 m high. **Distribution, habitat and status** Worldwide in deep water. Migrates between cold water summer feeding grounds and warm water winter and spring breeding grounds. Recorded from all Aust States except NT. Vulnerable. **Behaviour** More often in small groups (2–7) than other rorquals. Feeds by rolling onto right side with mouth open, mostly on krill in Antarctic waters. **Similar species** See under Sei Whale. Blue Whale is mottled blue-grey, has small, low fin set far back on tailstock, a more U-shaped rostrum, and taller, thicker blow.

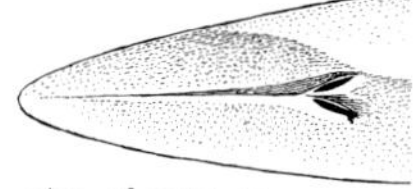

plan of rostrum

Blue Whale *Balaenoptera musculus*

tl males to 30 (25) m, females to 33 (26) m; **wt** to 160 tonnes
Largest mammal ever to live. Long, slender, strongly laterally-compressed body; from above rostrum U-shaped, not sharply pointed, pronounced ridges around blowholes merging into indistinct central ridge to tip of snout; small, variably-shaped fin located 3/4 of way back from snout; flippers slender, long for a rorqual (14% of tl); flukes wide (to 25% of tl), pointed, trailing edge slightly concave; throat pleats extend back to navel. Upperparts blue-grey or slate-grey mottled with paler blotches; paler below, whitish throat may appear yellowish when coated with diatoms. Blow very tall and vertical (to 9 m); fin usually not visible until head has submerged; flukes may show when diving. The Pygmy Blue Whale subspecies, *B. m. brevicauda*, is 21–24 m long, has proportionately shorter tail-stock. **Distribution, habitat and status** Worldwide. Recorded from all Aust states, with concentration off w. Vic and se. SA in summer and autumn. Most records in s. Aust waters are Pygmy Blue Whales. Endangered. **Behaviour** Usually occurs singly or as widely spaced small groups. Feeds on krill at surface by ingesting huge gulps of water, then expelling the water through sides of mouth to sieve out krill and other plankton. **Similar species** See under Fin Whale.

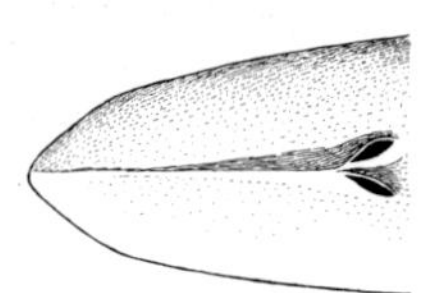

plan of rostrum

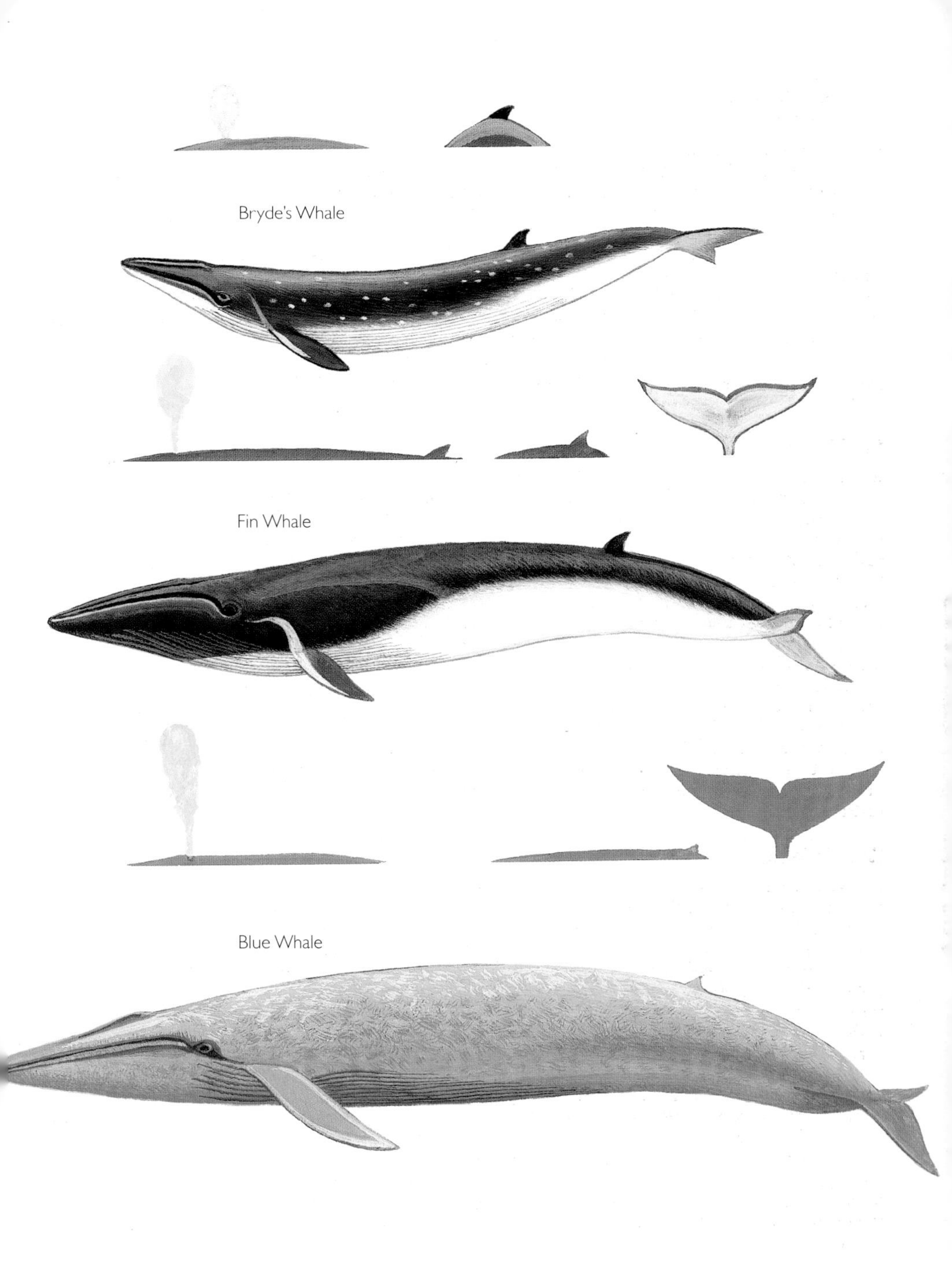

PLATE 99

Andrew's Beaked Whale *Mesoplodon bowdoini*

tl to 5 m; **wt** 1.0–1.5 tonnes
Body heavy with flat dorsal profile; small head, slight melon, thick shortish beak; mouthline arched upwards with one pair of flattened teeth set in raised sockets about halfway along beak; teeth erupt only in adult males and curve slightly outwards to lie outside upper jaw. Tailstock narrow; belly deep; fin low, round-tipped; flippers small, paddle-like; flukes all dark, without notch but slightly fluted trailing edge. Mostly blue-black except for forward $^2/_3$ of beak, which is whitish; belly may also be whitish. Prominent pale linear scars over upperparts. **Distribution, habitat and status** Confined to temperate waters of s. hemisphere. Rarely seen at sea; stranding records concentrated around Aust and NZ, including s. WA, SA, Vic, NSW. **Behaviour** Almost nothing known. Presumably prefers deep waters and avoids boats. Feeds on deepwater squid and fish. **Similar species** Blainville's Beaked Whale also has a tooth on each side about halfway along beak of males, but has very highly arched jawline, almost enclosing upper jaw, proportionately longer beak; females have extensive pale underparts. Ginkgo-toothed Beaked Whale indistinguishable at sea, but teeth of males barely protrude above gum. Goose-beaked Whale lacks obvious beak, shows flukes when diving, and occurs in larger groups.

Blainville's Beaked Whale *Mesoplodon densirostris*

Dense-beaked Whale

tl males to 5.8 m, females to 4.8 m; **wt** 1.0–1.5 tonnes
Similar build to Andrew's Beaked Whale but more evident dip in neck behind blowhole, taller, triangular backward-curved fin; long, tubular beak; lower jaw arched high over rear half of upper jaw; adult males have large, forward-pointing tooth at crest of the arch, protruding above upper jaw. Adults dark blue-grey, often with pale spots and streaks and darker eye-patch. Females develop whitish beak and extensive pale belly; calves are paler than adults. When surfacing, beak and melon appear first, followed by an indistinct blow and curved back with prominent fin as the animal dives; flukes are not exposed. **Distribution, habitat and status** Worldwide in tropical to temperate seas, mostly in deep water. Strandings recorded in all Aust states except SA and NT, s. to n. coast of Tas. **Behaviour** Seldom seen, avoids boats; usually in small groups; eats mostly squid and deep sea fish; deep dives may last 45 min. **Similar species** Probably only males can be identified at sea if close enough to see highly arched lower jaw with erupted tooth at its crest; colour pattern and prominent fin also useful. See also under Andrew's Beaked Whale.

Ginkgo-toothed Beaked Whale *Mesoplodon ginkgodens*

tl to 5 m; **wt** est. to 1.5 tonnes
Similar to Andrew's Beaked Whale, but beak is relatively long; single pair of teeth in males barely break through gums. Not as heavily scarred as other beaked whales. Males uniformly dark blue-grey except for beak, which is pale grey, and scattered small whitish spots on underside; females are paler grey, specially on underside. **Distribution, habitat and status** Few records, no confirmed sightings of live animals at sea. Presumably far from land in cool temperate to tropical waters. Known in Aust from 4 strandings of single animals: 3 in NSW, 1 w. Vic. **Behaviour** Unknown, probably similar to others in genus. **Similar species** Probably indistinguishable from other *Mesoplodon* species at sea; dark colour, comparative lack of scarring and position of teeth in males are useful clues. See also under Andrew's Beaked Whale.

Andrew's Beaked Whale

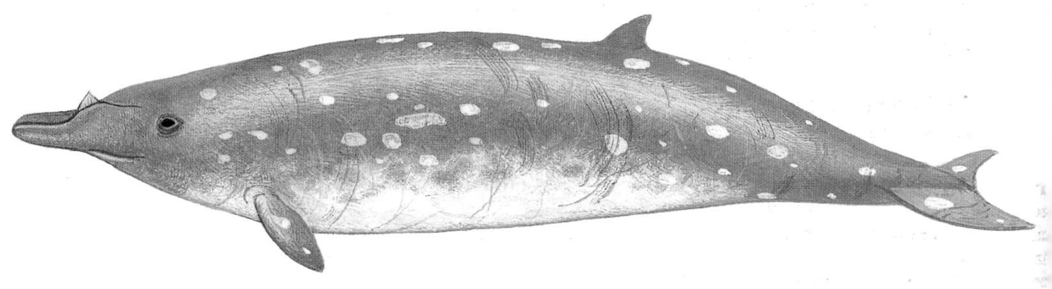

Blainville's Beaked Whale

Ginkgo-toothed Beaked Whale

PLATE 100

Strap-toothed Beaked Whale *Mesoplodon layardi*

tl to 6.3 (5) m; **wt** est. to 2 tonnes
The largest mesoplodont; body slender, laterally compressed, marked keel along tailstock, fin small, back-swept; beak long, slender; forehead steep; no notch in flukes. Adult males have a pair of long, strap-shaped tusks rising from halfway along lower jaw and curving back and over upper jaw, almost enclosing it. Adults blackish with broad areas of white or pale grey across back, forward of fin (most developed in older animals), down onto throat and lower jaw; oval patch around vent; distinct dark mask encloses eyes but not blowhole. Older animals heavily scarred. Immatures have colour pattern reversed, with less contrast. **Distribution, habitat and status** Circumglobal in mid latitudes of s. hemisphere (25–55°S). Strandings recorded in all Aust states but not NT, mostly along e. and s. coasts, and mostly Dec–Mar. Inhabits deep water but may move to inshore waters to calve. **Behaviour** Inconspicuous, wary of boats; when approached sinks imperceptibly, or may roll to one side, raising one flipper but not flukes, as it dives. Eats mostly squid, deep sea fish, crustaceans. Shows beak when surfacing. **Similar species** Colour pattern distinctive; strap-like teeth of males often visible when surfacing, but may be obscured by barnacles.

True's Beaked Whale *Mesoplodon mirus*

Wonderful Beaked Whale

tl to 5.4 (4.6) m; **wt** to 1.5 tonnes
Head small, melon distinct with depression behind blowhole, beak short, cone-shaped; males have pair of forward-pointing teeth at tip of lower jaw which lie outside the closed upper jaw; fin small, moderately backswept and pointed; flippers small, paddle-shaped, placed well forward and low; flukes broad, pointed, darker than tailstock; tailstock narrow, compressed with strong ridge. Upperparts dark grey forward of fin, paler behind fin and on melon and face; tip of beak whitish; dark eye-patch; underparts pale grey or cream, sometimes with darker spotting on flanks. Males heavily scarred. **Distribution, habitat and status** In Aust waters known only from strandings in WA, Vic and Tas. Probably prefers deep oceanic waters in cool temperate zone. **Behaviour** Unknown; shy and difficult to approach. **Similar species** Probably cannot be positively identified at sea. Hector's Beaked Whale very similar, fin more broadly based, leaves back at lower angle; teeth of adult males 2–3 cm behind tip of lower jaw, upright and triangular.

Hector's Beaked Whale *Mesoplodon hectori*

tl to 4.5 m; **wt** est. 1 tonne
The smallest beaked whale. Similar in shape to True's Beaked Whale. Fin low but long, leading edge joins back at sharp angle, tip rounded. Adult males have a pair of flat, triangular teeth protruding vertically from near tip of lower jaw. Stranded animals dark grey-brown above, paler below, palest on lower jaw and chin. In males the beak, chin and underside of flukes may be whitish. **Distribution, habitat and status** Circumpolar in cool temperate waters of s. hemisphere, mostly in deep oceanic waters. Known in Aust from strandings of single animals in SA and Tas. **Behaviour** Almost nothing known. Diet assumed to be deep-water squid and fish. Occasionally allows boats to approach. **Similar species** True's Beaked Whale probably indistinguishable at sea, has more upright fin; males have forward-pointing teeth at tip of lower jaw.

Strap-toothed Beaked Whale

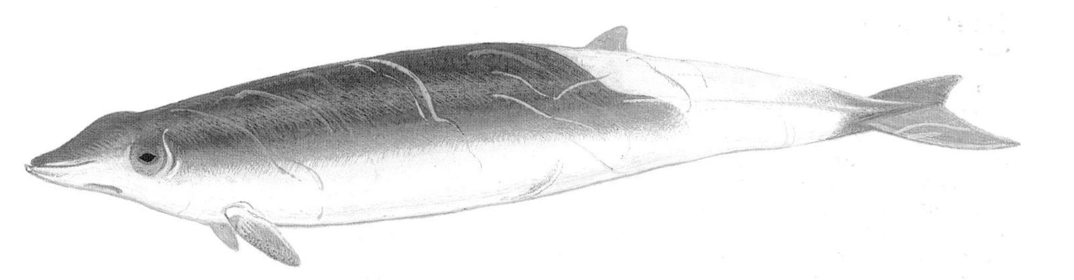
True's Beaked Whale

Hector's Beaked Whale

PLATE 101

Gray's Beaked Whale *Mesoplodon grayi*

Scamperdown Whale

tl males to 4.8 m, females to 5.6 m; **wt** 1.0–1.5 tonnes
Body spindle-shaped, slender; head small, flat with long, cylindrical beak and straight mouthline. Fin low, hooked; tailstock ridged along upper surface. Adults dark purplish grey except for white beak and whitish or yellowish patches on underparts, particularly in genital area. In males a single pair of teeth, 6 cm wide with serrated upper edge, may erupt about 20 cm from tip of lower jaw. White beak may be exposed at 45° when surfacing. **Distribution, habitat and status** Circumglobal in s. hemisphere at mid to high latitudes. The second most commonly stranded beaked whale. Aust records extend around s. coast from near Bunbury in WA to s. NSW, including Tas; mostly Dec–Apr, suggesting seasonal movement inshore to calve. **Behaviour** Little known; probably more gregarious than other beaked whales, often in groups of 6 or more; up to 28 have stranded together. Spends more time at surface than usual for beaked whales, known to breach, also 'porpoise' when travelling rapidly. Presumed to eat deep-sea squid and fish. **Similar species** Long, narrow white beak and small flat head diagnostic, if good view obtained.

Goose-beaked Whale *Ziphius cavirostris*

Cuvier's Beaked Whale

tl males to 7 m, females to 6.5 m; **wt** to 3.5 tonnes
Robust, spindle-shaped body with small head, indistinct melon, forehead sloping gently to stubby, poorly defined beak, short mouthline curves upwards at rear giving impression of smile. Flippers small, rounded tips; fin low (to 40 cm), somewhat backswept, $^2/_3$ of tl from beak; flukes broad (to 25% of tl). Males have 2 conical teeth at tip of lower jaw protruding beyond upper jaw. Colour variable; adult males mostly mid-grey, variable area of white on head and back forward of fin, pale linear scarring; females can be browner, even tan. Darker swirls and creamy blotches on underside. Calves blackish above, paler below. At surface, swims with lurching motion, exposing top of head; shows flukes when diving deep; blow low, inconspicuous. **Distribution, habitat and status** Throughout temperate to tropical seas in deep water, mostly >1000 m. Strandings not uncommon, recorded from all Aust states and NT. **Behaviour** Inconspicuous and rarely seen, but approaches boats at times. Usually alone or in groups up to 15. Eats deep sea squid and fish. **Similar species** Sloping forehead, broad stubby beak, combined with pale head and exposed teeth of males are diagnostic.

Shepherd's Beaked Whale *Tasmacetus shepherdi*

Tasman Whale

tl males to 7 m, females to 6.5 m; **wt** 2–2.7 tonnes
Robust, spindle-shaped body; elongated narrow beak, steep forehead and moderate melon; flippers small; fin low, backswept; flukes broad with fairly straight trailing edge. Unlike all other beaked whales, adults of both sexes have numerous teeth: 17–29 in each row, with elongated pair at tip of lower jaws in males. Few sightings of live animals, so colour patterns poorly known; brownish grey above with large dark cape extending onto flanks forward of fin; dark band from behind head to flipper insertion separates two broad whitish patches—whitish throat extends to level of eye forward of flippers, and behind flippers pale belly extends above eye level. **Distribution, habitat and status** Known from a small number of strandings from mid to low latitudes in s. hemisphere, mostly in NZ but also WA and SA. Unconfirmed sightings from NZ. **Behaviour** Unknown, probably eats mostly fish taken at depth, probably shy of boats. **Similar species** Forehead steeper, beak more pointed than other beaked whales. Only beaked whale with numerous functional teeth.

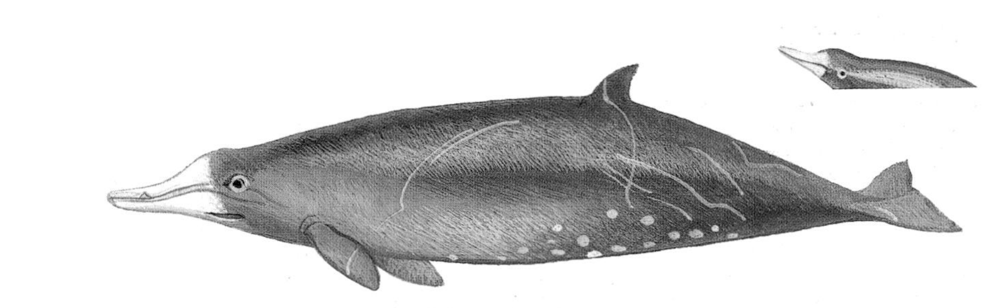

Gray's Beaked Whale

Goose-beaked Whale

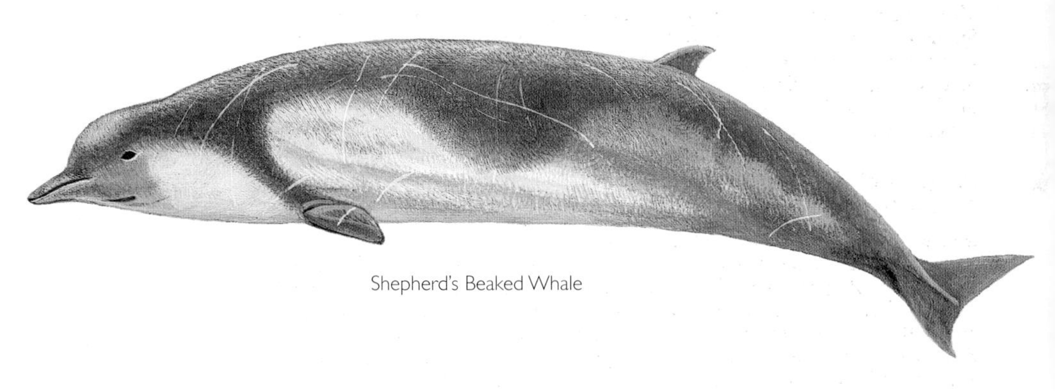

Shepherd's Beaked Whale

PLATE 102

Arnoux's Beaked Whale *Beradius arnuxii*

tl to 10 m; **wt** to 9 tonnes
The largest beaked whale in Southern Hemisphere. Spindle-shaped body; small head with steep forehead and prominent melon; distinct short tubular beak, lower jaw extending beyond upper jaw on all sides and bearing 2 pairs of conical teeth in all adults, the outer pair exposed outside the closed mouth; mouthline straight. Flippers very short and rounded; fin very small, backward pointing. Most animals distinctly brown; paler, sometimes whitish, on crown and belly; males often heavily scarred with pale slashes of 2 parallel incisions. Blow low, bushy, indistinct; sometimes shows flukes before a deep dive. **Distribution, habitat and status** Circumpolar s. of about 30°S, mostly s. of 40°, in deep cool temperate and subpolar waters. In Aust stranded animals recorded in summer from s. WA, SA and Tas; possible sightings off SA and s. NSW. Many sightings in Cook Strait, NZ, in spring–summer. **Behaviour** Often in tight groups of 6–10, aggregations up to 50 reported. Shy and difficult to observe. Can remain submerged for 1 hour. Probably eats mostly deepwater squid. **Similar species** Southern Bottlenose Whale may be indistinguishable at sea; smaller with shorter beak; well-developed melon that overhangs base of beak, forming indent between beak and melon; taller fin. Tropical Bottlenose Whale has distinctly larger backswept fin.

Southern Bottlenose Whale *Hyperoodon planifrons*

tl to 7.8 (6.5) m; **wt** to about 6 tonnes
Thick, robust body; very bulbous head rising perpendicularly from short, thick beak, in mature animals can overhang, forming indent between beak and melon; rear part of mouthline curves upward giving smiling appearance. Fin tall for beaked whale (to 40 cm), about 2/3 tl from snout; flukes broad, appear disproportionately large. Colour variable; upperparts often brownish with white patches on melon and beak; flanks and belly paler with irregular whitish patches; can have white flashes on undersides of flukes. Immature darker, more greyish. When surfacing shows beak and melon first; shows flukes when diving deep; blow low (1–2 m), dense, projects slightly forward. **Distribution, habitat and status** Temperate to Antarctic waters, mostly s. of 30°S; rarely inshore. Stranding records from s. Aust. **Behaviour** Occurs in small, close-knit groups; dives deep and remains submerged for up to 1 hour; probably eats mostly deep sea squid. **Similar species** See under Arnoux's Beaked Whale.

Tropical Bottlenose Whale *Indopacetus pacificus*

Longman's Beaked Whale

tl est. 6–8 m
The least-known of all whales. Large, with distinct whitish melon and tall backswept fin; melon not as bulging as in Southern Bottlenose Whale but rises perpendicularly from beak in some animals. Females generally tan or grey-brown; males grey, fin and area around base darker; becoming paler and more scarred with age until almost entirely white; juveniles darker grey-brown above with distinct whitish melon and white sides; dark transverse band separates pale melon and pale flanks, with embedded white 'ear' patch. Two forwardly inclined teeth at tip of lower jaw (presumably only in males and not visible at sea). **Distribution, habitat and status** Sightings of bottlenose whales in tropical Pacific and Indian Oceans now believed to be this species. Known with certainty from only 5 specimens, including a skull from Mackay, Qld, in 1882. **Behaviour** Pelagic in tight groups of 5–20, up to 100. Often exposes beak and melon when travelling fast. **Similar species** Southern Bottlenose Whale has more pronounced melon and constriction of rear margin of melon, pale colour on melon extends back beyond blowhole; immatures lack pale flanks, transverse dark band and whitish ear patch.

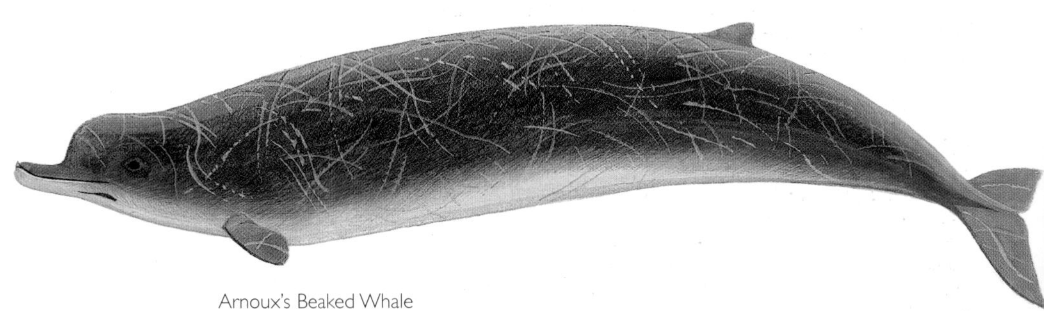

Arnoux's Beaked Whale

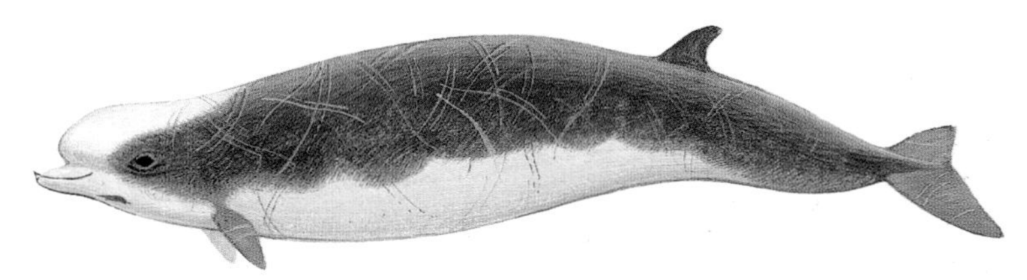

Southern Bottlenose Whale

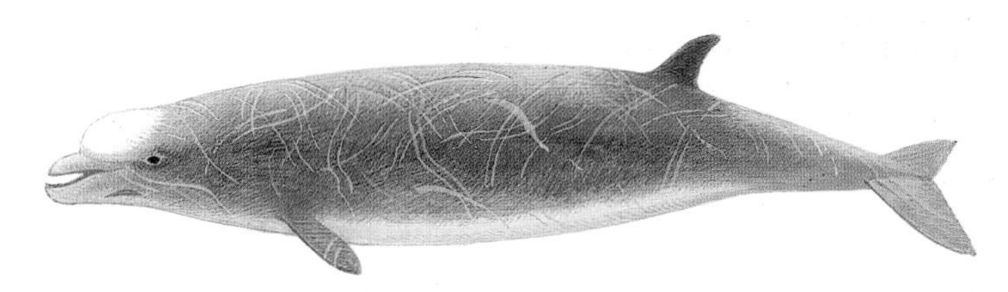

Tropical Bottlenose Whale

PLATE 103

Dugong *Dugong dugon*

tl to 3 m; **wt** to 420 kg
Unmistakable; rotund body; large triangular tail fluke with central notch; no dorsal fin; small flippers; muzzle deflected downwards, ending in large grasping disk with stiff, backward-pointing bristles. Head small; eyes and ears small, indistinct; nostrils on top of muzzle tip, with valve-like flap. Skin smooth, slate-grey, olive-grey or bronze, underparts pinkish white. **Distribution, habitat and status** Sheltered coastal seas and estuaries of n. Aust and Torres Strait, from Shark Bay in WA to Moreton Bay, se. Qld. Vagrants recorded s. to se. NSW. Dependent on sea-grass beds. Declining in e. Aust, probably stable in w. **Behaviour** Entirely marine, gregarious in herds of 5 to over 100. Slow-moving, surfaces discreetly; alert; grazes sea-grass beds. Slow growth rate and low reproductive capacity: females produce one calf every 3–7 years, usually in Sept–Apr. May live to 70 years. **Similar species** Could be confused only with small cetacean lacking dorsal fin, but has nostrils not blowhole, deep triangular flukes.

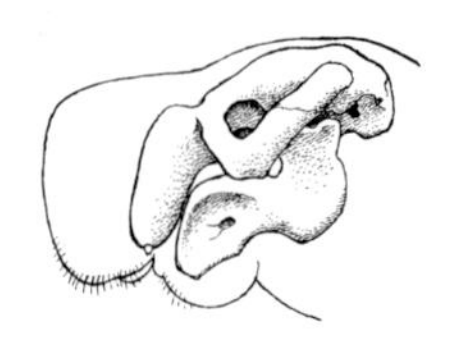

mouthparts

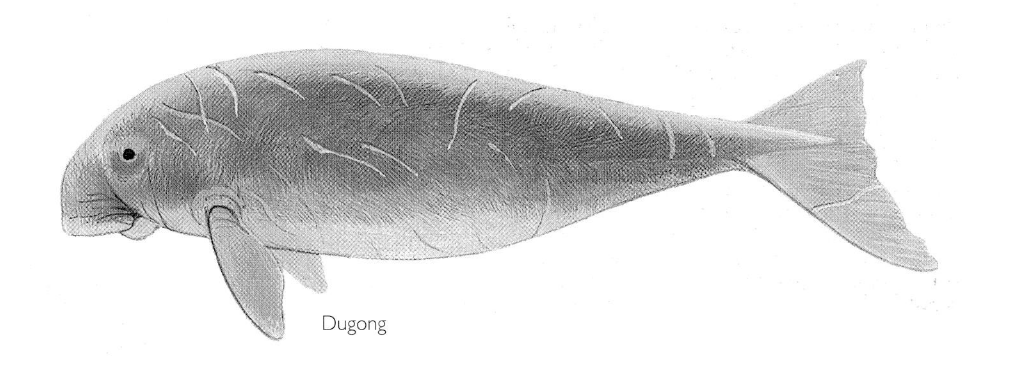
Dugong

PLATE 104

Addendum

Rusty Antechinus *Antechinus adustus*

hb 90–119 mm; **t** 89–103 mm; **wt** males 30–42 g, females 21–34 g; **teats** 6
Similar to Brown Antechinus but fur darker and longer. Upperparts uniformly sepia, tipped black. Underparts, shoulders, flanks and thighs cinnamon buff. Tail thin, length < hb, uniformly pale brown with blackish tip. **Distribution, habitat and status** Restricted to dense, upland (>800 m) tropical vine forest from Paluma (nw. of Townsville) to Mt Spurgeon (near Mossman), a north–south distance of 300 km. Most common in dense tangled edge vegetation and where there are rotting logs. **Behaviour** Terrestrial insectivore. Mates late Jun – late July after which all males die (by first week of August); pouch young first observed early Aug; remain in pouch for 4–5 weeks and are suckled until late Nov. **Similar species** Overlaps with Atherton Antechinus and Yellow-footed Antechinus (subspecies *rubeculus*) in the Ravenshoe district. Yellow-footed has gradation in dorsal colour from greyish on head and shoulders to russet on rump. Atherton is larger, predominantly ginger and has a near-naked tail.

Subtropical Antechinus *Antechinus subtropicus*

hb 94–136 mm; **t** 64–106 mm; **wt** males 52–67 g, females 24–32 g; **teats** 8
The largest of the Brown Antechinus complex but difficult to separate. Muzzle long and narrow. Upperparts mid brown, warmer on hindquarters. Underparts including chin, flanks and thighs pale fawn. Forefeet and hindfeet sparsely furred with pale buff hairs. Tail < hb, thin, olive-brown with weak ventral crest. **Distribution, habitat and status** On or e. of GDR from s. of Gympie, Qld to far ne. NSW, in subtropical vine forests from sea-level to 1000 m. Rarely in wet sclerophyll forest where restricted to fern gullies or dense riparian vegetation. **Behaviour** Terrestrial and arboreal insectivore. Mates in Sept after which all males die. Gestation lasts 25–26 days and pouch life a further 5 weeks. **Similar species** Brown Antechinus is generally smaller but the tail is proportionately longer, muzzle is shorter. Extremely large palatal vacuities are diagnostic.

Tan False Antechinus *Pseudantechinus roryi*

hb 77–90 mm; **t** 66–88 mm; **hf** 12–15 mm; **el** 13–19 mm; **teats** 6
Upperparts reddish brown, bright tan on back and shoulders; guard hairs dark brown. Underparts whitish—hairs are dark grey for basal third then whitish. Hairs of face and cheeks pale tan tipped grey, giving grizzled appearance. Bright orange patch behind ear. Upperside of feet white. Tail distinctly bicoloured, tan above, paler (sometimes whitish) below. Terminal pads smooth, other pads striated, remainder of sole granulated. Penis with elongated ventral process only slightly shorter than the penis itself. P_3 absent, P^2 larger and higher than the other upper premolars. **Distribution, habitat and status** Known from the Pilbara Uplands, including the Cape Range on North West Cape, eastwards and northwards into the Great Sandy and Gibson Deserts. False antechinus on Barrow I. may also belong to this species. Mostly found in rocky outcrops and breakaways. **Behaviour** Nothing known; presumably similar to others in genus. **Similar species** Woolley's False Antechinus is larger in all external measurements; retains P_3; males lack penile appendage. Potential zone of overlap with Fat-tailed False Antechinus in the Gibson Desert s. of Clutterbuck Hills. Fat-tailed is larger in all external measurements except hb, has less reddish fur and more uniform tail colour.

Further reading

General

Cronin, L. 1991. *Key Guide to Australian Mammals.* Reed Books, Sydney.
Hyett, J. and Shaw, N. 1980. *Australian Mammals: A Field Guide for New South Wales, Victoria, South Australia and Tasmania.* Thomas Nelson, Melbourne.
Menkhorst. P. (ed.) 1996. *Mammals of Victoria: Distribution, Ecology and Conservation.* Revised edition. Oxford University Press, Melbourne.
Ride, W.D.L. 1970. *A Guide to the Native Mammals of Australia.* Oxford University Press, Melbourne.
Strahan, R. (ed.) 1995. *The Mammals of Australia.* Reed Books, Sydney.
Taylor, J.M. 1984. *The Oxford Guide to Mammals of Australia.* Oxford University Press, Melbourne.
Triggs, B. 1996. *Scats, Tracks and Other Traces: A Field Guide to Australian Mammals.* Oxford University Press, Melbourne.
Walton, D.W. and Richardson, B.J. (eds) 1989. *Fauna of Australia. Volume 1B: Mammalia.* Australian Government Publishing Service, Canberra.
Watts, D. 1993. *Tasmanian Mammals: A Field Guide.* Revised edition. Peregrine Press, Kettering.
Wood Jones, F. 1923–25. *The Mammals of South Australia.* Government Printer, Adelaide.

Monotremes

Augee, M. (ed.) 1992. *Platypus and Echidnas.* Royal Zoological Society of New South Wales, Sydney.
Augee, M. and Gooden, B. 1993. *Echidnas of Australia and New Guinea.* University of New South Wales Press, Sydney.
Grant, T. 1995. *The Platypus: A Unique Mammal.* University of New South Wales Press, Sydney.
Rissmiller, P. 1999. *The Echidna: Australia's Enigma.* Hugh Lauter Levin Associates.

Marsupials

Archer, M. (ed). 1982. *Carnivorous Marsupials.* Surrey Beatty & Sons, Sydney.
Archer, M., Flannery, T. and Grigg, G. 1985. *The Kangaroo.* Wheldon, Sydney.
Beresford, Q. and Bailey, G. 1981. *Search for the Tasmanian Tiger.* Blubber Head Press, Hobart.
Dawson, T. 1995. *Kangaroos: Biology of the Largest Marsupials.* University of New South Wales Press, Sydney.
Frith, H.J. and Calaby, J.H. 1969. *Kangaroos.* F. W. Cheshire, Melbourne.
Grigg, G., Jarman, P. and Hume, I. (eds) 1989. *Kangaroos, Wallabies and Rat-kangaroos.* Surrey Beatty & Sons, Sydney.
Guiler, E. 1985. *Thylacine: The Tragedy of the Tasmanian Tiger.* Oxford University Press, Melbourne.
Mansergh, I. and Broome, L. 1994. *The Mountain Pygmy-possum of the Australian Alps.* University of New South Wales Press, Sydney.
Martin, R. and Handasyde, K. 1999. *The Koala: Natural History, Conservation and Management.* University of New South Wales Press, Sydney.
Maxwell, S., Burbidge, A.A. and Morris, K. (eds) 1996. *The 1996 Action Plan for Australian Marsupials and Monotremes.* Environment Australia, Canberra.
Paddle, R. 2000. *The Last Tasmanian Tiger: The History and Extinction of the Thylacine.* Cambridge University Press, Melbourne.
Russell, R. 1980. *Spotlight on Possums.* University of Queensland Press, Brisbane.
Seebeck, J.H., Brown, P.R., Wallis, R.L. and Kemper, C.M. (eds) 1990. *Bandicoots and Bilbies.* Surrey Beatty & Sons, Sydney.
Smith, A. and Hume, I. (eds) 1984. *Possums and Gliders.* Surrey Beatty & Sons, Sydney.
Triggs, B. 1996. *The Wombat: Common Wombats in Australia.* Revised edition. University of New South Wales Press, Sydney.

Bats

Churchill, S. 1998. *Australian Bats.* New Holland Publishers, Sydney.

Duncan, A., Baker, G.B. and Montgomery, N. (eds) 1999. *The Action Plan for Australian Bats.* Environment Australia, Canberra.

Hall, L.S. and Richards, G.C. 1979. Bats of Eastern Australia. Booklet No. 12. Queensland Museum, Brisbane.

Hall, L. and Richards, G. 2000. *Flying Foxes: Fruit and Blossum Bats of Australia.* University of New South Wales Press, Sydney.

Parnaby, H. 1992. An Interim Guide to Identification of Insectivorous Bats of South-eastern Australia. Technical Reports of the Australian Museum No. 8. Australian Museum, Sydney.

Reardon, T. and Flavel, S. 1987. *A Guide to the Bats of South Australia.* South Australian Museum, Adelaide.

Thomson, B.G. 1991. *A Field Guide to Bats of the Northern Territory.* Conservation Commission of the Northern Territory, Darwin.

Rodents

Covacevich, J. and Easton, A. 1974. Rats and Mice in Queensland. Booklet No. 9. Queensland Museum, Brisbane.

Lee, A.K. 1995. *The Action Plan for Australian Rodents.* Australian Nature Conservation Agency, Canberra.

Watts, C.H.S. and Aslin, H.J. 1981. *The Rodents of Australia.* Angus & Robertson, Sydney.

Carnivores

Breckwoldt, R. 1988. *The Dingo: A Very Elegant Animal.* Angus & Robertson, Sydney.

Corbert, L. 1995. *The Dingo in Australia and Asia.* University of New South Wales Press, Sydney.

Shaughnessy, P. 1999. *The Action Plan for Australian Seals.* Environment Australia, Canberra.

Deer

Bentley, A. 1998. *An Introduction to the Deer of Australia, With Special Reference to Victoria.* Second (revised) edition. Australian Deer Research Foundation, Croydon.

Harrison, M. 1998. *Wild Deer of Australia.* Australian Deer Research Foundation, Croydon.

Mayze, R. and Moore, G. 1990. *The Hog Deer.* Australian Deer Research Foundation, Croydon.

Cetaceans

Baker, A. 1998. *Whales and Dolphins of Australia and New Zealand: An Identification Guide.* Victoria University Press, Wellington.

Bannister, J.L., Kemper, C.M. and Warneke, R.M. 1996. *The Action Plan for Australian Cetaceans.* Australian Nature Conservation Agency, Canberra.

Bryden, M., Marsh, H. and Shaughnessy, P. 1998. *Dugongs, Whales, Dolphins and Seals, With Special Reference to Australasia.* Allen & Unwin, Sydney.

Carwardine, M. 1995. *Whales, Dolphins and Porpoises: The Visual Guide to All the World's Cetaceans.* Harper Collins, Sydney.

Carwardine, M., Hoyt, E., Fordyce, R.E., and Gill, P. 1998. *Whales, Dolphins and Porpoises.* Readers' Digest, Sydney.

Gill, P. and Burke, C. 1999. *Whale Watching in Australian and New Zealand Waters.* New Holland Publishers, Sydney.

Watson, L. 1981. *Sea Guide to Whales of the World.* Hutchison, London.

Glossary

abdomen the belly.

allogrooming cooperative grooming of fur and skin.

alpine above the treeline where snow normally lies continuously for more than four months of the year.

apical granule a large granule at the front of an interdigital pad.

arboreal living in trees.

arthropod a member of the phylum Arthropoda: animals with segmented bodies, hard exoskeleton and many, jointed legs, e.g. insects, spiders, crustaceans, millipedes, centipedes, scorpions.

baleen plates comb-like fibrous plates that hang from the upper jaw of baleen whales (Mysticeti). Used to sieve planktonic organisms from the water.

beak (of cetaceans) the elongated forward portion of the head of some dolphins and some toothed whales (Odontoceti).

benthic living on or near the bottom of a water body.

bipedal moving on only the two hind legs.

blow (of cetaceans) the column of water vapour forcefully exhaled when a cetacean surfaces to breathe.

blowhole nasal opening(s) on the top of the head of a cetacean, through which the animal breathes. Baleen whales have two blowholes, toothed whales have one.

bow-riding riding on the pressure wave created by the bow of a boat.

breaching (of cetaceans) leaping out of the water.

brigalow a woodland community dominated by *Acacia harpophylla*, formerly widespread in central east Qld.

brindled tawny or grey, streaked with darker colour.

browse to feed on leaves of shrubs, trees, or ferns (compare with *graze*).

calcar in bats, a cartilaginous spur that extends from the ankle towards the tail, supporting the trailing edge of the tail membrane.

callosities patches of rough, lumpy, hardened skin on the head of right whales, infested with whale lice and barnacles.

canine the 'eye' tooth of mammals. Normally the longest and most distinctly pointed tooth, sitting behind the incisors and separating them from the cheek teeth (premolars and molars).

canopy the cover of foliage of a plant, or of the tallest layer of vegetation.

canter an easy gait between a trot and a gallop, in which 3 feet are simultaneously off the ground.

cape (of cetaceans) dark area over the shoulders and back of some cetaceans, sometimes extending onto the flanks.

carnivore (adj. carnivorous) a flesh-eating animal, feeding on other animals (see also *insectivore*).

carrion dead animal flesh.

cephalopod a member of the Cephalopoda, a class of marine molluscs with tentacles attached to their heads; includes squids, octopuses, cuttlefishes and nautiluses.

cetacean a member of the order Cetacea, the whales, dolphins and porpoises.

chenopod a member of the plant family Chenopodiaceae, the saltbushes.

chevron stripes meeting at an angle, like a V.

circumglobal extending around the globe.

circumpolar extending around the globe at a high latitude.

commensal relating to two species that share resources or habitat without disadvantage to either. In the context of this book, it refers to rodents living in association with humans.

continental shelf the submerged and generally flat part of a continental mass, extending from the shore to the edge of the continental slope.

crepuscular active around dawn and dusk, in 'twilight'.

crustacean a member of the Crustacea, a subphylum of arthropods that have two pairs of antennae in front of the mouth, breathe through gills, and have a hardened extrusion (carapace) covering the thorax.

cryptic difficult to see.

dasyurid a member of the family Dasyuridae (carnivorous marsupials).

diagnostic (of a character) confirming the identification of a species.

diprotodont A member of the order Diprotodonta, marsupials having only one functional pair of lower incisors; e.g. possums, kangaroos, wombats, koalas.

digit a finger or toe.

distal towards the free end of a structure such as a digit or tail, farthest from point of attachment. Opposite of *proximal.*

diurnal active during daylight.

dorsal pertaining to the upper surface or back. Opposite of *ventral.*

drey globular nest constructed in the branches of a tree or shrub, not in a hollow or other caivity.

echolocation use of reflected high-frequency sound to determine the location of an object.

ecotone a boundary or mixing zone between vegetation formations or habitats.

embryonic diapause temporary halt to embryonic development.

endangered at risk of extinction.

epiphyte a plant growing upon another plant, without roots reaching into the soil.

erectile capable of being raised to an upright position.

extinct having died out completely; no longer present in the world population.

extralimital outside the limits of the region under consideration.

falcate (of a cetacean fin) backswept in a curve.

family the primary level of classification above genus, consisting of genera with a number of features in common.

feral pertaining to domesticated animals which have established wild populations.

fin (of cetaceans) the fin located on the midline of the back.

flanks the lower sides, between the forelimbs and hindlimbs.

flippers (of seals) the paddle-shaped forelimbs and hindlimbs; (of cetaceans) the paddle-shaped forelimbs.

fluke (of cetaceans or dugong) the horizontally flattened tail fin used for propulsion, not a modified hindlimb.

forb herbaceous plant other than a grass.

forearm that part of the forelimb between the elbow and the wrist.

form a scrape or hollow in the soil where certain mammals rest and sleep.

gallery forest tall complex riparian forest.

gape the corner of the mouth, or the arc through which the lower jaw rotates.

genus the primary level of classification above species, a group of closely related species (pl. genera).

gestation pregnancy; the period from fertilisation to birth.

gibber an arid plain with surface composed of densely packed, small, rounded rocks.

glans penis the terminal portion of the penis, covered by the foreskin.

glean to pick food items from the surface of plants or other substrates.

granulated surface covered in small hemispheres or granules.

graze to eat grass or forbs, not shrubs or trees (compare with *browse*).

Great Dividing Range (GDR) the range of mountains running close to the east coast of Australia from northern Cape York Peninsula to western Victoria.

gregarious (of animals) living in groups, opposite of solitary.

grizzled intermixed with grey; evenly sprinkled with fine pale markings.

guard hairs long straight hairs, frequently black tipped, which project above the general fur level on the back and sides of mammals, give colour and texture to the pelage.

habitat the environment in which an organism lives. Strictly the components of the environment (physical and biological) required by a particular species to maintain life, but often reduced to/summarised as vegetation types utilised.

hallucal pertaining to the hallux, e.g. hallucal pad—footpad immediately behind the insertion of the hallux and the second toe.

hallux first, inner ('big') toe of hindfoot.

herb a plant, other than a grass, that does not produce a woody stem.

herbivorous feeding on plants.

hibernate to reduce activity and metabolism by lowering body temperature for a long period during winter.

home range the area normally utilised by an individual animal, but usually not the total range of an individual. May overlap with the home range of other individuals.

honeydew a sugary liquid excreted by sap-sucking bugs.

hummock grass grasses of the genera *Triodia* and *Plectrachne* which form spiny rounded hummocks (spinifex).

incisor the front teeth, located forward of the canines and used for cutting.

inguinal in the region of the groin.

insectivore (adj. **insectivorous**) animal that eats insects or other arthropods.

interdigital pad raised pad on the sole of the hindfoot located between the bases of the toes.

invertebrate any animal that does not possess a backbone.

keel (of cetacean) prominent ridge along the top and/or bottom of the tailstock.

krill planktonic shrimp-like crustaceans of the order Euphausiacea, a major part of the diet of baleen whales and the Crab-eater Seal.

larva pre-adult form of many animals, usually morphologically different to adults, usually sexually immature, often the dispersal stage.

lobe a rounded, fleshy projection.

lob-tailing (of cetaceans) raising the flukes and smacking them down onto the water's surface, to make a loud splash.

macropod a member of the family Macropodidae: kangaroos and wallabies.

mallee small, multistemmed eucalypts with underground woody tuberous root systems; a plant community dominated by these plants; (Mallee) the geographical area where mallee communities are prevalent.

mane long hair growing on the back and sides of the neck.

mangroves trees which grow in the intertidal zone and obtain oxygen partly through pneumatophores.

manna a sugary secretion from certain insects that feed on the sap of plants, especially eucalypts.

mantle the upper back between the shoulders.

marsupial a mammal belonging to the subclass Marsupialia, characterised by young being born at a very early developmental stage. The young marsupial then attaches permanently to a teat (with or without a protective pouch) until almost weaned.

melon the bulge on the forehead of some toothed whales and dolphins. Believed to be used to focus sounds in echolocation.

midline a line of contrasting colour running along the middle of the neck and back.

molars the most posterior teeth, have a large grinding surface with a complex pattern of ridges and grooves.

monsoonal relating to regions experiencing wet summers and dry winters, associated with a reversal of seasonal winds.

montane pertaining to mountainous environments below the alpine zone.

moult to shed feathers, hair, skin or cuticle of birds, mammals, reptiles and arthropods.

mulga Acacia woodland or tall shrubland in arid parts of Australia.

muzzle the snout or nose; the head forward of the eyes.

nasal exfoliation small, poorly developed noseleaf around the nostrils of bats in certain genera, consisting of small flaps and ridges.

nectar sugar-rich liquid secretion of many flowers.

nocturnal active during the night.

nomadic undertaking movements which are without an obvious seasonal pattern.

noseleaf a series of fleshy plates and processes surrounding the nostrils and mouth of some bats, used in echolocation.

nostril external opening of the nose leading into the nasal cavity (pl. nares).

occlusal the surfaces of the teeth that contact those in the opposite jaw.

oceanic occurring in the sea where the depth exceeds 200 metres.

omnivore (adj. **omnivorous**) organism feeding on both animals and plants.

opportunistic adapted to take swift advantage of temporary resources.

owl-pellet a mass of undigested material (fur, feathers, bones) regurgitated by an owl.

palatal foramina a pair of small passages through the palatine bone of skull (palate).

palatal vacuities large paired holes in the palatine bone.

pectoral pertaining to the chest.

pelage covering of fur or hair.

pelagic living in the upper waters of the open seas, not close to land.

pendulous suspended, hanging down.

phalanx bone of the finger or toe (pl. phalanges).

planktonic living in open water and moving mostly by the action of currents.

plantar pad a raised pad on the outside of the sole of the hindfoot roughly midway between the ball of the foot and the heel.

pod (of cetaceans) long-term social group.

pollen powdery male propagules discharged from the anthers of a flower.

polyprotodont a marsupial having more than one pair of functional lower incisors, as in dasyurids and bandicoots.

porpoise (verb) to move rapidly by repeatedly leaping out of the water; (noun) a cetacean belonging to the family Phocoenidae, comprising small cetaceans without a beak and with laterally compressed teeth used for cutting.

post-hallucal pad a raised pad situated on the sole of the hindfoot heelwards from the insertion of the big toe or hallux.

post-interdigital pad any pad on the sole of the hindfoot further towards the heel than the interdigital pads.

post-nasal ridge a ridge of tissue running transversely across the snout of bats of the genus *Nyctophilus*, the long-eared bats.

prehensile able to grasp.

premolar a tooth lying between the canine and the molars, usually having a shearing or cutting function.

proboscis an elongated snout or mouthpiece.

pupae inactive stage between larval and adult insect.

quadrupedal using four legs for locomotion.

rainforest closed-forest dominated by soft-leaved species, often with vines and epiphytes prominent.

rhinarium area of bare skin around and between the nostrils of a mammal.

riparian pertaining to the banks of a stream.

riverine pertaining to the banks or flood plain of a stream.

roan colour of fur where the predominant colour, usually but not exclusively reddish-brown, is thickly mixed with another, usually white or grey.

rodent any member of the mammalian order Rodentia: gnawing mammals with one pair of chisel-like incisors in the upper and lower jaw, and lacking canines.

rorqual a whale belonging to the family Balaenopteridae.

rostrum (of rorquals) the top of the head forward of the blowhole.

rump the lower back and hindquarters.

savannah grassy plain with widely spaced trees.

scat faecal pellet.

scavenger an animal that feeds on dead animals.

sclerophyll pertaining to plants with hard, stiff or leathery leaves. Here mostly referring to eucalypt or heath species.

seral stage one of a sequence of plant communities that replace one another through time.

spatulate broad and laterally compressed; spade-shaped.

species the basic unit of classification, consisting of a group of genetically similar organisms potentially capable of interbreeding to produce fertile offspring.

spinifex spiny hummock grass of the genera *Triodia* or *Plectrachne*.

sporangia structures containing the spores of ferns, mosses, etc.

spy-hopping (of cetaceans) holding the head vertically out of the water and observing the surroundings above water.

sthenurine a member of the macropod subfamily Sthenurinae, the broad-faced kangaroos.

striated (of footpads) having transverse ridges.

sub-antarctic the region around the Antarctic Convergence.

subspecies a subgrouping within a species which is a geographically isolated variant.

succulent thick and juicy.

supratragus a small flap above the ear aperture.

swale the valley between sand dunes.

syndactylous having the second and third toes of the hindfoot contained within a single sheath up to the terminal joints and claw.

tailstock (of cetaceans) the region of the body between the vent and the flukes.

taxon a taxonomic category of any rank, e.g. family, genus, species, subspecies, etc. (pl. taxa).

taxonomy the theory and practice of describing, classifying and naming organisms by arranging them in groups of related forms.

teat a nipple, the external opening of a mammary gland.

temperate the climatic zone between the subtropics and subantarctic.

terminal pad a raised pad on the underside of the distal end of a digit.

terrestrial land-based.

territorial defending an area against other animals of the same or other species.

throat pleats long grooves on the underside of baleen whales, extending back from the chin which allow the throat to distend to take in water when feeding.

throat pouch a pouch of skin on the throat.

tibia the main bone of the lower leg; the shinbone.

tines the points on a deer's antlers.

torpor a short period of dormancy, a state of reduced body temperature and metabolism to save energy; similar to hibernation, but not a response to regular seasonal climatic changes.

tragus a small upright, cartilage-based lobe, often elongate, originating from the base of the ear opening in most bats.

tropical relating to the hot climatic region of the Earth between the Tropic of Capricorn (23.5°S) and the Tropic of Cancer (23.5°N).

tubercle a small granular nodule.

vegetarian eating only plant matter.

vent the cloacal aperture.

ventral relating to the underside of an animal. Opposite of *dorsal.*

vertebrate an animal with a backbone, a member of the phylum Vertebrata.

vestigial currently of little or no utility, but ancestrally well-developed.

vine forest/thicket vegetation dominated by vines.

vulnerable a category of 'threatened' species that are in less imminent danger of extinction than endangered species.

wallum sandy coastal heath in Qld and NSW, frequently containing Wallum (*Banksia aemula*).

warren a network of burrows.

wean to learn to cease suckling and begin eating solid food.

wing pouch a flap of skin extending from the forearm to the fifth finger of some emballonurid bats.

Index to common names

Agile Antechinus 56
Agile Wallaby 110
Alice Springs Mouse, *see* Djoongari
Allied Rock Wallaby 130
Ampurta 52
Andrew's Beaked Whale 242
Antarctic Fur Seal 222
Antarctic Minke Whale 238
Anteater, Banded, *see* Numbat
Anteater, Spiny, *see* Short-beaked Echidna
Antechinus,
  Agile 56
  Atherton 54
  Brown 56
  Cinnamon 54
  Dusky 56
  Fat-tailed, *see* Fat-tailed False Antechinus
  Fawn 54
  Little Red, *see* Kaluta
  Red-eared, *see* Fat-tailed False Antechinus
  Rusty 252
  Sandstone, *see* Sandstone False Antechinus
  Spinifex, *see* Kaluta
  Subtropical 252
  Swainson's, *see* Dusky Antechinus
  Swamp 56
  Yellow-footed 54
    see also False Antechinus
Antilopine Wallaroo 110
Arnhem Long-eared Bat 168
Arnhem Rock Rat 182
Arnhem Sheathtail Bat 148
Arnoux's Beaked Whale 248
Ash-grey Dunnart, *see* White-tailed Dunnart
Ash-grey Mouse 190
Atherton Antechinus 54
Australian Fur Seal 218
Australian Sea-lion 220
Bali Banteng 212
Bali Cattle, *see* Bali Banteng
Banded Hare Wallaby 122
Bandicoot,
  Brindled, *see* Northern Brown Bandicoot
  Desert 78
  Eastern Barred 78
  Golden 76
  Long-nosed 78
  Northern Brown 76
  Pig-footed 80
  Rabbit, *see* Bilby
  Rabbit-eared, *see* Bilby
  Rufous Spiny 80
  Rufescent, *see* Rufous Spiny Bandicoot
  Short-nosed, *see* Southern Brown
  Southern Brown 76
  Western Barred 78
Banteng, Bali 212
Bare-backed Fruit Bat 138
Bare-rumped Sheathtail Bat 146
Barrow Island Wallaroo, *see* Euro
Bat,
  Arnhem Long-eared 168
  Arnhem Sheathtail 148
  Bare-backed Fruit 138
  Bare-rumped Sheathtail 146
  Beccari's Freetail 150
  Broad-nosed (undescribed) 166
  Cape York Sheathtail, *see* Papuan Sheathtail Bat
  Chocolate Wattled 154
  Coastal Sheathtail 148
  Common Bentwing 162
  Common Blossom, *see* Eastern Blossom Bat
  Common Sheathtail 148
  Diadem Leaf-nosed 146
  Dusky Leaf-nosed 144
  East-coast Freetail 150
  Eastern Blossom 136
  Eastern Broad-nosed 166
  Eastern Cave 158
  Eastern Forest 160
  Eastern Freetail 152
  Eastern Horseshoe 142
  Eastern Tube-nosed 136
  Fawn Leaf-nosed 144
  Flute-nosed 156
  Ghost 142
  Golden-tipped 156
  Gould's Long-eared 170
  Gould's Wattled 154
  Greater Broad-nosed 164
  Greater Long-eared 170
  Hairy-nosed Freetail 152
  Hill's Sheathtail 148
  Hoary Wattled 154
  Inland Broad-nosed 166
  Inland Cave 158
  Inland Forest 160
  Inland Freetail 152
  Kimberley Cave 158
  Large Bentwing, *see* Common Bentwing Bat
  Large Forest 160
  Large-eared Horseshoe 142
  Large-eared Pied 154
  Least Blossom 136
  Lesser Long-eared 170
  Little Bentwing 162
  Little Broad-nosed 166
  Little Forest 160
  Little Northern Freetail 152
  Little Pied 154
  Mangrove Freetail 152
  Northern Blossom, *see* Least Blossom Bat
  Northern Broad-nosed 166
  Northern Cave 158
  Northern Freetail 150
  Northern Leaf-nosed 144
  Northern Long-eared 168
  Northern Mastiff, *see* Northern Freetail Bat
  Orange Leaf-nosed 142
  Papuan Sheathtail 146
  Pygmy Long-eared 168
  Semon's Leaf-nosed 144
  Southern Forest 160
  Southern Freetail 152
  Torresian Tube-nosed 136
  Troughton's Sheathtail 148
  Tube-nosed Insectivorous, *see* Flute-nosed Bat
  Western Broad-nosed, *see* Inland Broad-nosed Bat
  Western Freetail 152
  White-striped, *see* White-striped Freetail Bat
  White-striped Mastiff, *see* White-striped Freetail Bat
  White-striped Freetail 150
  Yellow-bellied Sheathtail 146
  Yellow-lipped, *see* Kimberley Cave Bat
Beaked Whale,
  Andrew's 242
  Arnoux's 248
  Blainville's 242
  Cuvier's, *see* Goose-beaked Whale
  Dense, *see* Blainville's Beaked Whale

Ginkgo-toothed 242
Gray's 246
Hector's 244
Longman's, *see* Tropical Bottlenose Whale
Shepherd's 246
Strap-toothed 244
Tasman, *see* Shepherd's Beaked Whale
True's 244
Wonderful, *see* True's
Beccari's Freetail Bat 150
Bennett's Tree Kangaroo 106
Bennett's Wallaby, *see* Red-necked Wallaby
Bettong,
Burrowing 102
Brush-tailed, *see* Woylie
Northern 100
Rufous 100
Southern 100
Tasmanian, *see* Southern Bettong
Tropical, *see* Northern Bettong
Big-eared Hopping Mouse 180
Bilby 80
Bilby,
Greater, *see* Bilby
Lesser 80
White-tailed, *see* Lesser Bilby
Black Flying-fox 138
Black Rat 200
Black Wallaby 122
Black Wallaroo 118
Black-flanked Rock Wallaby 128
Black-footed Rock Wallaby, *see* Black-flanked Rock Wallaby
Black-footed Tree Rat 174
Black-gloved Wallaby, *see* Western Brush Wallaby
Black-striped Wallaby 120
Black-tailed Wallaby, *see* Black Wallaby
Blossom Bat,
Common, *see* Eastern Blossom Bat
Eastern 136
Least 136
Northern, *see* Least Blossom Bat
Blainville's Beaked Whale 242
Blue Flier, *see* Red Kangaroo
Blue Whale 240
Blue-grey Mouse 192
Bobuck, *see* Mountain Brushtail Possum
Bolam's Mouse 188
Boodie, *see* Burrowing Bettong
Bottlenose Dolphin 230
Bottlenose Whale,
Southern 248
Tropical 248
Bramble Cay Melomys 172
Bridled Nailtail Wallaby 124
Brindled Bandicoot, *see* Northern Brown Bandicoot
Broad-cheeked Hopping Mouse 180
Broad-faced Potoroo 104
Broad-toothed Rat 198
Brown Antechinus 56
Brown Hare 208
Brumby, *see* Horse
Brush-tailed Bettong *see* Woylie
Brush-tailed Rabbit Rat 176
Brush-tailed Rock Wallaby 132
Brush-tailed Tree Rat, *see* Brush-tailed Rabbit Rat
Bryde's Whale 240
Buffalo,
Swamp 212
Water, *see* Swamp Buffalo
Burramys, *see* Mountain Pygmy Possum
Burrowing Bettong 102
Bush Rat 198
Butler's Dunnart 70
Calaby's Pebble-mound Mouse 186
Camel, One-humped 210
Canefield Rat 202
Cape York Melomys 172
Cape York Pipistrelle 156
Cape York Rat 202
Cape York Rock Wallaby 130
Cape York Sheathtail Bat, *see* Papuan Sheathtail Bat
Carpentarian False Antechinus 60
Carpentarian Rock Rat 182
Cat, House 206
Central Hare Wallaby 108
Central Pebble-mound Mouse 188
Central Rock Rat 184
Cinnamon Antechinus 54
Chestnut Dunnart 70
Chital 216
Chocolate Wattled Bat 154
Chudich, *see* Western Quoll
Coastal Sheathtail Bat 148
Common Bentwing Bat 162
Common Blossom Bat, *see* Eastern Blossom Bat
Common Brushtail Possum 86
Common Dolphin 230
Common Dunnart 66
Common Planigale 64
Common Ringtail Possum 96
Common Rock Rat 182
Common Sheathtail Bat 148
Common Spotted Cuscus 84
Common Striped Possum 90
Common Wallaroo, *see* Euro
Common Wombat 82
Crab-eater Seal 222
Crescent Nailtail Wallaby 124
Cuscus,
Common Spotted 84
Grey, *see* Southern Common
Southern Common 84
Cuvier's Beaked Whale, *see* Goose-beaked Whale
Daintree River Ringtail Possum 98
Dalgyte, *see* Bilby
Dama Wallaby, *see* Tammar Wallaby
Darling Downs Hopping Mouse 180
Deer,
Chital 216
Fallow 216
Hog 216
Red 214
Rusa 214
Sambar 214
Delicate Mouse 186
Dense-beaked Whale, *see* Blainville's Beaked Whale
Desert Bandicoot 78
Desert Rat-kangaroo 102
Desert Mouse 188
Desert Short-tailed Mouse 184
Devil, Tasmanian 46
Diadem Leaf-nosed Bat 146
Dibbler 58
Dingo 206
Djoongari 190
Dolphin,
Bottlenose 230
Common 230
Fraser's 228
Dusky 226
Indo-pacific Humpback 224
Irrawaddy 224
Long-beaked Bottlenose 230
Pantropical Spotted 228
Risso's 226
Rough-toothed 224
Southern Rightwhale 226
Spinner 228
Striped 228
Shortfin, *see* Fraser's Dolphin
Snubfin, *see* Irrawaddy Dolphin
see also Spectacled Porpoise
Donkey 210
Dromedary, *see* One-humped Camel
Dugong 250
Dunnart,
Ash-grey, *see* White-tailed Dunnart
Butler's 70
Chestnut 70
Common 66
Fat-tailed 66
Gilbert's 68
Grey-bellied 68
Hairy-footed 72
Julia Creek 70
Kakadu 70
Kangaroo Island 74
Lesser Hairy-footed 72
Little Long-tailed 68
Long-tailed 72
Mouse, *see* Common Dunnart
Ooldea 72
Red-cheeked 74
Sandhill 74
Sooty, *see* Kangaroo Island Dunnart
Stripe-faced 66

White-footed 66
White-tailed 68
Dusky Antechinus 56
Dusky Dolphin 226
Dusky Flying-fox 138
Dusky Hopping Mouse 178
Dusky Leaf-nosed Bat 144
Dusky Rat 202
Dwarf Minke Whale 238
Dwarf Sperm Whale 234
Echidna, Short-beaked 44
East-coast Freetail Bat 150
Eastern Barred Bandicoot 78
Eastern Blossom Bat 136
Eastern Broad-nosed Bat 166
Eastern Cave Bat 158
Eastern Chestnut Mouse 194
Eastern False Pipistrelle 164
Eastern Fre-tail Bat 152
Eastern Forest Bat 160
Eastern Hare Wallaby 108
Eastern Horseshoe Bat 142
Eastern Pebble-mound Mouse 194
Eastern Pygmy Possum 88
Eastern Quoll 48
Eastern Tube-nosed Bat 136
Eastern Wallaroo, *see* Euro
Euro 118
European Rabbit 208
Fallow Deer 216
False Antechinus,
Carpentarian 60
Fat-tailed 60
Ningbing 62
Sandstone 60
Tan 252
Woolley's 60
False Killer Whale 234
False Pipistrelle,
Eastern 164
Western 164
False Water Rat, *see* Water Mouse
Fat-tailed Antechinus, *see* Fat-tailed False Antechinus
Fat-tailed False Antechinus 60
Fat-tailed Dunnart 66
Fawn Antechinus 54
Fawn-footed Melomys 172
Fawn Hopping Mouse 178
Fawn Leaf-nosed Bat 144
Feathertail Glider 90
Fin Whale 240
Five-lined Palm Squirrel 204
Flash Jack, *see* Bridled Nailtail Wallaby
Fluffy Glider, *see* Yellow-bellied Glider
Flute-nosed Bat 156
Flying-fox,
Black 138
Dusky 138
Grey-headed 140
Large-eared 140
Little Red 140
Spectacled 138
Torresian 140
Percy Island, *see* Dusky Flying-fox
Forester, *see* Eastern Grey Kangaroo
Forrest's Mouse, *see* Desert Short-tailed Mouse
Fox, Red 206
Fraser's Dolphin 228
Fruit Bat, Bare-backed 138
see also Flying-fox
Fur Seal,
Antarctic 222
Australian 218
New Zealand 218
Sub-antarctic 218
Ghost Bat 142
Giant White-tailed Rat 174
Gilbert's Dunnart 68
Gilbert's Potoroo 104
Gile's Planigale 64
Ginkgo-toothed Beaked Whale 242
Glider,
Feathertail 90
Fluffy, *see* Yellow-bellied Glider
Greater 94
Mahogany 94
Squirrel 92
Sugar 92
Yellow-bellied 94
Goat 208
Godman's Rock Wallaby 130
Golden Bandicoot 76
Golden-backed Tree Rat 174
Golden-tipped Bat 156
Goose-beaked Whale 246
Gould's Long-eared Bat 170
Gould's Mouse 196
Gould's Wattled Bat 154
Grassland Melomys 172
Gray's Beaked Whale 246
Great Hopping Mouse, *see* Broad-cheeked Hopping Mouse
Great Pipistrelle, *see* Eastern False Pipistrelle
Greater Broad-nosed Bat 164
Greater Glider 94
Greater Long-eared Bat 170
Greater Stick-nest Rat 204
Green Ringtail Possum 98
Grey-bellied Dunnart 68
Grey-headed Flying-fox 140
Grey Cuscus, *see* Southern Common Cuscus
Hairy-footed Dunnart 72
Hairy-nosed Freetail Bat 152
Hare, Brown 208
Hare Wallaby,
Banded 122
Central 108
Eastern 108
Rufous, *see* Mala
Spectacled 108
Hastings River Mouse 194
Heath Mouse 192
Heath Rat, *see* Heath Mouse
Hector's Beaked Whale 244
Herbert River Ringtail Possum 98
Herbert's Rock Wallaby 130
Hill's Sheathtail Bat 148
Hoary Wattled Bat 154
Hog Deer 216
Honey Possum 90
Hopping Mouse,
Big-eared 180
Broad-cheeked 180
Darling Downs 180
Dusky 178
Fawn 178
Great, *see* Broad-cheeked Hopping Mouse
Long-tailed 180
Mitchell's 178
Northern 180
Short-tailed 180
Spinifex 178
Horse 210
House Cat 206
House Mouse 184
Humpback Whale 236
Inland Broad-nosed Bat 166
Inland Cave Bat 158
Inland Freetail Bat 152
Inland Forest Bat 160
Indo-pacific Humpback Dolphin 224
Irrawaddy Dolphin 224
Julia Creek Dunnart 70
Kakadu Dunnart 70
Kakadu Pebble-mound Mouse, *see* Calaby's Pebble-mound Mouse
Kaluta 58
Kangaroo,
Black-faced, *see* Western Grey Kangaroo
Eastern Grey 114
Great Grey, *see* Eastern Grey Kangaroo
Mallee, *see* Western Grey Kangaroo
Red 116
Western Grey 114
Kangaroo Island Dunnart 74
Killer Whale 230
Kimberley Cave Bat 158
Kimberley Pebble-mound Mouse 186
Kimberley Rock Rat 182
Koala 82
Kowari 52
Kultarr 58
Lakeland Downs Mouse, *see* Tropical Short-tailed Mouse
Large Bentwing Bat, *see* Common Bentwing Bat
Large-eared Flying-fox 140
Large-eared Horseshoe Bat 142
Large-eared Pied Bat 154
Large-footed Myotis 162
Large Forest Bat 160
Leadbeater's Possum 92
Least Blossom Bat 136
Lemur-like Ringtail Possum, *see* Lemuroid Ringtail Possum
Lemuroid Ringtail Possum 98

Leopard Seal 220
Lesser Hairy-footed Dunnart 72
Lesser Long-eared Bat 170
Lesser Stick-nest Rat 204
Little Bentwing Bat 162
Little Broad-nosed Bat 166
Little Forest Bat 160
Little Long-tailed Dunnart 68
Little Northern Freetail Bat 152
Little Pied Bat 154
Little Pygmy Possum 88
Little Red Antechinus, *see* Kaluta
Little Red Flying-fox 140
Little Red Kaluta, *see* Kaluta
Little Rock Wallaby, *see* Nabarlek
Lesser Bilby 80
Long-beaked Bottlenose Dolphin 230
Long-flippered Pilot Whale 232
Long-footed Potoroo 104
Long-haired Rat 200
Longman's Beaked Whale, *see* Tropical Bottlenose Whale
Long-nosed Bandicoot 78
Long-nosed Potoroo 104
Long-tailed Dunnart 72
Long-tailed Hopping Mouse 180
Long-tailed Planigale 64
Long-tailed Pygmy Possum 88
Long-tailed Mouse 196
Lumholtz's Tree Kangaroo 106
Mala 108
Mahogany Glider 94
Mallee Ningaui 62
Mangrove Freetail Bat 152
Mangrove Pipistrelle 156
Mardo, *see* Yellow-footed Antechinus
Marl, *see* Western Barred Bandicoot
Mareeba Rock Wallaby 130
Marsupial Mole,
  Northern 52
  Southern 52
Masked White-tailed Rat 174
Melomys,
  Bramble Cay 172
  Cape York 172
  Fawn-footed 172
  Grassland 172
Melon-headed Whale 232
Merrin, *see* Bridled Nailtail Wallaby
Mitchell's Hopping Mouse 178
Mole, *see* Marsupial Mole
Monjon 126
Mountain Brushtail Possum 86
Mountain Pygmy Possum 90
Mouse,
  Alice Springs, *see* Djoongari
  Ash-grey 190
  Blue-grey 192
  Bolam's 188
  Calaby's Pebble-mound 186
  Central Pebble-mound 188
  Delicate 186
  Desert 188
  Desert Short-tailed 184
  Eastern Chestnut 194
  Eastern Pebble-mound 194
  Forrest's, *see* Desert Short-tailed Mouse
  Gould's 196
  Hastings River 194
  Heath 192
  House 184
  Kakadu Pebble-mound, *see* Calaby's Pebble-mound Mouse
  Kimberley, *see* Kimberley Pebble-mound Mouse
  Kimberley Pebble-mound 186
  Lakeland Downs, *see* Tropical Short-tailed Mouse
  Long-tailed 196
  New Holland 196
  Pilliga 194
  Pilbara Pebble-mound 190
  Plains 196
  Sandy Inland 188
  Shark Bay, *see* Djoongari
  Silky 192
  Smoky 192
  Tropical Short-tailed 184
  Water 176
  Western 190
  Western Chestnut 186
  Western Pebble-mound, *see* Pilbara Pebble-mound Mouse
  see also Djoongari, Hopping Mouse
Mouse Dunnart, *see* Common Dunnart
Mulgara 52
Mundarda, *see* Western Pygmy Possum
Musky Rat-kangaroo 106
Myotis, Large-footed 162
Nabarlek 126
Narrow-nosed Planigale 64
Native Cat, *see* Quoll
New Holland Mouse 196
New Zealand Fur Seal 218
Ningbing False Antechinus 62
Ningaui,
  Mallee 62
  Pilbara 62
  Ride's, *see* Wongai Ningaui
  Southern, *see* Mallee Ningaui
  Wongai 62
Noolbenger, *see* Honey Possum
Northern Bettong 100
Northern Blossom Bat, *see* Least Blossom Bat
Northern Broad-nosed Bat 166
Northern Brown Bandicoot 76
Northern Cave Bat 158
Northern Freetail Bat 150
Northern Hairy-nosed Wombat 82
Northern Hopping Mouse 180
Northern Leaf-nosed Bat 144
Northern Long-eared Bat 168
Northern Marsupial Mole 52
Northern Nailtail Wallaby 124
Northern Pipistrelle, *see* Mangrove Pipistrelle
Northern Planigale, *see* Long-tailed Planigale
Numbat 50
One-humped Camel 210
Ooldea Dunnart 72
Orange Leaf-nosed Bat 142
Orca, *see* Killer Whale
Pacific Rat 200
Pademelon,
  Red-bellied, *see* Rufous-bellied Pademelon
  Red-legged 134
  Red-necked 134
  Tasmanian, *see* Rufous-bellied Pademelon
  Rufous-bellied 134
Pale Field Rat 202
Pantropical Spotted Dolphin 228
Papuan Sheathtail Bat 146
Parma Wallaby 112
Paucident Planigale, *see* Gile's Planigale
Pebble-mound Mouse,
  Calaby's 186
  Central 188
  Eastern 194
  Kakadu, *see* Calaby's Pebble-mound Mouse
  Kimberley 186
  Pilbara 190
  Western, *see* Pilbara Pebble-mound Mouse
Percy Island Flying-fox, *see* Dusky Flying-fox
Phascogale,
  Brush-tailed 50
  Red-tailed 50
Pig 212
Pig-footed Bandicoot 80
Pilbara Ningaui 62
Pilbara Pebble-mound Mouse 190
Pilliga Mouse 194
Pilot whale,
  Long-flippered 232
  Short-flippered 232
Pipistrelle,
  Cape York 156
  Mangrove 156
  Northern, *see* Mangrove Pipistrelle
  Great, *see* Eastern False Pipistrelle
  Tasmanian, *see* Eastern False Pipistrelle
Plague Rat, *see* Long-haired Rat
Plains Mouse 196
Planigale,
  Common 64
  Gile's 64

Narrow-nosed 64
Long-tailed 64
Northern, *see* Long-tailed Planigale
Paucident, *see* Gile's Planigale
Platypus 44
Polynesian Rat, *see* Pacific Rat
Porpoise, Spectacled 226
Possum,
Bushy-tailed, *see* Common Brushtail Possum
Common Brushtail 86
Common Ringtail 96
Common Striped 90
Daintree River Ringtail 98
Green Ringtail 98
Herbert River Ringtail 98
Honey 90
Mountain Brushtail 86
Leadbeater's 92
Lemur-like Ringtail, *see* Lemuroid Ringtail
Lemuroid Ringtail 98
Rock Ringtail 96
Scaly-tailed 84
Silver-grey, *see* Common Brushtail Possum
Western Ringtail 96
Potoroo,
Broad-faced 104
Gilbert's 104
Long-footed 104
Long-nosed 104
Prehensile-tailed Rat 204
Pretty-face Wallaby, *see* Whiptail Wallaby
Proserpine Rock Wallaby 132
Purple-necked Rock Wallaby 132
Pygmy Long-eared Bat 168
Pygmy Killer Whale 232
Pygmy Possum,
Eastern 88
Little 88
Long-tailed 88
Western 88
Mountain 90
Pygmy Right Whale 238
Pygmy Sperm Whale 234
Quenda, *see* Southern Brown Bandicoot
Quokka 122
Quoll,
Eastern 48
Northern 48
Spot-tailed 48
Tiger, *see* Spot-tailed
Western 48
Rabbit, European 208
Rabbit Bandicoot, *see* Bilby
Rabbit-eared Bandicoot, *see* Bilby
Rabbit Rat,
Brush-tailed 176
White-footed 176
Rat,
Black 200
Black-footed Tree 174
Broad-toothed 198
Brown 200
Brush-tailed Rabbit 176
Canefield 202
Cape York 202
Bush 198
Dusky 202
False Water, *see* Water Mouse
Giant White-tailed 174
Golden-backed Tree 174
Greater Stick-nest 204
Heath, *see* Heath Mouse
Lesser Stick-nest 204
Long-haired 200
Masked White-tailed 174
Pacific 200
Pale Field 202
Plague, *see* Long-haired Rat
Polynesian, *see* Pacific Rat
Prehensile-tailed 204
Roof, *see* Black Rat
Sewer, *see* Brown Rat
Ships, *see* Black Rat
Swamp 198
Water 176
White-footed Rabbit 176
Rat-kangaroo,
Desert 102
Musky 106
Red Deer 214
Red Fox 206
Red-bellied Pademelon, *see* Rufous-bellied Pademelon
Red-cheeked Dunnart 74
Red-eared Antechinus, *see* Fat-tailed False Antechinus
Red-legged Pademelon 134
Red-necked Pademelon 134
Red-necked Wallaby 120
Red-tailed Phascogale 50
Red-tailed Wambenger, *see* Red-tailed Phascogale
Ride's Ningaui, *see* Wongai Ningaui
Right Whale,
Pygmy 238
Southern 236
Risso's Dolphin 226
Rock Rat,
Arnhem 182
Carpentarian 182
Central 184
Common 182
Kimberley 182
Rock Ringtail Possum 96
Rock Wallaby,
Allied 130
Black-flanked 128
Black-footed, *see* Black-flanked Rock Wallaby
Brush-tailed 132
Cape York 130
Godman's 130
Herbert's 130
Little, *see* Nabarlek
Mareeba 130
Purple-necked 132
Proserpine 132
Rothschild's 128
Sharman's 130
Short-eared 126
Unadorned 130
Yellow-footed 128
Rothschild's Rock Wallaby 128
Roof Rat, *see* Black Rat
Ross Seal 222
Rough-toothed Dolphin 224
Rufescent Bandicoot, *see* Rufous Spiny Bandicoot
Rufous-bellied Pademelon 134
Rufous Bettong 100
Rufous Hare Wallaby, *see* Mala
Rufous Rat-kangaroo, *see* Rufous Bettong
Rufous Spiny Bandicoot 80
Rusa 214
Rusty Antechinus 252
Sambar 214
Sandhill Dunnart 74
Sandstone Antechinus, *see* Sandstone False Antechinus
Sandstone False Antechinus 60
Sandy Inland Mouse 188
Sandy Nailtail Wallaby, *see* Northern Nailtail Wallaby
Scaly-tailed Possum 84
Scamperdown Whale, *see* Gray's Beaked Whale
Scrub Wallaby, *see* Black-striped Wallaby
Seal,
Antarctic Fur 222
Australian Fur 218
Crab-eater 222
New Zealand Fur 218
Sub-antarctic Fur 218
Leopard 220
Ross 222
Southern Elephant 220
Weddell 222
Sea-lion, Australian 220
Sei Whale 238
Semon's Leaf-nosed Bat 144
Sewer Rat, *see* Brown Rat
Sooty Dunnart, *see* Kangaroo Island Dunnart
Shark Bay Mouse, *see* Djoongari
Sharman's Rock Wallaby 130
Shepherd's Beaked Whale 246
Ships Rat, *see* Black Rat
Short-beaked Echidna 44
Short-eared Rock Wallaby 126
Short-flippered Pilot Whale 232
Short-nosed Bandicoot, *see* Southern Brown Bandicoot
Short-tailed Hopping Mouse 180
Silky Mouse 192
Smoky Mouse 192
Snubfin Dolphin, *see* Irrawaddy Dolphin
Southern Bettong 100
Southern Bottlenose Whale 248
Southern Brown Bandicoot 76
Southern Common Cuscus 84
Southern Elephant Seal 220
Southern Dibbler, *see* Dibbler
Southern Forest Bat 160
Southern Freetail Bat 152

Southern Hairy-nosed Wombat 82
Southern Marsupial Mole 52
Southern Right Whale 236
Southern Rightwhale Dolphin 226
Spectacled Flying-fox 138
Spectacled Hare Wallaby 108
Spectacled Porpoise 226
Sperm Whale 236
Spinifex Hopping Mouse 178
Spinner Dolphin 228
Spiny Ant-eater, *see* Short-beaked Echidna
Spot-tailed Quoll 48
Squirrel, Five-lined Palm 204
Squirrel Glider 92
Sub-antarctic Fur Seal 218
Sugar Glider 92
Strap-toothed Beaked Whale 244
Striped Dolphin 228
Stripe-faced Dunnart 66
Swainson's Antechinus, *see* Dusky Antechinus
Swamp Antechinus 56
Swamp Buffalo 212
Swamp Rat 198
Swamp Wallaby, *see* Black Wallaby
Tammar Wallaby 112
Tan False Antechinus 252
Tasman Whale, *see* Shepherd's Beaked Whale
Tasmanian Bettong, *see* Southern Bettong
Tasmanian Devil 44
Tasmanian Pipistrelle, *see* Eastern False Pipistrelle
Tasmanian Tiger, *see* Thylacine
Thylacine 46
Tiger Cat, *see* Spot-tailed Quoll
Tiger Quoll, *see* Spot-tailed Quoll
Toolache 120
Torresian Flying-fox 140
Torresian Tube-nosed Bat 136
Tree Kangaroo,
  Bennett's 106
  Lumholtz's 106
Tree Rat,
  Black-footed 174
  Brush-tailed, *see* Brush-tailed Rabbit Rat
  Golden-backed 174
Tropical Whale, *see* Bryde's Whale
Tropical Bottlenose Whale 248
Tropical Bettong, *see* Northern Bettong
Tropical Short-tailed Mouse 184
Troughton's Sheathtail Bat 148
True's Beaked Whale 244
Tuan, *see* Brush-tailed Phascogale
Tube-nosed Bat,
  Eastern 136
  Torresian 136
Tube-nosed Insectivorous Bat, *see* Flute-nosed Bat
Unadorned Rock Wallaby 130
Wallaby,
  Agile 110
  Banded Hare 122
  Bennett's, *see* Red-necked Wallaby
  Black 122
  Black-gloved, *see* Western Brush Wallaby
  Black-striped 120
  Black-tailed, *see* Black Wallaby
  Bridled Nailtail 124
  Central Hare 108
  Crescent Nailtail 124
  Dama, *see* Tammar Wallaby
  Eastern Hare 108
  Northern Nailtail 124
  Parma 112
  Pretty-face, *see* Whiptail Wallaby
  Red-necked 120
  Rufous Hare, *see* Mala
  Sandy Nailtail, *see* Northern Nailtail Wallaby
  Scrub, *see* Black-striped Wallaby
  Spectacled Hare 108
  Swamp, *see* Black Wallaby
  Tammar 112
  Western Brush 112
  Whiptail 110
  White-throated, *see* Parma Wallaby
Wallaroo,
  Antilopine 110
  Barrow Island, *see* Euro
  Black 118
  Common, *see* Euro
  Eastern, *see* Euro
Wambenger, *see* Phascogale
Warabi, *see* Monjon
Water Buffalo, *see* SwampBuffalo
Water Mouse 176
Water Rat 176
Weddell Seal 222
Western Barred Bandicoot 78
Western Broad-nosed Bat, *see* Inland Broad-nosed Bat
Western Brush Wallaby 112
Western Chestnut Mouse 186
Western False Antechinus *see* Woolley's False Antechinus
Western False Pipistrelle 164
Western Freetail Bat 152
Western Mouse 190
Western Native Cat, *see* Western Quoll
Western Pebble-mound Mouse, *see* Pilbara Pebble-mound Mouse
Western Pygmy Possum 88
Western Quoll 48
Western Ringtail Possum 96
Whale,
  Andrew's Beaked 242
  Antarctic Minke 238
  Arnoux's Beaked 248
  Blainville's Beaked 242
  Blue 240
  Bryde's 240
  Cuvier's Beaked, *see* Goose-beaked Whale
  Dense-beaked, *see* Blainville's Beaked Whale
  Dwarf Minke 238
  Dwarf Sperm 234
  False Killer 234
  Fin 240
  Ginkgo-toothed Beaked 242
  Goose-beaked 246
  Gray's Beaked 246
  Hector's Beaked 244
  Humpback 236
  Killer 230
  Long-flippered Pilot 232
  Longman's Beaked, *see* Tropical Bottlenose Whale
  Melon-headed 232
  Pygmy Killer 232
  Pygmy Right 238
  Pygmy Sperm 234
  Scamperdown, *see* Gray's Beaked Whale
  Sei 238
  Shepherd's Beaked 246
  Short-flippered Pilot 232
  Southern Bottlenose 248
  Southern Right 236
  Sperm 236
  Strap-toothed Beaked 244
  Tasman, *see* Shepherd's Beaked Whale
  True's Beaked 244
  Tropical, *see* Bryde's Whale
  Tropical Bottlenose 248
  Wonderful Beaked, *see* True's Beaked Whale
Whiptail Wallaby 110
White-footed Dunnart 66
White-footed Rabbit Rat 176
White-striped Bat, *see* White-striped Freetail Bat
White-striped Freetail Bat 150
White-striped Mastiff Bat, *see* White-striped Freetail Bat
White-tailed Bilby, *see* Lesser Bilby
White-tailed Dunnart 68
White-throated Wallaby, *see* Parma Wallaby
Wombat,
  Common 82
  Northern Hairy-nosed 82
  Southern Hairy-nosed 82
Wonderful Beaked Whale, *see* True's Beaked Whale
Wongai Ningaui 62
Woolley's False Antechinus 60
Woylie 102
Yellow-bellied Glider 94
Yellow-bellied Sheathtail Bat 146
Yellow-footed Antechinus 54
Yellow-footed Rock Wallaby 128
Yellow-lipped Bat, *see* Kimberley Cave Bat

Index to scientific names

Acrobates pygmaeus 90
Aepyprymnus rufescens 100
Antechinomys laniger 58
Antechinus,
adustus 252
agilis 56
bellus 54
flavipes 54
godmani 54
leo 54
minimus 56
stuartii 56
subtropicus 252
swainsonii 56
Arctocephalus,
forsteri 218
gazella 222
pusillus doriferus 218
tropicalis 218
Australophocoena dioptrica 226
Balaenoptera,
acutorostrata 238
bonaerensis 238
borealis 238
edeni 240
musculus 240
physalus 240
Berardius arnuxii 248
Bettongia,
gaimardi 100
lesueur 102
penicillata 102
tropica 100
Bos javanicus 212
Bubalus bubalus 212
Burramys parvus 90
Caloprymnus campestris 102
Camelus dromedarius 210
Canis lupus dingo 206
Caperea marginata 238
Capra hircus 208
Cercartetus,
caudatus 88
concinnus 88
lepidus 88
nanus 88
Cervus,
axis 216
dama 216
elaphus 214
porcinus 216
timorensis 214
unicolor 214
Chalinolobus,
dwyeri 154
gouldii 154
morio 154
nigrogriseus 154
picatus 154
Chaerephon jobensis 150
Chaeropus ecaudatus 80
Conilurus,
albipes 176
penicillatus 176
Dactylopsila trivirgata 90
Dasycercus,
cristicaudata 52
hillieri 52
Dasykaluta rosamondae 58
Dasyuroides byrnei 52
Dasyurus,
geoffroii 48
hallucatus 48
maculatus 48
viverrinus 48
Delphinus delphis 230
Dendrolagus,
bennettianus 106
lumholtzi 106
Dobsonia magna 138
Dugong dugon 250
Echymipera rufescens 80
Equus,
asinus 210
caballus 210
Eubalaena australis 236
Falsistrellus,
mackenziei 164
tasmaniensis 164
Felis catus 206
Feresa attenuata 232
Funambulus pennanti 204
Globicephala,
macrorhynchus 232
melas 232
Grampus griseus 226
Gymnobelideus leadbeateri 92
Hemibelideus lemuroides 98
Hipposideros,
ater 144
cervinus 144
diadema 146
semoni 144
stenotis 144
Hydromys chrysogaster 176
Hydrurga leptonyx 220
Hyperoodon planifrons 248
Hypsiprymnodon moschatus 106
Indopacetus pacificus 248
Isoodon,
auratus 76
macrourus 76
obesulus 76
Kerivoula papuensis 156
Kogia,
breviceps 234
simus 234
Lagenodelphis hosei 228
Lagenorhynchus obscurus 226
Lagorchestes,
asomatus 108
conspicillatus 108
hirsutus 108
leporides 108
Lagostrophus fasciatus 122
Lasiorhinus,
krefftii 82
latifrons 82
Leggadina,
forresti 184
lakedownensis 184
Leporillus,
apicalis 204
conditor 204
Leptonychotes weddellii 222
Lepus capensis 208
Lissodelphis peroni 226
Lobodon carcinophagus 222
Macroderma gigas 142
Macroglossus minimus 136
Macropus,
agilis 110
antilopinus 110
bernardus 118
dorsalis 120
eugenii 112
fuliginosus 114
giganteus 114
greyi 120
irma 112
parma 112
parryi 110
robustus 118
rufogriseus 120
rufus 116
Macrotis,
lagotis 80
leucura 80
Mastacomys fuscus 198
Megaptera novaeangliae 236
Melomys,
burtoni 172
capensis 172
cervinipes 172
rubicola 172
Mesembriomys,
gouldi 174

macrurus 174
Mesoplodon,
bowdoini 242
densirostris 242
ginkgodens 242
grayi 246
hectori 244
layardii 244
mirus 244
pacificus, see *Indopacetus pacificus*
Miniopterus,
australis 162
schreibersii 162
Mirounga leonina 220
Mormopterus,
beccarii 150
loriae 152
norfolkensis 150
planiceps 152
spp. 152
Murina florium 156
Mus musculus 184
Myotis adversus 162
Myotis macropus, see *Myotis adversus*
Myrmecobius fasciatus 50
Neophoca cinerea 220
Ningaui,
ridei 62
timealyi 62
yvonneae 62
Notomys,
alexis 178
amplus 180
aquilo 180
cervinus 178
fuscus 178
longicaudatus 180
macrotis 180
mitchelli 178
sp 180
mordax 180
Notoryctes
caurinus 52
typhlops 52
Nyctimene,
robinsoni 136
cephalotes 136
Nyctophilus,
arnhemensis 168
bifax 168
geoffroyi 170
gouldi 170
timoriensis 170
walkeri 168
Ommatophoca rossii 222
Onychogalea,
fraenata 124
lunata 124
unguifera 124
Orcaella brevirostris 224
Orcinus orca 230
Ornithorhynchus anatinus 44
Oryctolagus cuniculus 208
Parantechinus apicalis 58
Peponocephala electra 232
Perameles,
bougainville 78
eremiana 78
gunnii 78
nasuta 78
Petauroides volans 92
Petaurus,
australis 92
breviceps 90
gracilis 92
norfolcensis 90
Petrogale,
assimilis 130
brachyotis 126
burbidgei 126
coenensis 130
concinna 126
godmani 130
herberti 130
inornata 130
lateralis 128
purpureicollis 132
mareeba 130
penicillata 132
persephone 132
rothschildi 128
sharmani 130
xanthopus 128
Petropseudes dahli 96
Phalanger intercastellanus 84
Phascogale,
calura 50
tapoatafa 50
Phascolarctos cinereus 82
Physeter macrocephalus 236
Pipistrellus,
adamsi 156
westralis 156
Planigale,
gilesi 64
ingrami 64
maculata 64
tenuirostris 64
Pogonomys mollipilosus 204
Potorous,
gilbertii 104
longipes 104
platyops 104
tridactylus 104
Pseudantechinus,
bilarni 60
macdonnellensis 60
mimulus 60
ningbing 62
roryi 252
woolleyae 60
Pseudocheirus,
occidentalis 96
peregrinus 96
Pseudochirops archeri 98
Pseudochirulus,
cinereus 98
herbetensis 98
Pseudomys,
albocinereus 190
apodemoides 192
australis 196
bolami 188
calabyi 186
chapmani 190
deliculatus 186
desertor 188
fieldi 190
fumeus 192
glaucus 192
gouldi 196
gracilicaudatus 194
hermannsbergensis 188
higginsi 196
johnsoni 188
laborifex 186
nanus 186
novaehollandiae 196
occidentalis 190
oralis 194
pilligaensis 194
patrius 194
shortridgei 192
Pseudorca crassidens 234
Pteropus,
alecto 138
banakrisi 140
brunneus 138
conspicillatus 138
macrotis 140
poliocephalus 140
scapulatus 140
Rattus,
colletti 202
exulans 200
fuscipes 198
leucopus 202
lutreolus 198
norvegicus 200
rattus 200
sordidus 202
tunneyi 202
villosisimus 200
Rhinolophus,
megaphyllus 142
philippinensis 142
Rhinonicteris aurantius 142
Saccolaimus,
flaviventris 146
mixtus 146
saccolaimus 146
Sarcophilus harrisi 46
Scoteanax rueppellii 164
Scotorepens,
balstoni 164
greyii 164
orion 164
sanborni 164
sp 164
Setonix brachyurus 122
Sminthopsis,
aitkeni 74
archeri 70
bindi 70
butleri 70
crassicaudata 66
dolichura 68
douglasi 70
gilberti 68
granulipes 68
griseoventor 68
hirtipes 72
leucopus 66
longicaudata 72
macroura 66

murina 66
ooldea 72
psammophila 74
virginae 74
youngsoni 72
Sousa chinensis 224
Spilocuscus maculatus 84
Stenella,
attenuata 228
coeruleoalba 228
longirostris 228
Steno bredanensis 224
Sus scrofa 212
Syconycteris australis 136
Tachyglossus aculeatus 44
Tadarida australis 150
Taphozous,
australis 148
georgianus 148
hilli 148
kapalgensis 148
troughtoni 148
Tarsipes rostratus 90
Tasmacetus shepherdi 246
Thylacinus cyanocephalus 46
Thylogale,
billardierii 134
stigmatica 134
thetis 134
Trichosurus,
caninus 86
vulpecula 86
Tursiops,
aduncus 230
truncatus 230
Uromys,
caudimaculatus 174
hadrourus 174
Vespadelus,
baverstocki 160
caurinus 158
darlingtoni 160
douglasorum 158
finlaysoni 158
pumilus 160
regulus 160
troughtoni 158
vulturnus 160
Vombatus ursinus 82
Vulpes vulpes 206
Wallabia bicolor 122
Wyulda squamicaudata 84
Xeromys myoides 176
Ziphius cavirostris 246
Zyzomys,
argurus 182
maini 182
palatalis 182
pedunculatus 184
woodwardi 182

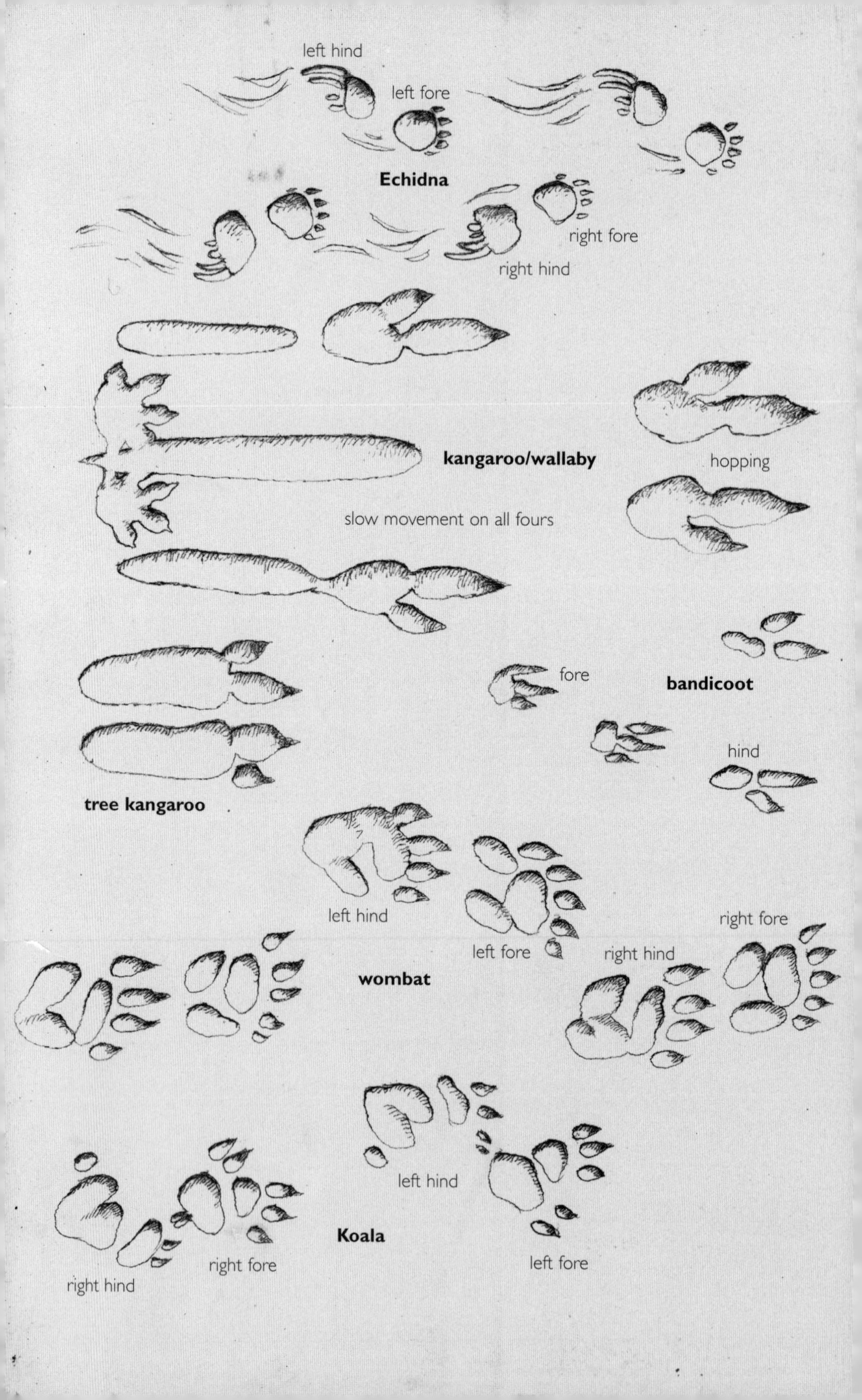

left hind
left fore
Echidna
right fore
right hind
kangaroo/wallaby
hopping
slow movement on all fours
fore
bandicoot
hind
tree kangaroo
left hind
left fore
right fore
right hind
wombat
left hind
Koala
right fore
left fore
right hind